The Science of Shakespeare *

ALSO BY DAN FALK

In Search of Time

Universe on a T-Shirt

THE SCIENCE OF SHAKESPEARE

A New Look at the Playwright's Universe

DAN FALK

Thomas Dunne Books
St. Martin's Press
New York

THOMAS DUNNE BOOKS.
An imprint of St. Martin's Press.

www.thomasdunnebooks.com
www.stmartins.com

Designed by Jonathan Bennett

Library of Congress Cataloging-in-Publication Data

Falk, Dan, author.
The Science of Shakespeare : a New Look at the Playwright's Universe / Dan Falk.—First Edition.
pages cm
Includes bibliographical references and index.
ISBN 978-1-250-00877-0 (hardcover)
ISBN 978-1-250-00878-7 (e-book)
1. Shakespeare, William, 1564–1616—Knowledge—Science.
2. Literature and science—England—History—17th century. I. Title.
PR3047.F35 2014
822.3'3—dc23

2013046842

St. Martin's Press books may be purchased for educational, business, or promotional use. For information on bulk purchases, please contact Macmillan Corporate and Premium Sales Department at 1-800-221-7945, extension 5442, or write specialmarkets@macmillan.com.

First Edition: April 2014

10 9 8 7 6 5 4 3 2

contents

list of illustrations

preface and acknowledgments

As it happens, my love of science and my fondness for Shakespeare date back to about the same time: I was ten or eleven when my parents took me to see *Macbeth* for the first time; and it was around then that they bought me a copy of H. A. Rey's *Know the Stars*, a wonderful children's astronomy book. (Rey was into monkeys as well as astronomy; with his wife, Margaret, he was the creator of Curious George.) I would later come to realize that Shakespeare understood a fair bit about astronomy. This was obvious from his frequent allusions to sunrises, eclipses, the pole star, and such; but for many years I gave the matter little thought. I also knew that Shakespeare and Galileo were born in the same year; but that fact is usually dismissed as little more than an item of trivia. (Sure, 1564 was a good year, but so what?)

A turning point came in January 1996, when Peter Usher, an astronomer, presented a paper titled "A New Reading of Shakespeare's *Hamlet*" at a meeting of the American Astronomical Society. (The meeting, as luck would have it, was held in my home city of Toronto.) But Usher's work was clearly controversial, and his paper made only a mild splash; soon I was back to writing articles about black holes and the big bang. But in roughly 2010, with Shakespeare's 450th birthday looming, I began to dig a little deeper—quickly realizing that I had only scratched the surface of a rich and under-explored topic. I soon discovered that a handful of respected Shakespeare scholars were beginning to investigate the playwright's knowledge of science, and astronomy in particular. Another bout of luck: One of the scholars at the forefront of this research,

Scott Maisano, was based in Boston, where I was living in 2011–12 while a Knight Science Journalism Fellow at the Massachusetts Institute of Technology.

Sensing that "Shakespeare and science" would make an intriguing radio documentary, I pitched it to CBC Radio's *Ideas*, and they gave me the green light. This book takes advantage of many of the interviews originally conducted for that project. As my research continued, the library became my second home; I audited university Shakespeare classes; and I went to as many Shakespeare productions as I could. I've lost track of the total number of performances, but they came in all varieties: indoor and outdoor; minimalist and lavish; no-budget, low-budget, and professional. I loved being a "groundling" for *Henry V* at Shakespeare's Globe in London; I saw multiple *As You Like It*s, *Twelfth Night*s, and *Richard III*s; I took in a sadomasochist production of *Antony and Cleopatra* (let's just say there was a lot of leather) and an "anarchist" production of *Measure for Measure* in which Barnardine was played by a hand puppet (which proved quite effective, though it restricted his movements somewhat).

In *The Science of Shakespeare*, I examine the playwright's world, taking a close look at the science of his day (bearing in mind that "science," as we think of it today, was only just coming into existence). This subject—the birth of modern science—is fascinating in its own right, and I hope that readers will enjoy this work as a history of ideas, focusing on this remarkable period of discovery. I also investigate how these discoveries are reflected in Shakespeare's work, and, more broadly, how they reshaped society at large. And so, even though much has been written on this period, and Shakespeare is one of the most-studied figures in history, I hope that, by exploring the connections between the playwright and this aspect of his world, *The Science of Shakespeare* offers something new.

This book could never have been written were it not for the scholars—too many to name—who have explored the subject matter within these pages in far greater depth than myself; their books and journal articles have been invaluable. I am particularly indebted to those researchers who allowed me to pester them (sometimes repeatedly) with all manner of Shakespeare questions, and especially those who allowed me to interview them, microphone in hand, for the CBC Radio documentary. This group includes Stephen Greenblatt at Harvard, John Pitcher at Oxford, Eric Mallin at the University of Texas–Austin, and Colin McGinn, recently retired from the University of Miami. Scott Maisano, who sat for more than one interview and answered numerous queries, is due a special note of appreciation. Peter Usher, whose work was one of the catalysts for this project, also deserves particular thanks. I am also grateful to numerous museum curators, tour guides, and librarians on both sides of the Atlantic. In London, meetings with Boris Jardine at the Science Museum, and Kevin Flude at the Old Operating Theater Museum and Herb Garret, were especially fruitful. Owen Gingerich and Donald Olson answered many of my questions on the history of astronomy, and Ray Jayawardhana pointed me in the right direction on the physics of supernovas. Many scholars have helped me without knowing it; David Levy's work, for example, led me to numerous astronomical references in early modern English literature. I also thank the professors who welcomed me into their classrooms—a list that includes Gordon Teskey at Harvard, Peter Donaldson at MIT, and Christopher Warley and Jeremy Lopez at the University of Toronto.

I am indebted to my tireless agent, Shaun Bradley of the Transatlantic Agency, and my very patient editor, Peter Joseph of St. Martin's Press, as well as production editor David Stanford Burr and copy editor Terry McGarry. Jessica Misfud was invaluable in helping to gather the images that illustrate this work, and I owe special thanks to Marina De Santis for her Italian-to-English translation skills. Dr. Maisano generously took the time to look over portions of the manuscript, as did Bill Lattanzi and the ever-supportive Amanda Gefter. (Nonetheless, the reader should not presume that any of the researchers mentioned in the book would necessarily agree with any of the particular conclusions that I draw; and of course any mistakes are purely my own.) My family and friends stood by me at every step along this journey, and I couldn't have succeeded without their love and support.

Finally, in spite of all my hours in libraries, classrooms, and theaters, I make no pretense of being a professional Shakespeare scholar: I'm merely a journalist who is fascinated by science, intrigued by history, and—like millions of people around the world—in awe of Shakespeare's achievement. I've used footnotes, endnotes, and a thorough bibliography to document my sources and to point the reader to further information; yet the book is aimed squarely at non-experts—those who marvel at the way that science has transformed our world, and those who enjoy reading and watching Shakespeare for the joy of it, as I do.

By the Favor of the Heavens

PROLOGUE

Stratford-upon-Avon, Warwickshire, England
November 19, 1572
6:05 p.m.

Father!"

A middle-aged man turns to greet his son, one of a dozen schoolboys making their way out of the King's New School and onto Chapel Lane. It's getting cold; the man pulls his cloak to his chest. He's thankful to be wearing his new fur cap rather than the felt one he'd had to make do with the previous winter. The boy, as full of energy as ever, doesn't seem to mind the cold.

"You don't need to walk me home, Father. I'm almost nine years old." The boy's breath is visible in the crisp winter air.

"Eight and a half is not 'almost nine.' But you're right, William, you are a young man now," the father replies. "It happens that I had some business at the church, and I was just on my way back. Let us make haste now, your mother and the children are waiting. I hope you didn't give Master Hunt any trouble today?"

"Master Hunt had to leave for Alveston, on account of his mother being sick."

The father is taken aback; usually he is the first to hear any news of that kind. "Is that so?"

"But another teacher took his place," the boy continues. "Master Jenkins. We still had to do all of that Latin grammar. But we also talked about the Bible, and the children in the upper form read a poem by Horace, and got to act out a scene from a Roman play."

"Horace was my favorite. Can you remember a few lines?"

"Let me think. . . . *There is nothing that the hands of the Claudii will not accomplish*—"

"Not in *English*. Horace isn't meant to be read in English. In Latin, William, please."

"Oh, Father, school is out. And I don't *like* Latin."

"Whether you like it is hardly the point. You must learn it to be a gentleman—and, for the next few years, to escape the birch. Now continue. In Latin."

"Um . . . *nil Claudiae non perficient manus, quas et* . . . um . . . *benignus numine Iuppiter*—"

"*Benigno numine*," his father interrupted, correcting the boy's grammar. "It means 'by the favor of the heavens.' That's enough for now. You did very well, William."

The pair turn from Chapel Street onto High Street. It is now growing dark; the long winter's night stretches ahead. The full moon will provide some relief, but it is only just creeping above the eastern horizon. It has been a cloudy day—a little snow fell earlier—but as the wind blows, the clouds finally begin to part. In the southeast shines mighty Jupiter—the same Jupiter the Romans had put their faith in as they marched into battle; the same Jupiter that Horace had rhapsodized over. As they reach Henley Street, William stops and gazes upward.

"What are you looking at, son?"

"It's something Master Jenkins told us about. He said there was a new star in the sky. He said he had been in Oxford yesterday, and everyone was talking about it."

The father lets out a hearty laugh. "Don't be silly, William. I heard some talk of it also at the guild, but the reverend said it couldn't be, and of course he is right. It could be a comet perhaps."

"But Father, Master Jenkins said it was a star. In the constellation of—the queen with the funny shape. The queen shaped like an 'M.'"

"Cassiopeia," the father replies. In spite of himself, he turns northward to see what may be there. His son turns to follow his gaze. "The Lord doesn't just create new stars, the way Mr. Smith hammers out horseshoes. God created the world thousands of years ago, and he doesn't need to make improvements."

A pause.

"I think that's it!" William points to a bright star, eastward from the pole, just visible now that the clouds have passed. It stands just to the left of the unmistakable "M" of Cassiopeia.

The father has to admit there is *something* there. Whatever it is, it's even brighter than Jupiter. Brighter even than Venus had been that morning, as far as he could recall.

"Father—what does it mean?"

"I don't know, son. And I don't know that it really is what it appears to be. It could just as well be the devil's work as the Lord's. And now we really must carry on, or supper will be cold. Not to mention my fingers."

"I'm coming, Father." But the boy lingers for one last look as his father heads off down the street. "It's beautiful," he says, and then runs to catch up. "I don't think it's a comet, Father, because comets have tails."

"More nonsense from Master Jenkins? Well, cats have tails too, but Mrs. Olden's cat doesn't have one, and it's still a cat."

The boy pauses, seemingly deep in thought. "Why doesn't Mrs. Olden's cat have a tail?"

"They say Mr. Olden's dog bit it off," his father replies.

"Well, maybe a dog bit off the new star's tail," the boy offers.

"That's quite an imagination you have, son. And how many dogs are there in the sky, William?"

Another pause—and then a wide smile. "Two, Father! You showed them to me last winter—the big dog and the little dog!"

The father laughs. "You do have quite a wit, don't you, son? Now say their names in Latin, please."

"Oh, Father! *Canis . . . Canis Major* and *Canis Minor.*"

"Very good, son. By Jove, I swear you'll make a fine lawyer one day."

The Science of Shakespeare *

Introduction

*"The poet's eye, in a fine frenzy rolling
Doth glance from heaven to earth,
from earth to heaven . . ."*

I'm sitting in a large, airy room on the ground floor of the Houghton Library, a small, elegant neoclassical building in the shadows of Harvard University's gargantuan Widener Library. The semester is winding down, and there are only eight or nine people in the room, leafing through dusty books or clicking away on their laptops. Portraits of forgotten scholars peer down at us, while a giant clock with gold hands looms above the doorway. As a gloomy drizzle falls outside, I stare at two books on the table in front of me.

They're both old—four centuries, give or take—though the one on the left had an eighty-year head start. I gently pick up the first book. Its pale beige cover is made from a pig's skin stretched over wood, and may be nearly as old as the pages themselves. (Back then, customers who bought a "book" were actually buying a bundle of pages from the bookseller; one could then pay a bookbinder to put it all together in an attractive package.) Scenes from the Bible, barely discernible, have been pressed onto the front and back covers; the process was called "blind stamping," the librarian tells me. On the spine, the author's name has been nearly obliterated with the passage of time.

Two slim metal clasps, probably brass, hold the covers shut. I gently release them, and lift the front cover. The pages are stiff and warped, as though they had been damp at one time—who knows how many years ago—and then left to dry. A blank inside page has a few scribbles from a previous owner, as well as a sticker indicating the name of the man—a graduate of the class of 1922—who donated the book to Harvard. Then I come to the title page. Here the author's name is very legible indeed—though the typesetter apparently had trouble fitting it all on one line:

NICOLAI CO-
PERNICI TORINENSIS
DE REVOLUTIONIBUS ORBI-
um coeleſtium, Libri VI.

It's Latin, of course. And in those days, an "s" looked like an "f," so it's actually "*coelestium*"—"celestial," or, perhaps more accurately, "heavenly." The author's name, in the genitive, is given matter-of-factly as Nicolaus Copernicus of Toruń. The full title is *On the Revolutions of the Heavenly Spheres, in Six Books*, often shortened to *On the Revolutions* or *De revolutionibus*, or even "*De rev*"; I figure a tiny bit of Latin won't hurt us, so I'll stick to calling it *De revolutionibus*. However we label it, this is the book that turned the universe inside out. At the bottom of the page is the publisher's name (Johannes Petreius), the city where the book was printed (Nuremberg), and the year (1543):

Norimbergae apud Ioh. Petreium,
Anno M. D. XLIII.

Each page makes a peculiarly satisfying sound as it is turned. I soon come to Copernicus's famous diagram, located on page 10 verso (meaning "left-hand page"), nestled between two chunks of Latin text (see figure 0.1). There are more than 140 other diagrams in the book, most of them very technical and now of interest only to historians of science—but this one has become iconic. It may be the most important diagram in the history of Western thought.

The diagram shows a series of concentric circles; at their center is a very small circle with a dot in the middle, labeled "Sol." (My rudimentary Latin is enough to know that "*sol*" is "sun.") The larger circles mark the paths of the planets as they revolve around the sun in their orbits. And there *we* are: the third rock from the sun, a mere dot, labeled "Terra"—Earth. Circling this diminutive dot is another tiny object, a little crescent moon. It is a diagram that I had drawn countless times as a

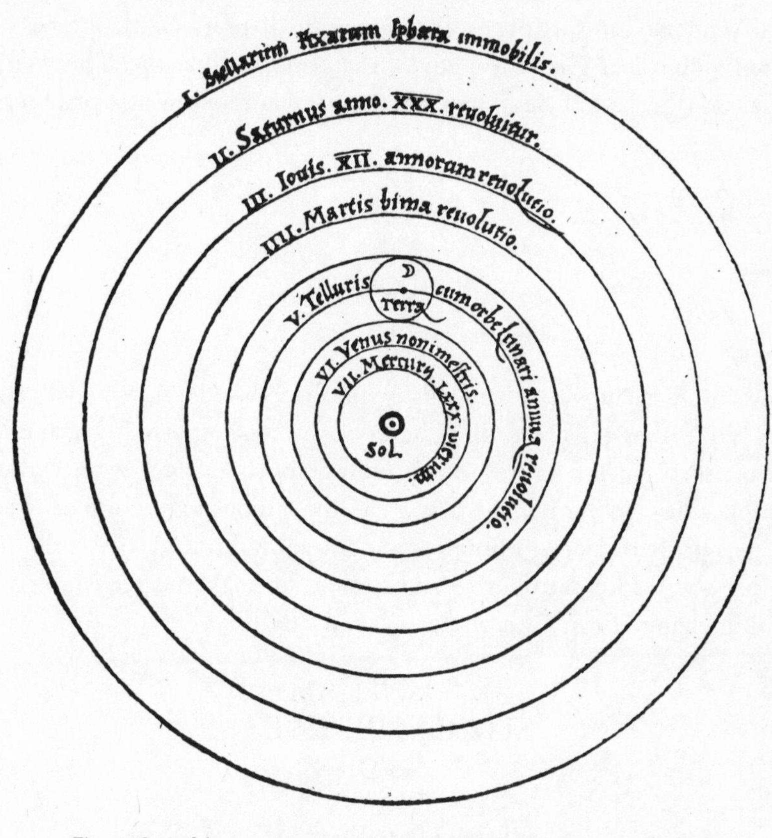

Fig. 0.1 One of the most important diagrams in the history of Western thought: Copernicus's illustration of a sun-centered universe, from *De revolutionibus* (*On the Revolutions*), published in 1543. Image Select/Art Resource, NY

---

nerdy, science-minded kid,* but you don't get any brownie points for being the hundred thousandth child to do so. You *do* get credit for being the person who came up with the picture in the first place.

The book on my right is a bit larger, measuring about nine-by-thirteen inches; it must weigh three or four pounds. Its cover, like that of its companion, is made from wood, this time covered in dark brown leather. The binding is too good to be original; it must have been re-bound a

---

* Of course, my version included nine planets—the five known since antiquity, plus the four discovered in modern times. Today it would be eight, assuming that Pluto's demotion in 2006 has taken hold in the minds of twenty-first-century children.

century or two after its pages first came off the press. Someone has also applied a gold-leaf gilt to the edges; the book still gleams. The writing on the spine is crystal clear, naming both the author and the printers:

SHAKESPEARE
I. JAGGARD
AND
E. BLOUNT
1623

Just inside the cover, a previous owner had pasted a newspaper article from November 11, 1848, on "The Folios of Shakespeare." This collection of thirty-six of Shakespeare's most important plays went through multiple editions, but it is the first one—the famous First Folio of 1623—that gets pride of place in libraries and museums around the world. The title page will look familiar to any student of Shakespeare. And—a refreshing change from Copernicus—it's in English:

MR. WILLIAM
SHAKESPEARES
COMEDIES,
HISTORIES, &
TRAGEDIES.

Published according to the True Originall Copies.

Don't be alarmed by the spelling: Scholars of early modern English assure us that spellings had not yet been standardized, and Shakespeare himself was known to mix it up even when signing his own name.* Just below is the familiar black-and-white engraving by Martin Droeshout—one of only two known depictions of the playwright that have a fighting chance of being accurate likenesses (see figure 0.2). (The other is the funerary monument in Holy Trinity Church in Stratford-upon-Avon, which dates from sometime between the playwright's death, in 1616, and the publication of the First Folio, seven years later.)

An introductory note from Shakespeare's friend and fellow playwright,

* Six undisputed examples of the playwright's signatures have survived, all on legal documents, and they vary from "Shakp" and "Shakspe" to "Shaksper" and "Shakspere." The more familiar "Shakespeare" was, however, used in the first printed works to bear his name—his two narrative poems, *Venus and Adonis* and *The Rape of Lucrece*, published in 1593 and 1594.

Mr. WILLIAM
SHAKESPEARES
COMEDIES,
HISTORIES, and
TRAGEDIES.
Publiſhed according to the True Originall Copies.

LONDON
Printed by Iſaac Iaggard, and Ed. Blount. 1623.

Fig. 0.2 The frontispiece from Shakespeare's First Folio, a collection of thirty-six of his plays, compiled by his colleagues John Heminges and Henry Condell. It was published in 1623, seven years after the playwright's death. The Bridgeman Art Library, London

Ben Jonson, asks the reader not to spend too much time staring at the portrait; it is Shakespeare's words that will bring him immortality. The note urges us to "looke Not on his Picture, but his Booke."

Is there any connection between these two books? Did Shakespeare know about Copernicus's revolutionary idea? Did he care? History is so much clearer in hindsight: Looking back after four centuries, it's obvious *to us* that Shakespeare lived in a remarkable time. The medieval world—a world of magic, astrology, witchcraft, and superstition of all kinds—was just beginning to give way to more modern ways of thinking. Shakespeare and Galileo were born in the same year, and new ideas about the human body, the Earth, and the universe at large were just starting to transform Western thought. The first modern anatomy book, by the Flemish-born physician Andreas Vesalius, was published in 1543, the same year as *De revolutionibus.* Is it possible that Shakespeare was unaware of these developments—or that he was vaguely conscious of them, but uninterested?

For some literary figures, the impact of this new picture of the world is obvious: In a famous passage from *An Anatomy of the World* (1611), John Donne laments that "the new philosophy calls all in doubt. . . . The sun is lost, and th'earth, and no man's wit / Can well direct him where to look for it." A half century later, John Milton would devote lengthy passages in *Paradise Lost* to a debate over the structure of the cosmos; indeed, he refers to Galileo three times in the poem (once by name; the astronomer is the only living figure to warrant such a mention). Milton is even said to have met the Italian scientist in person, when Galileo was in his final years, under house arrest, in his villa outside Florence. By Milton's time, as one scholar puts it, the Copernican system was "a scientific force with which all thinking men had to reckon." But Milton went to Cambridge, and Donne studied at *both* Oxford and Cambridge. Shakespeare flourished a little bit earlier, and had only the benefit of his local grammar school; as Jonson famously quipped, his colleague had only "small Latin and less Greek."

The traditional view is that Shakespeare was unconscious, or barely conscious, of the "new philosophy." It's not that Shakespeare scholars, or historians of early modern science, have neglected to look at possible connections between Shakespeare's works and the ideas and discoveries that mark what we now call the Scientific Revolution: They've looked, and concluded—wrongly, I believe—that no such connections exist. As recently as 2005, John Cartwright and Brian Baker, in *Literature and Science: Social Impact and Interaction*, find that ". . . the greatest poet of the age, William Shakespeare, shows little awareness or interest in the achievements or concerns of the astronomers." A few years earlier, William Burns declared in *The Scientific Revolution: An Encyclopedia* (2001) that "William Shakespeare . . . took almost no interest in science." Thomas McAlindon, meanwhile, believes that Shakespeare, in spite of being deeply concerned with cosmological matters, showed "no sign of [the Copernican] revolution" in his plays. Why is it so easy to read Shakespeare as a wholly prescientific figure? One reason is that Shakespeare's plays are littered with references to the medieval worldview. He frequently mentions the stars and the heavens, typically in a manner consistent with the thinking of the ancient Greek astronomer Ptolemy, dead for fourteen centuries. Shakespeare couldn't have known much about the new way of thinking, the theory goes, because ideas circulated slowly in those days, and Copernicanism took many decades to reach England, which at any rate was an intellectual backwater. Moreover, Copernicus's novel conception of the cosmos didn't really gain intellectual currency until Galileo's telescopic observations lent it some measure of observational support—and that came only in 1610, just as Shakespeare was packing his bags for a well-earned retirement in his hometown of Stratford.

But perhaps we shouldn't be so hasty. First of all, while acceptance of the Copernican theory came slowly, finding pockets of enthusiasm in a handful of university towns in central Europe, the theory did attract a number of early adherents in England, where a spirit of intellectual freedom and rational inquiry was in the air (arguably nurtured by the Protestant national faith, in contrast to the more repressive atmosphere in Catholic Europe).* Copernicus's groundbreaking book had been published in 1543, twenty-one years before Shakespeare's birth; by 1556 it was already mentioned favorably in an English book, Robert Recorde's *The Castle of Knowledge.* The first full account of the theory by an Englishman came from astronomer Thomas Digges in 1576 (when Shakespeare was twelve). Digges's book included a diagram of the solar system

---

* This is a simplification, of course—the Jesuits, a Catholic religious order, were establishing some of Europe's best schools—but the two religions' differing view of the miraculous is worth noting. According to Church doctrine, Catholics were obliged to believe in continuing divine intervention in human affairs, while Protestants held the opinion that, as Shakespeare's Lafeu puts it in *All's Well That Ends Well*, "miracles are past" (2.3.3). See Dear, *"Miracles, Experiments, and the Ordinary Course of Nature"*; Kocher, p. 191; Johnson, *Astronomical Thought in Renaissance England*, pp. 149–50.

in which the stars were seen to extend outward without limit, a vision of an infinite cosmos that surpassed even Copernicus in its daring.

As we will see, Shakespeare had multiple connections to the Digges family. (For a time he and Digges's son, Leonard, lived less than three blocks apart in their north London neighborhood. Leonard, a poet, was an early Shakespeare "fan" who contributed an introductory verse at the start of the First Folio.) Shakespeare may have encountered England's other great men of science of the day, from Thomas Harriot to Queen Elizabeth's own "science advisor," John Dee—the man often put forward as the model for Prospero in *The Tempest*. And then there was the Italian philosopher and mystic Giordano Bruno, who traveled to England in the 1580s, lecturing on Copernicanism and other provocative notions. Shakespeare is unlikely to have met Bruno, but may well have encountered his ideas.

Moreover, Shakespeare could have seen at least some of the evidence for the "new astronomy" with his own eyes, as hinted at in the fictionalized prologue. In November 1572, a bright new star lit up the night sky, appearing in the constellation Cassiopeia. (Today we know such an event as a supernova, the explosive death of a massive star.) It was so bright that for several months it outshone even Venus, making it the brightest object in the sky apart from the sun and the moon. (Indeed, it could be spotted even in daylight.) It was observed by Digges in England, and watched even more closely in Denmark by astronomer Tycho Brahe, whose published account of the new star was making waves even before the object had faded from view. The strange and wonderful apparition—today we call it simply "Tycho's star"—dealt a shattering blow to the cosmology of the ancients, refuting the idea of immutable heavenly spheres.

Amazingly, another new star blazed forth thirty-two years later, in 1604, and was studied by the German mathematician Johannes Kepler.* Shakespeare was forty, and at the height of his career, when Kepler's star illuminated the skies of Europe. Even if he somehow failed to see Tycho's star, he could not have missed Kepler's. It was a dazzling sight, one that could not be ignored. In fact, Shakespeare lived during a remarkably eventful period in terms of celestial drama: A dazzling comet in 1577 displayed a tail stretching one-eighth of the way across the sky, and two more comets appeared in 1582 and 1607; and a solar eclipse darkened

* I say "amazingly" because, in general, supernovae are quite rare. Kepler's star of 1604 was the last known star to explode in our Milky Way galaxy.

the skies over Europe in the autumn of 1605. There were ample reasons for taking an interest in cosmic happenings.

We should also note that England, and in particular London, was hardly a backwater. The city was teeming with tradesmen, merchants, and sailors who took a keen interest in what we would now call "science," and in particular in the latest technological advances, especially those connected to the art of navigation. The curriculum at Gresham College in London, founded in 1597, included astronomy, geometry, and medicine. Francis Bacon's groundbreaking work, *The Advancement of Learning*, championing the importance of observation and empirical knowledge, was published in 1605, around the time Shakespeare was working on *King Lear.* And the bold ideas penned by the French statesman and essayist Michel de Montaigne had appeared in English translation two years earlier. (Although Shakespeare scholars routinely discuss Montaigne's influence on the playwright—several of the plays contain passages lifted almost verbatim from the *Essays*—the fact that Montaigne specifically mentions the Copernican theory is often overlooked.)

But a reassessment may finally be at hand. In the last few years, a handful of scholars have begun to look more closely at Shakespeare's interest in the scientific discoveries of his time—asking what he knew, when he knew it, and how that knowledge might be reflected in his work. Scott Maisano at the University of Massachusetts–Boston, for example, has written extensively on the evidence for Shakespeare's awareness of the science of his day, and for its influence on his plays, especially the late romances. Other scholars, like John Pitcher and Jonathan Bate, both at Oxford, have acknowledged Shakespeare's interest in contemporary science, discussing it in popular biographies and in scholarly editions of the plays. One result of this reassessment is that it allows for a familiar passage to be read in a new light. Consider Ulysses's speech in *Troilus and Cressida*, in which he refers to "the glorious planet Sol / In noble eminence enthroned and sphered . . ." (1.3.89–90). The reference to "spheres" sounds at first like straight-ahead medieval cosmology, including the reference to the sun as a "planet." In the 1940s, this passage served as the backbone for E. M. W. Tillyard's thesis that Shakespeare's time ought to be seen as medieval rather than modern, a case he argued in his influential book *The Elizabethan World Picture*. Some current scholars continue to follow in Tillyard's footsteps; in the Arden edition, David Bevington tags the line simply as "a Ptolemaic conception." But as Bate points out, by emphasizing the role of the sun, the passage "may hint at the new heliocentric astronomy." James Shapiro, meanwhile, concedes that Shakespeare knew that Ptolemaic science "was already discredited by the Copernican revolution."

And Shakespeare wasn't quite ready to retire in 1610; he had a few years to go, and would produce five more plays in that time (two on his own, including *The Tempest*, and three more in collaboration with colleagues). It is from this period that we find *Cymbeline*—and an even more tantalizing hint that the playwright may have been conscious of the new cosmology. This admittedly weird play, combining elements of ancient Britain and ancient Rome, seems to have been written in 1610—just late enough that Shakespeare could have read Galileo's account of his telescopic discoveries, published in the spring of that year. Both Maisano and Pitcher have written in support of this hypothesis. "Jupiter" himself appears near the end of the play, while a stage direction calls for four ghosts to dance in a circle; could this be an allusion to the planet's four newly discovered moons, described by Galileo?

We will also take a look at the work of a more controversial figure, the astronomer Peter Usher, recently retired from Pennsylvania State University. Like Maisano and Pitcher, Usher sees the Jupiter scene in *Cymbeline* as a response to Galileo's discovery—but he takes "Shakespearian science" much further, arguing that examples of the playwright's scientific knowledge can be found in works spanning his entire career. Usher has taken a particular interest in *Hamlet*, which he sees as an allegory about competing cosmological worldviews. According to Usher, the play references not only Copernicus, but also Ptolemy, as well as Tycho Brahe, who pushed for a hybrid model of the solar system (a compromise that preserved elements of the ancient Ptolemaic system as well as the new Copernican model). Digges, too, is central to Usher's theory. When Hamlet envisions himself as "a king of infinite space" (2.2.255), could he be alluding to the new, infinite universe described—for the first time—by his countryman Thomas Digges?

Usher's proposal may sound far-fetched—but even skeptics do a double take when they look at Tycho Brahe's coat of arms, noticing that two of Tycho's relatives were named "Rosencrans" and "Guildensteren." And Usher isn't quite alone; several mainstream Shakespeare scholars are at least willing to admit that the playwright was influenced by Tycho's astronomy.

Shakespeare's characters were connected to the cosmos in a way that seems quite foreign to the modern reader. They have, to use Thomas

McAlindon's phrase, "cosmic imagination": Whether crying for joy or shedding tears of anguish, they look to the heavens for confirmation, calling out to "Jupiter" or "the gods" or "the heavens" as they struggle to make sense of their lives.

And so we find, not surprisingly, a multitude of references to astrology. But some of Shakespeare's characters also speak out *against* such superstitions, as when Cassius declares, "The fault, dear Brutus, is not in our stars, but in ourselves, that we are underlings" (*Julius Caesar* 1.2.139–40), or when Edmond, in *King Lear*, ridicules those who blame their misfortune on the heavens, dismissing such astrological conceit as "the excellent foppery of the world" (1.2.104). As for religion, though Shakespeare often alludes to biblical stories, he never once uses the word "bible." Nor do his characters put much faith in life continuing beyond death. He lived in an age of belief, yet a streak of skepticism runs through his work, especially toward the end of his career; in *King Lear* it reaches an almost euphoric nihilism. His characters often call upon the gods to help them, but their desperate pleas are rarely answered. Was Shakespeare a closet atheist, like his colleague Christopher Marlowe?

Of course, one has to tread carefully. Shakespeare is not only the most beloved writer in the English language, but also the most closely scrutinized. There is an enormous amount of very good scholarship on the playwright's life and work, and a significant amount of not-so-good scholarship. One reason there is so much to potentially be said about Shakespeare is that *he* said so much: He was prolific, with an output in the ballpark of 885,000 words. Yet there are only a few scraps of documentation to illuminate his personal life, and we can make only educated guesses regarding his private thoughts and beliefs. With no diaries, no letters, and no manuscripts, we have to rely on Shakespeare's published works. There is the perpetual danger of attributing the beliefs of his characters to the playwright himself. And, since Shakespeare uses language that is often challenging for the modern reader, his precise meanings are sometimes elusive (indeed, sometimes they are intentionally ambiguous). There is always the temptation to bend the facts to fit one's pet theory. (As with the Bible, one can find anything in Shakespeare if one looks hard enough.) We will consider a variety of opinions—mainly from established Shakespeare scholars, but occasionally from those whose expertise lies in another field but who nonetheless have something to contribute to our understanding of Shakespeare's world. I will do my best, however, to always indicate how widely accepted—or not—the various viewpoints are.

In the chapters ahead, we will examine the science of Shakespeare's time, beginning with a detailed look at the astronomical knowledge of his day, and then broadening our canvas to include the physical and life sciences more generally—along with the astrology, alchemy, and magic with which they were so deeply intertwined. Throughout the journey we will stop to ask what Shakespeare knew, and how it may have influenced his work. Obviously, Shakespeare was not the Carl Sagan of the Elizabethan Age—his first commitment was to his stagecraft, not to philosophy or science.* But I would argue that a close reading of his works reveals the depth of his interest in the natural world, and I hope to show that he was more conscious of the changing conception of the cosmos than we usually imagine. Shakespeare's writing often reflects the scientific ideas of his time—and the philosophical problems they were raising—and the more carefully we look at those ideas the better we can appreciate the scope of his achievement.

---

* The word "science" did not acquire something close to its modern meaning until roughly 1700, while the word "scientist" entered the language only in the 1830s. In the Renaissance, the study of the natural world was often called "natural philosophy," though, as we will see, this covered a broader domain of inquiry than present-day science. (To complicate matters, the word "science" *was* in use—meaning, roughly, "knowledge.") In the interest of readability, however, I will use the word "science," anachronistic as it may be, to refer to those endeavors that would today be seen as scientific pursuits.

1. "Arise, fair sun . . ."

A BRIEF HISTORY OF COSMOLOGY

Shakespeare's audience did not have to look far to see the stars: A wooden canopy projected out over the stage, and its underside—known as "the heavens"—was decorated with brightly painted stars and constellations. It served its purpose in *Hamlet*, for example, when the prince refers to "this brave o'erhanging firmament, this majestical roof fretted with golden fire" (2.2.283–5) or when Caesar declares that "the skies are painted with unnumbered sparks" (*Julius Caesar* 3.1.63).

The view of the universe engendered by this simple theatrical device wasn't so far off from how our ancestors had envisioned the cosmos for thousands of years: We look up at night, and we see an uncountable number of stars, brilliant pinpoints of light, seemingly painted on the vast dark canvas of the night sky.* And back then, before the light pollution brought by electrical lighting, the sky *really was* black. In *Antony and Cleopatra*, when Lepidus says to Caesar, "Let all the number of the stars give light / To thy fair way!", we might imagine that the stars truly shone brightly enough for the purpose (3.2.65–66). (In practice, a bit of moonlight would probably help.) The stars were intimately familiar, yet at the same time deeply mysterious. They were certainly far away—climbing the highest hills did not seem to bring them any closer—but how far away, one couldn't say. Perhaps they lay just out of reach; a

* Or at least, they *seem* uncountable. Today we know that only about two thousand stars—a bit more for a person with perfect eyesight under ideal conditions—can be seen with the unaided eye at any one time.

little farther, perhaps, than the great oceans or the highest mountain peaks.

The sun was more familiar, its presence more intimate: the brightest of lights; the giver of life. Everyone knew that it rose in the east and set in the west, but they also knew the subtle variation in that pattern over the course of a year: In the winter, the sun makes only a low arc across the southern sky, while summer brings longer days in which the sun takes a much higher path across the sky. The cycle repeats, with perfect dependability, year after year. A farmer had to know the sun's movement—but so, too, did a playwright; for the action to be visible, one had to contend with the harsh sunlight of midsummer as well as the long shadows of autumn and the all-too-early darkness of the winter months. Sophisticated stagecraft and spectacular costumes mean nothing if audience members have to squint to see them. As Peter Ackroyd writes, Shakespeare was "aware of the passage of time and of daylight across the open stage, so that he wrote shadowy scenes for the hour when the shadows begin to deepen across London itself." Stage directions calling for a character to enter "with a torch" or "with a light" tend to come in a play's final act. (There is also some evidence that the Globe was constructed in alignment with the position of the rising sun on the summer solstice.) Of course, one might misread a signal: In *Romeo and Juliet*, the two lovers famously quibble over the signs of the coming dawn. A bird cries—but was it the lark, or the nightingale? "Night's candles are burnt out," Romeo declares, "and jocund day / Stands tiptoe on the misty mountain tops." Juliet has heard and seen the same signals, but her wishful thinking interprets them quite differently: "Yon light is not day-light, I know it, I: / It is some meteor that the sun exhales." (The physics of meteors was not yet understood; a common guess was that they were vapors "exhaled" by the earth under the sun's influence.) Eventually, Romeo gives in; if Juliet says it is night, so be it:

> I'll say yon grey is not the morning's eye,
> 'Tis but the pale reflex of Cynthia's brow.
> Nor that is not the lark whose notes do beat
> The vaulty heaven so high above our heads.
>
> (3.5.19–22)

The only tricky part for a modern reader is perhaps the reference to "Cynthia"; in a good scholarly edition, a footnote will explain that Cynthia was a name for the moon goddess in Greek mythology. As Romeo notes, a cloud reflecting the light of the moon could indeed be mistaken for the coming dawn.

The rising sun intrudes on the young lovers in *Romeo and Juliet*; it intrudes, too, on the conspirators in *Julius Caesar.* They gather for a nighttime meeting in Brutus's garden to plot their next move—but take time out of their scheming to argue about where, exactly, the sun will rise:

> DECIUS
>
> Here lies the east. Doth not the day break here?
>
> CASKA
>
> No.
>
> CINNA
>
> O pardon, sir, it doth, and yon grey lines
> That fret the clouds are messengers of day.
>
> CASKA
>
> You shall confess that you are both deceived.
> Here, as I point my sword, the sun arises,
> Which is a great way growing on the south,
> Weighing the youthful season of the year.
> Some two months hence, up higher toward the north
> He first presents his fire, and the high east
> Stands as the Capitol, directly here.
>
> (2.1.100–110)

There are murders to plan, ambitions to thwart, and nations to rebuild—but first, *let's argue about the position on the horizon where the sun will rise!* Nothing will happen, it seems, until this point can be agreed upon. Interestingly, Shakespeare gets it *almost* right. We know that it's mid-March (the "ides" and all that), which means it's almost the equinox—and therefore the sun will rise almost due east, not "a great way growing on the south," as Caska proclaims. But he is right that, as the weeks pass, the sun's position as it rises will advance to the north. (But the *time* is a problem:

Later in the scene we are told that it's three o'clock—too soon for the sunrise, or even the dawn's early light, at any time of year.)*

The moon's appearance and movement is every bit as familiar as that of the sun: It, too, rises in the east and sets in the west, though its appearance changes dramatically as it goes through its familiar phases, waxing and waning in its monthly cycle. For a few days each month it disappears completely, only to reappear as a thin crescent in the western sky, where it shines for a short time after sunset. About a week later it reaches "first quarter," shining like a capital "D" in the southern sky. Another week passes, and it becomes a majestic full moon, rising opposite the setting sun and shining all night long. The lunar cycle repeats as dependably as its solar counterpart.

And then there were the stars—"these blessed candles of the night," as Bassanio poetically describes them in *The Merchant of Venice* (5.1.219). They move as well—not haphazardly, but in unison, also from east to west. If you face north, they appear to revolve in a counterclockwise direction, as if attached to a giant pinwheel. Only the north star, or "pole star," seems to remain fixed at the center of this pinwheel. (Known as "Polaris" since the seventeenth century, the north star happens to lie close to the north celestial pole, the imaginary spot that the Earth's axis points toward.) This basic astronomical fact was, of course, well known to Shakespeare. In *Julius Caesar*, the general compares himself to the pole star: ". . . I am constant as the northern star, / Of whose true-fixed and resting quality / There is no fellow in the firmament" (3.1.60–62). Because the other stars move around the north star in a smooth circle and at a steady rate, one can use the sky itself as a clock. Telling time by the stars is a straightforward task for Shakespeare's characters, as it must have been for his audience. In *Henry IV, Part 1*, a farmer tracks the time by noting the position of the Big Dipper, known in Britain today as "the Plough" but in Elizabethan times as "Charles's Wain," that is, "Charles's Wagon": "Charles's Wain is over the new chimney, and yet our horse not packed" (2.1.2–3).

Although the distance to the stars was unknown, it was convenient to

* As you might imagine, this seemingly light-hearted interlude has been subject to much scholarly analysis. Regarding the too-early sunrise, and its not-quite-right location, Arthur Humphreys (in the Oxford edition, p. 135) urges the reader not to worry about such "minor inconsistencies" which, after all, "pass unremarked on the stage." The main function of the scene, says Humphreys, is to relieve tension; it also "creates the local atmosphere, marks the significant progress of the hours, and fixes attention on the Capitol."

imagine them lying at some fixed distance from the Earth, attached to the inner surface of a vast, transparent sphere. The sphere turned about the Earth, carrying the stars with it; one lived at the center of this arrangement, watching the heavens' endless procession.

The stars also display a second kind of motion. Along with the daily rising and setting, the entire pinwheel seems to shift slightly from night to night. As the weeks pass, the shift becomes more noticeable. Consider Orion, the mighty hunter. In autumn, it rises about midnight. By Christmas, however, it rises much earlier, around the time of sunset. By the following autumn, Orion once again rises at midnight. This cycle, like that of the seasons, lasts one year. These motions are straightforward and predictable. A shepherd would have known which constellations were visible in which season, and in which direction one would have to gaze.

THE WANDERERS

But there were certain objects in the night sky whose behavior wasn't quite so simple. From ancient times, skywatchers noticed that there were a handful of starlike objects that did not quite move in unison with the other stars; they changed their position against the constellations from night to night and from season to season. Today we call them planets; the word derives from the Greek term for "wanderer." Five of these wandering stars were known in ancient times—Mercury, Venus, Mars, Jupiter, and Saturn.

Although the planets wandered, they did not run amok: One could always depend on finding them within a narrow band that circles the night sky, the belt defined by the twelve constellations of the zodiac. In this respect, they are like the sun and moon, which also keep within the zodiac, and so one sometimes spoke of seven (rather than five) wandering bodies. But there were some intriguing differences in the paths that these wandering stars seemed to take: Mercury, looking like a dim reddish star, moved swiftly—only the moon seemed to move faster—and yet it always appeared close to the sun. Brilliant white Venus moved a little slower, and strayed a little farther from the sun, but not too far (neither planet was ever seen *opposite* the sun). Mars, with its distinctive reddish color, moved more slowly than the sun; so, too, did Jupiter and Saturn, both creamy white in color. Their paths took them across the entire sky, so that sometimes they were near the sun, sometimes opposite the sun. Of these, Saturn was the slowest of all; it took weeks before its motion against the background stars was perceptible, and required almost thirty years to complete a full circle relative to the background stars.

To the title character in Christopher Marlowe's *Doctor Faustus*, such motions were elementary. The doctor distinguishes "the double motion of the planets"—referring to their daily rising and setting, and also to their more complicated motion against the stars of the zodiac. Saturn's motion, he says, is completed "in thirty years, Jupiter in twelve, Mars in four, the sun, Venus, and Mercury in a year, the moon in twenty-eight days. Tush, these are freshmen's suppositions" (7.51–56). Actually, the period for Mars is closer to two years than four, but close enough: For Marlowe, and for his learned doctor, this basic comprehension of the heavens—knowing which objects were visible, in which part of the sky, and for how long—was everyday knowledge.

The planets, however, were more than just points of light in the night sky: They were also associated with gods. Each had its own powers, its own domain of influence. For both the Greeks and the Romans, Venus was the goddess of love; Mars was the god of war. Saturn was a god of agriculture and of time, while Mercury was a kind of messenger, a god of travel—which makes sense, given Saturn's plodding pace and Mercury's swiftness. Jupiter, often the brightest of the planets, was the king of the gods.*

The movement of the planets showed many regularities—but also some downright peculiar behavior. From night to night, the planets *usually* edged a little bit to the east; as the weeks passed, this was easily observed. Eventually, they completed a full circle against the backdrop of the stars. But for several weeks or months each year they would reverse their direction, moving westward from night to night, before resuming their usual eastward motion. Astronomers refer to this backtracking as "retrograde" motion, in contrast to the more usual "direct" motion. Again, these were familiar terms in Elizabethan times—as much for their use in astrology as in astronomy. In *All's Well That Ends Well*, Helena plays with this idea, poking fun at Parolles's skills on the battlefield:

---

* Venus usually outshines Jupiter; however, Venus can only be visible for, at most, a few hours after sunset or a few hours before sunrise. Jupiter, depending on its position in its orbit, can shine at any time of night.

HELENA

Monsieur Parolles, you were born under a charitable star.

PAROLLES

Under Mars, I.

HELENA

I especially think under Mars.

PAROLLES

Why under Mars?

HELENA

The wars have so kept you under that you must needs be born under Mars.

PAROLLES

When he was predominant.

HELENA

When he was retrograde, I think rather.

PAROLLES

Why think you so?

HELENA

You go so much backward when you fight.

(I.I.190–200)

Mars, aside from being the god of war, was also the most perplexing of the planets. The magnitude of its retrograde movement was greater than that of the other planets, making it the most readily visible example of backward motion in the heavens and, at the same time, the object whose movement was most urgently in need of explanation. As the French king points out early in *Henry VI, Part 1*, "Mars his true moving, even as in the heavens / So in the earth, to this day is not known" (1.2.191–92). As familiar as retrograde motion was, it proved baffling to astronomers, who struggled to tweak their models of the heavens to explain this odd feature of planetary motion.

THE SPHERES ABOVE

Imagining the sun, moon, and planets affixed to a giant, transparent sphere was a promising start, but it was not quite enough: At the very least, each planet had to have *its own* sphere, so that it could move independently of the other wanderers; these nested spheres—think of the layers of an onion—could then rotate at different speeds, with the Earth at rest in the center. The innermost sphere carried the moon, which moved a significant distance from night to night; next was Mercury, then Venus. After that came the sun itself. Beyond the sun lay the spheres of Mars, Jupiter, and Saturn; and finally the sphere containing the stars themselves, sometimes called the "firmament" (as Prince Hamlet referred to it in the passage quoted at the start of the chapter). And so one would not speak of a single giant sphere, but of a system of spheres—a system like that imagined in figure 1.1. Perhaps the spheres were composed of some kind of crystal; they needed to be rigid and yet perfectly transparent.

Although this model had evolved significantly by the sixteenth century, the ancient picture just described was more or less how ordinary people imagined the universe in the time of Shakespeare's youth. When Hamlet, after seeing his father's ghost, says the vision threatens to make his eyes "like stars, start from their spheres" (1.5.22), his audience would have had no trouble catching the metaphor. Similar turns of phrase can be found throughout the canon. In *A Midsummer Night's Dream*, Oberon describes a mermaid's song—music so lovely that "certain stars shot madly from their spheres" in order to hear it better (2.1.153). And if you've ever seen a Western in which one character says to another that "this town isn't big enough for the both of us," remember that Shakespeare was there first—though entire planets, rather than towns, were at issue. In *Henry IV, Part 1*, Prince Henry says to his archenemy, Harry Percy, "Two stars keep not their motion in one sphere, / Nor can one England brook a double reign / Of Harry Percy and the Prince of Wales" (5.4.64–66).

Fig. 1.1 In ancient Greece, the universe was earth-centered, with a system of concentric spheres carrying the stars, sun, and planets— including the sun and moon—across the sky. Only in the terrestrial realm do we find the four elements: earth, water, air, and fire. (In this fanciful 1599 engraving, Atlas carries the whole affair on his back.) This ancient model—with various tweaks—remained the dominant view for nearly 2,000 years. The Granger Collection, New York

What we've described here is, roughly, how ancient civilizations across the Near East imagined the heavens for thousands of years: The cosmos was pictured as an intricate system of nested, transparent spheres, carrying the sun, moon, planets, and stars across the sky in their daily and yearly cycles. It was also the way the great thinker Aristotle imagined the universe in the fourth century B.C. By Aristotle's time, it was accepted that the Earth itself was spherical; but it was thought to be immobile, fixed at the center of the universe, surrounded by this intricate array of translucent spheres, carrying the five planets—or seven, if we count the sun and moon among these "wanderers."

Aristotle also noticed a profound difference between what happened down here on the Earth, and what transpired in the heavens. The terrestrial realm—the "sublunar" world—was marked by continuous change; it was subject to corruption and decay. This stood in stark contrast to the perfection of the sun, moon, and planets, whose movements were as predictable and regular as a well-oiled machine. (The metaphor is less of an anachronism if we think of the cosmic machine as a divine creation rather than something constructed in a blacksmith's workshop, but either way we have an artifact bearing witness to the talent of its creator.) Here on Earth, everything was thought to be composed of the four elements—earth, air, fire, and water—described by the Greeks even before Aristotle. All that we see around us, from mice to mountains, can be thought of as a particular arrangement of these elements, as they move and combine in different forms. As Christopher Marlowe's Tamburlaine observes, "Nature that framed us of four elements, / Warring within our breasts for regiment . . ." (*Tamburlaine the Great, Part 1* 2.6.58–59). Even Sir Toby Belch, in *Twelfth Night*, asks: "Does not our life consist of the four elements?" (2.3.9).

In a world of hierarchies, it is not surprising that the elements themselves were ranked according to their presumed nobility. Fire was the most worthy; next was air. Water, being heavier, filled the oceans below. Earth, the basest of the elements, lay at the bottom. However quaint such a system may seem today, it basically worked: When flames were observed to rise, it could be seen as an attempt to reach the heavenly spheres, their natural home; the fall of rain to the sea, or a thrown rock to the ground, could be similarly accounted for.

These elements, confined to the sublunar world, were constantly in flux. But the "superlunar" world—the heavens—showed no such signs of change. To Aristotle, this heavenly realm, with its various spheres, was composed of a kind of quintessence—literally a "fifth element." Sometimes an additional sphere was added beyond the sphere of the fixed stars; this was the *primum mobile* ("that which moves first"), which was believed to set the whole system in motion.

In considering the motion of the heavenly bodies, Aristotle was influenced by Plato, who had in turn been influenced by the followers of Pythagoras, an early Greek thinker who saw the universe as inherently mathematical, its creator a kind of divine geometer. Among the many shapes pondered by the geometers, one was seen as more perfect than any other. This was the circle (or, in three dimensions, the sphere). As a medieval astronomer named Sacrobosco noted, there were three reasons why the heavens must be spherical: First, a sphere has no beginning and no end, and is therefore "eternal." Second, a sphere encloses a larger volume than any other shape having the same surface area. And third, any other shape would seem to leave "unused" space. The first of these reasons, in particular, permeated Greek mathematical thought. And so Aristotle imagined the planets as moving in perfect circles. This was a little bit tricky, since it was well known that the planets do, in fact, display irregularities in their motion, as seen from Earth. But surely, he reasoned, this was an illusion: Aristotle and his followers were confident that, from the correct perspective, all heavenly motion was indeed perfectly uniform and perfectly circular.

CIRCLES UPON CIRCLES

This system of nested crystalline spheres was immensely appealing—but anyone who followed the movements of the planets closely came to realize that it was not quite enough; the motion of the planets was too complex. For example, it was still unclear how the circular movement of those spheres could account for retrograde motion. The best guess was that each planet required two such spheres: a large one, to account for the basic eastward motion; and a smaller one, to account for the "loops" that the planet traces out when moving in retrograde. These smaller circles were known as *epicycles* (from a Greek term meaning "a cycle displaced from the center").

The most detailed account of such a system comes from the Greek mathematician and astronomer Claudius Ptolemy (ca. A.D. 90–168).*

* Ptolemy lived in Alexandria, in Egypt, which at that time was a province of the Roman Empire. Ptolemy wrote in Greek.

Ptolemy's system was intricate and sophisticated, employing geometrical contrivances that today sound unfamiliar to anyone except for historians of astronomy. We will not wade into Ptolemaic astronomy any more than we have to, but it is worth looking at its main elements. As in Aristotle's system, the Earth lies at the center of the universe. Each planet, as mentioned, has two motions: it moves in a small epicycle, with the center of the epicycle revolving around the Earth in a larger circle called a *deferent*. The deferent, meanwhile, is not centered precisely on the Earth, but on a nearby point called the *eccentric*. One more aspect of Ptolemy's astronomy merits our attention: Ptolemy had imagined not only that the heavenly bodies moved in perfect circles, but that they did so at a constant speed. This was problematic, because, as measured from Earth, the speed would not be constant in the system as described. But the speed *would* be constant relative to an imaginary point on the "other side" of the eccentric, displaced from the center by the same amount as the Earth. That imaginary point was called an *equant*.

If you're thinking that all of this is frighteningly complex, you're not alone. In the thirteenth century, the king of León, Alfonso X, commissioned a new set of astronomical tables to be drawn up; the calculations were carried out using the Ptolemaic system, which still reigned supreme in celestial matters. When one of his aides explained the system to him, the king is said to have remarked, "If the Lord Almighty had consulted me before embarking upon the creation, I should have recommended something simpler."

Three centuries later, this apparent complexity would trouble the poet John Milton. In *Paradise Lost*, Adam inquires about the structure of the heavens; the angel Raphael replies that God must surely be laughing at man's desperate efforts to explain the cosmos:

> . . . when they come to model heav'n
> And calculate the stars, how they will wield
> The mighty frame, how build, unbuild, contrive
> To save appearances, how gird the Sphere
> With centric and eccentric scibbled o'er,
> Cycle and epicycle, Orb in Orb . . .
>
> (8.79–84)

Remarkably, as complicated as the Ptolemaic system sounds, it worked: It allowed astronomers (and astrologers) to predict the positions of the planets with reasonable accuracy, allowing them to "save the appearances" of the wandering lights in the night sky. (That phrase, derived from a Greek expression, had long been in common use when Milton borrowed it for use in his poem.) And it worked in spite of a fairly serious glitch. It's not just that Ptolemy had placed the Earth, rather than the sun, at the center; that, by itself, would not affect the predicted positions, as the two schemes are mathematically equivalent. But his estimates of the sizes of the spheres were all quite far off. They were based on the "best guess" at the distance between the Earth and the sun—which, it turns out, Ptolemy had underestimated by a factor of twenty; this, in turn, threw off all of the other estimates of distance.

Ptolemy's vision was laid out in his hefty book, which, thankfully, is no longer known by its Greek title (translated roughly as "Mathematical Systematic Treatise") but by the name it took on centuries later, the *Almagest*—derived from an Arabic phrase meaning "the majestic" or "the great." The *Almagest* is divided into thirteen sections, or "books," each crammed full of diagrams, charts, and equations. (And remember, for the first thirteen hundred years of its existence, it could be copied only by hand.) Far more thorough and authoritative than any previous astronomical text, it would dominate cosmological thinking and teaching for the next fourteen centuries.

In medieval Europe, Christian theology adopted some aspects (though not all) of the ancient Greek description of the cosmos. What endured, both in Catholic nations and in the newly Protestant lands, was a kind of "Christianized Aristotelianism." It was a worldview that embraced the structure of the heavens and the Earth as described by Aristotle, and the basic elements of the various celestial movements as described by Ptolemy, with all of those deferents, eccentrics, and epicycles. It was an ingenious synthesis—and a remarkably cohesive picture of the world.

MICROCOSM AND MACROCOSM

What this discussion of the motion of the heavenly spheres leaves out is just how *intimate* the medieval universe was—at least, the version of the universe that emerged once Christianity and Greek philosophy had completed their merger. It was not a complete unification, however. Some of the key ideas of Greek thought—those seen to be compatible with the Christian faith—were embraced by the early Church; others were discarded. (For example, the idea of the primum mobile was absorbed rather easily; for

Christians, this realm could simply be associated with God himself, who could act as a "first cause," giving the spheres their initial motion.)

The picture that emerged was one of profound unity: A sublime order was seen to underlie the arrangement of the natural world, from the lowest rocks to the loftiest stars—with man occupying a unique position in the middle, noble in reason but frail in body. Everything, and every person, had its place in this grand cosmic hierarchy, sometimes called the "Great Chain of Being." Kings ruled over men; men ruled over their households. It was an interconnected web, with a just and omnipotent God supervising from above. With this hierarchy in mind, we can see why cosmology had a political dimension—or, if you prefer, why politics had a cosmic dimension. The king was next to God, and God ruled the heavens.

It was a small step to imagine a connection between the monarch and the heavens themselves—an idea illustrated rather vividly on the frontispiece of *Sphaera Civitatis* (*The Sphere of State*), a commentary on Aristotle's *Politics* by the writer John Case, published in 1588 (see figure 1.2). As a queen without an heir, Elizabeth could be forgiven for fearing disorder above all. But the engraving goes beyond simply equating the sovereign with divine order; it places her in the realm of the heavens themselves. The diagram is solidly Ptolemaic, with the Earth lying as the center. But as Jonathan Bate points out, it wouldn't be all that disruptive had it been presented as Copernican, with the sun, symbolizing the monarch, placed at the center; either way, the queen "presides over the whole scheme [with] implacable authority." (Much later, in the second half of the seventeenth century, Louis XIV of France would push the metaphor as far as one might reasonably expect to, declaring himself the "sun king.") Royalty need not be compared to the sun; a star might suffice. In one of Ben Jonson's masques, a prince declares,

> I, thy Arthur, am
> Translated to a star; and of that frame
> Or constellation that was called of me
> So long before, as showing what I should be,
> Arcturus, once thy king, and now thy star.

Elizabeth couldn't be compared to Arcturus, since she was a woman; instead, as Alastair Fowler points out, she was often compared to Astraea, the "star maiden" of Greek mythology. Astraea was associated

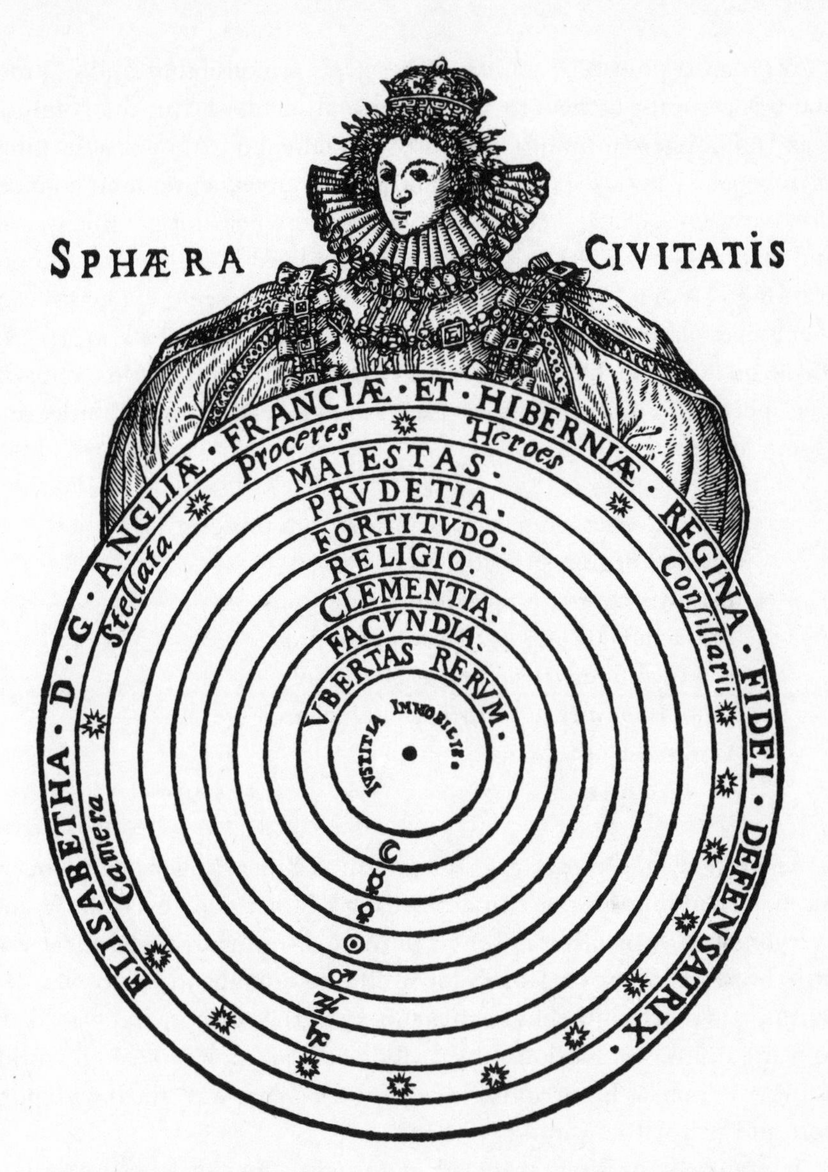

Fig. 1.2 With the monarch imagined to have divine properties, it was reasonable to depict her in the heavens, preserving the very order of the cosmos—an idea illustrated vividly in the frontispiece of a 1588 book of political commentary. (Note that Queen Elizabeth presides over an Aristotelian, earth-centered universe.) The Granger Collection, New York

---

with justice as well as innocence and purity. Born a human, she was repulsed by the wickedness of mankind and ascended to the sky to become the constellation Virgo (the Virgin)—rather appropriate for the Virgin Queen.

With great power, of course, comes great responsibility, and so kings and princes must be held to a higher moral standard than the common man. As Imogen notes in *Cymbeline*, ". . . falsehood / Is worse in kings than beggars" (3.6.13–14). Few would have doubted a profound connection between social and celestial order, the inherent unity of microcosm and macrocosm—a way of thinking encapsulated in Calpurnia's famous warning: "When beggars die, there are no comets seen; / The heavens themselves blaze forth the death of princes" (*Julius Caesar* 2.2.30–31).* In *Troilus and Cressida*, Ulysses takes the analogy much further. In a remarkable speech, he describes an intricate parallel between social order and cosmic order:

> The heavens themselves, the planets and this centre
> Observe degree, priority and place,
> Insisture, course, proportion, season, form,
> Office and custom, in all the line of order.
> And therefore is the glorious planet Sol
> In noble eminence enthroned and sphered
> Amidst the other . . .
>
> (1.3.85–91)

As we've seen, this passage can be taken as either Ptolemaic or Copernican, depending on one's interpretation.† Either way, everything and everyone had their place and their purpose. Not surprisingly, there was little hope for improving one's lot in life; to attempt to do so was like putting a wrench in the divine machinery of the cosmos, and was likely to bring divine retribution. It was, above all, an interconnected world; its every corner, as historian Lawrence Principe says, was "filled with purpose and rich with meaning."

The ancient writings were taken seriously. By Shakespeare's time, Plato and Aristotle had been dead for nineteen centuries, yet were deemed

---

* Such comparisons have more or less disappeared today in the West, but note that in North Korea, the birth of Kim Jong-il was said to have been heralded by the appearance of a new star in the heavens.

† Kirstin Olsen writes, "Ulysses' speech is sometimes taken to be a profession of Shakespeare's affinity for the Copernican system . . . But this passage is ambiguous at best; 'this centre' could just as easily be the earth as the sun, and the sun is described as 'Amidst the other' planets, which could mean at the center of all their orbits or in its traditional Aristotelian/Ptolemaic position between Venus and Mars." (Kirstin Olsen, *All Things Shakespeare: An Encyclopedia of Shakespeare's World*, vol. 1, pp. 69–70)

more authoritative than any living thinker. For the natural sciences in particular, Aristotle was *the* authority. But it was Plato who spoke of the link between man and the cosmos—between microcosm and macrocosm ("little ordered world" and "large ordered world"). As we struggled to understand our lives here on Earth, we could look to the heavens for guidance: Their orderly structure was a model, a blueprint, for living a rational, meaningful life. Every branch of learning, from astrology to medicine, flowed from this simple assertion. We can see why natural philosophy, though roughly equivalent to what we now call "science," was broader in scope: It encompassed not only the observational sciences, but also theology and metaphysics. And we can understand why as late a figure as Sir Isaac Newton, in the latter part of the seventeenth century, was able to carry out scientific experiments one day, dabble in alchemy the next, and study obscure biblical passages the day after that.

To study nature was to study God's creation. That sentiment was commonplace in Renaissance Europe, but its most compact and eloquent expression is found in Psalm 19: "The heavens declare the glory of God; and the firmament sheweth his handywork." (On this point, Protestants and Catholics were in full agreement. As Calvin writes, "The skillful ordering of the universe is for us a sort of mirror in which we can contemplate God, who is otherwise invisible.") To see God's handiwork is one thing; to comprehend it is another. The Creator worked in mysterious ways, and no mere mortal could grasp his plan for humankind in its entirety—but one could glimpse a small part of it, perhaps, by studying God's creation. One could come to know God through either of the "two books"—the book of nature or the book of scripture. His Word or his Work.

By Shakespeare's day, the metaphor was ubiquitous: Nature was seen as a book that could be read by someone with the right training. We have some idea of the texts that the playwright perused, and he almost certainly had access to an encyclopedia written by a Frenchman named Pierre de la Primaudaye, who declares that we must consult both "books" in order to know God: "We must lay before our eyes two bookes which God hath given unto us to instruct us by, and lead us to the knowledge of himselfe, namely the booke of nature, and the booke of his world." Mind you, Shakespeare wasn't shy about projecting the metaphor back to ancient Rome. In *Antony and Cleopatra*, the soothsayer says of his abilities: "In nature's infinite book of secrecy / A little I can read" (1.2.10–11).

Two books, but a common purpose: to know the mind of God, and through God to understand the meaning and purpose of life. As a new era dawned, Lawrence Principe writes, the greatest of thinkers "looked out on a world of connections and a world full of purpose and meaning as well as of mystery, wonder, and promise."

FROM MEDIEVAL TO MODERN

A profound change in this way of seeing the world was on the horizon—though of course no one at the time would have recognized its first stirrings. The period that we now think of as the Scientific Revolution—roughly 1500 to 1700—was seen as nothing of the sort at that time. Moreover, the discoveries that we celebrate in science museums today probably had little impact on ordinary men and women at the time. As Peter Dear observes, it is "unclear how much difference the classic 'Scientific Revolution' of the sixteenth and seventeenth centuries made to ordinary people." The innovations that it brought "left most features of their everyday lives unchanged." As Steven Shapin points out, the term "Scientific Revolution" saw widespread use beginning only in the late 1930s.* (It has been fashionable in recent years to quote the first sentence of Shapin's book *The Scientific Revolution*: "There was no such thing as the Scientific Revolution. And this is a book about it.") And yet it was, undeniably, a time of unprecedented inquiry, investigation, and discovery.

Whatever we may call this period, something rather important happened, even if it was more gradual, and constituted less of a break with past traditions, than the name "revolution" might suggest. But it did not come out of the blue; rather, it was built on a foundation established in the latter part of the Middle Ages. And it did not happen everywhere at the same time; what we would now recognize as "modern" developments in medicine, engineering, and commerce, as well as in the visual arts and literature, could be seen in Italy many decades before they reached more remote parts of the Continent. It was, as Principe puts it, "a rich tapestry of interwoven ideas and currents, a noisy marketplace of competing systems and concepts, a busy laboratory of experimentation in all areas of thought and practice." The printing press, a fifteenth-century invention, fostered the spread of ideas at a new, accelerated pace,

---

* Shapin also provides a sobering reminder of the Eurocentric nature of much historical investigation: ". . . the overwhelming majority of seventeenth-century people did not live in Europe, did not know that they lived in 'the seventeenth century,' and were not aware that a Scientific Revolution was happening." (Steven Shapin, *The Scientific Revolution*, p. 8)

while voyages of discovery were opening up new worlds for colonization and exploitation. And the rediscovery of classical texts, via Arabic translations, triggered a new wave of learning across Europe. Those works included the writings of Aristotle and Ptolemy, which we've touched on, as well as the geometry of Euclid, the medical writings of Galen, and much more.

This wave of learning was closely linked to the activities of the Roman Catholic Church. The best medieval schools had been those associated with the monasteries and the great cathedrals. By the late Middle Ages these had also become centers of what we would now call science (as mentioned, at that time such pursuits would have fallen under the umbrella of natural philosophy). There were also the universities, the earliest having been founded around 1200; these, too, functioned largely as religious institutions. The highest degree offered was in theology—though to obtain it, the student also had to master mathematics, logic, and natural philosophy.

This close connection between science and faith may seem strange to the modern reader, living at a time when Western society, and Western science in particular, has become a secular endeavor—and with best-selling authors declaring religion to be antithetical to science and an obstacle to progress. The evolving relationship between science and religion is a large and complex subject, but one thing is clear: Whatever that relationship may be like today, it was very different four hundred years ago. There was no "conflict" between science and religion for the simple reason that no such division between the two pursuits existed. For one thing, religion was simply part of the fabric of society; all of the key figures of the Scientific Revolution were men of faith of one kind or another. (These days Francis Bacon is hailed as one of the founders of modern science—but a first-time reader of his *Advancement of Learning* [1605] might be surprised to find that he expounds at some length on God and the Bible.) Moreover, the very thing being investigated—the heavens—was seen as indisputable evidence of God's craftsmanship. Robert Recorde, writing in the middle of the sixteenth century, is typical in his enthusiasm. His *Castle of Knowledge* is at once a textbook on astronomy and an expression of religious devotion. He begins:

> Oh worthy temple of Gods magnificence. Oh throne of glory and seate of the Lord: thy substance most pure what tongue

> can describe? Thy beauty with stares so garnished and glittering . . . Oh marvellous Maker, oh God of good governance: thy works are all wonderous, thy cunning unknowen: yet seeds of all knowledge in that booke are sowen. . . .

To study astronomy was to study "the book of nature"; it could not fail to lead one to a greater understanding of God. As Recorde assured his readers, "there was never any good astronomer that denied the majestie and providence of God." As historian Paul Kocher puts it, early modern science "was more often cited as proving God's existence than disproving it, under the staple argument that study of the marvellous structure of the universe led man's mind to see that it must have had a Creator." And as Principe writes, the study of nature was seen as "an inherently religious activity." The twenty-first-century notion that scientific investigation requires checking one's faith at the door is a much more modern idea, one that would have been incomprehensible to the great thinkers of this time.

But the link between science and faith, as Principe stresses, is deeper than this. For the thinkers of early modern Europe, he writes, "the doctrines of Christianity were not personal choices. They had the status of natural or historical facts":

> Never was theology demoted to the status of "personal belief"; it constituted, like science today, both a body of agreed-upon facts and a continuing search for truths about existence. . . . Thus theological ideas played a major part in scientific study and speculation—not as external "influences," but rather as serious and integral parts of the world the natural philosopher was studying.

This is a useful reminder, because it is commonplace today to pick up a weekly newsmagazine and read of religious thinkers being "influenced" by science or, less frequently, of a scientific idea having been influenced by religious thought (the big-bang-model theory of modern cosmology comes to mind). The presumption is that science is a kind of monolithic, ever-expanding pool of wisdom, with religion (and philosophy) merely coming along for the ride. Whatever we think of this view today, it would have been meaningless in early modern Europe. The so-called war between science and religion was largely an invention of the late nineteenth century, and, while it may have some relevance today, Principe is correct to remind us that it "does not portray the historical situation."*

Perhaps the clearest example of the interdependence of science and faith during the late Middle Ages, and into the Renaissance, comes from the field we have been discussing: astronomy. Religious leaders relied on the work of astronomers to determine the date of Easter, the holiest day in the Christian calendar. The date of Easter was calculated by means of a complex algorithm that depended on the date of the vernal equinox—the day when the hours of daylight exactly equal the hours of darkness, marking the first day of spring. But determining the date of the equinox is itself a difficult problem, and can only be worked out through careful observations of the heavens. To aid in this work, dozens of churches and cathedrals across Christian Europe also served as observatories; many were equipped with strategically placed apertures in walls or ceilings that allowed a beam of sunlight to strike a north-south "meridian line" on the floor. The resulting measurements helped establish the dates of the solstices and equinoxes, on which the Easter calculations depended. And so the Roman Catholic Church was, for many centuries, the largest sponsor of astronomical research in Europe.†

It is hardly a surprise, then, that many of the natural philosophers of the early modern era were also churchmen of one kind or another. That is certainly the case with the pious Catholic cleric from the remote eastern edge of Europe who, early in the sixteenth century, decided that studying the heavens might be just as rewarding as studying the intricacies of ecclesiastical law. He would soon turn the universe on its head.

---

* Two books were particularly influential in initiating the "conflict" paradigm—John William Draper's *History of the Conflict between Religion and Science* (1874) and Andrew Dickson White's *A History of the Warfare of Science with Theology in Christendom* (1896). The relationship between science and faith has been endlessly scrutinized since the publication of Darwin's *Origin of Species* in 1859, and remains a fascinating and complex subject. For those interested in a historical perspective, Ronald L. Numbers's introduction to *Galileo Goes to Jail and Other Myths about Science and Religion* (2009) is a good starting point.

† For a more thorough look at the "problem of Easter," leading to the Gregorian calendar reform, see Chapter 2 of my earlier book *In Search of Time*.

2. *"He that is giddy thinks the world turns round . . ."*

NICOLAUS COPERNICUS, THE RELUCTANT REFORMER

We know precious little about the man who overturned a millennium and a half of astronomical thinking. The portrait that hangs in the town hall in Toruń, the city of his birth, shows Nicolaus Copernicus as a slim, hollow-cheeked man sporting a red jerkin and thick, wavy, jet-black hair. His dark, piercing eyes gaze not ahead but sharply to his left, as though someone or something out of the frame demanded his attention. It reveals little of his personality. We do know that he was born in 1473, the son of a wealthy merchant. The city of Toruń was, at that time, part of Royal Prussia (encompassing those Prussian districts which fell under the control of the king of Poland). His family, however, spoke German, and in his youth he went by the name Niklas Koppernigk.* Copernicus did much of his studying abroad, which explains his decision to adopt a Latinized name. He studied first in Cracow and later in Bologna, where he focused on canon law (the laws governing the rights and responsibilities of Church leaders). He also became a committed humanist.† After the death of his father, his education and travels were supported by his maternal uncle, a powerful bishop. Copernicus

* The question of whether Copernicus should be considered German or Polish is a nonstarter, as the concept of "nationality" as we know it today did not exist in the sixteenth century. He likely would have considered himself a "Prussian." (See Sobel, p. 5; Davies, p. 20.)

† The meaning of "humanist" has evolved over the centuries—today we think of "secular humanism," but in the sixteenth century it referred to an engagement with civic life, devotion to learning, and the quest to live a virtuous life; the art and literature of antiquity were held up as models of what could be achieved. It is from this usage that we refer to certain streams of higher education as "the humanities."

returned to his homeland in 1497, where his uncle appointed him as a canon (more than an administrator but less than a priest) at Frauenburg Cathedral (today Frombork Cathedral), in the province of Warmia on the Baltic Sea. Copernicus's ecclesiastical ambitions seem to have stopped short of full-fledged priesthood, and it is likely that he was never ordained. He would later return to Italy, enrolling as a medical student at Padua, returning to Frauenburg after his graduation, in 1510.

Of his personal life, however, we remain ignorant; the documentation is simply absent. Owen Gingerich, the astronomer and historian of science who probably knows Copernicus better than any scholar today, admits defeat on the question of Nicolaus's personality:

> What sort of person was Copernicus? Did he like puns? Did he ever play jokes on his classmates or his fellow canons? Did he enjoy music? . . . Did he ever have a girlfriend? Did he like children? Alas, these are unanswerable questions.

We do know that Copernicus became captivated by astronomy at an early age, though we can't say exactly when this fascination began. He was likely exposed to the Ptolemaic model during his time at Cracow. In Bologna, he assisted a well-known astronomer named Novara, observing an occultation of the star Aldebaran by the moon in the spring of 1497 (when he was twenty-four); three years later he observed a partial lunar eclipse from Rome.*

SEEKING HIDDEN CAUSES

By the time of his studies in Italy, Copernicus was likely familiar with the shortcomings of the Ptolemaic system. Once he settled in Frauenburg, where he would remain for the final four decades of his life, he devoted nearly all of his energy to studying the heavens and improving the description of how celestial bodies moved. He had an observing tower added to the wall surrounding the cathedral; visitors can see it to this day. With damp fog often rolling in from the mouth of the Vistula River, however, it was not an ideal location. ("The ancients had the advantage of a clearer sky," he admitted.) At first, he told only his closest confidants of his ideas, but to those who knew him, it was clear that he was up to

* In an occultation, the moon appears to pass in front of a star, causing the star to disappear for up to several hours.

something. As a colleague observed, "He discusses the swift course of the Moon and [the sun] as well as the stars together with the wandering planets . . . he knows how to seek out hidden causes of phenomena by the aid of wonderful principles." It is also worth noting that in an age when most people believed in astrology, Copernicus apparently did not.

As we've seen, the Ptolemaic system did not lack for accuracy; and its picture of a stationary Earth, with the heavens whirling around, was compatible with the commonsense view of the cosmos. So what drove Copernicus to challenge the established model of the heavens? There is probably more than one answer—but an important element is Copernicus's commitment to the Platonic ideal of circular motion, which he saw as the only conceivable motion that a heavenly body would execute; and his belief that a planet must move at a constant speed along these circles. Recall that in Ptolemy's system, the planets were seen to move at a constant speed only relative to an imaginary point, the equant. To Copernicus, this seemingly arbitrary invention was at odds with the spirit of Plato's vision; it was "neither sufficiently absolute nor sufficiently pleasing to the mind." Copernicus offered an alternative model: Perhaps the sun, not the Earth, lay at the center of the observed motions. As early as 1510, he realized that, with this simple switch, he could construct a system in which the planets moved with truly uniform, circular motion. This seems to have been his first concern; the new structure—*heliocentric* (sun-centered) rather than *geocentric* (Earth-centered)—was perhaps secondary. Yet once he had made the switch, he became enamored with it. "All the spheres revolve about the sun as their mid-point," he would write, "and therefore the sun is the center of the universe." There was no going back.*

As it happens, Copernicus was not the first to propose a heliocentric model; it had been put forward by Aristarchus of Samos (ca. 310–230 B.C.) and by a handful of other ancient Greek astronomers. However, no one had worked out the details of the theory and the idea seems to have been abandoned. (Indeed, there is no evidence that Copernicus was aware of Aristarchus's writings on heliocentrism, which had not yet appeared in Latin translation.) And so the idea began its revival when Copernicus penned a short manuscript outlining the "new" theory. This text, from about 1510, is known as the *Commentariolus* (*The Little Commentary*).

---

* For an interesting analysis of why it took fourteen hundred years for anyone (beyond a handful of ancient Greek thinkers) to take this simple step, see Margolis, pp. 91–102.

This was the first ripple in the slow but inevitable spread of Copernicus's theory. Nearly half a century would pass before anyone in England took note; this would come with the publication of Robert Recorde's *The Castle of Knowledge* in 1556. Shakespeare, born just eight years later, entered a world in which the heliocentric model was still young and tentative; as we will see, however, it became less tentative with each passing decade. Yet its first mention was more like a whisper: The *Commentariolus* was not published in Copernicus's lifetime, and only a handful of manuscript copies are believed to have circulated. In fact, if word of the new theory hadn't reached a young German astronomer named Georg Joachim Rheticus, history might have unfolded quite differently. Rheticus was so taken by the heliocentric model that he set off to Frauenburg to meet its author in person. He would become Copernicus's sole pupil.

Copernicus, meanwhile, was becoming more and more convinced that Ptolemy's system was too inelegant to represent the true structure of the heavens. With its hodgepodge of epicycles, Ptolemy's depiction seemed downright ugly. The astronomers who supported it "have been like someone attempting a portrait by assembling hands, feet, head, and other parts from different sources," he wrote. "These several bits may be well painted, but they do not fit together to make a single body. Bearing no genuine relationship to each other, such components, joined together, would compose a monster, not a man."

Contrary to popular myth, Copernicus's model was not "simpler" than that of Ptolemy in any objective sense, as we shall see. And yet, a sun-centered system did manage to solve many of the problems that had challenged astronomers from the beginning. For starters, it presented a simple explanation for the observed retrograde motion the planets sometimes displayed. Consider the case of Mars: As the faster-moving Earth passes or "laps" the slower-moving red planet, it appears to temporarily reverse its direction, as seen against the background stars. (This instantly made the largest set of epicycles in the Ptolemaic system redundant.) Secondly, the new model gave a straightforward explanation for why Mercury and Venus always seem to lie close to the sun in the sky (namely: they *really are* close to the sun). It solved a third problem as well: The planets were seen to vary in brightness over a period of weeks and months. In the new system, the Earth was in motion; as it moved in its orbit, it was sometimes closer to a particular planet, and sometimes farther away. The changing brightness of the planets now made perfect sense. By simplifying these issues, the Copernican model could be seen as a simpler way of describing the motion of the planets; it required fewer hypotheses and fewer arbitrary assumptions. As Copernicus noted, he chose to "follow the wisdom of nature, which, as it takes very great care not to have produced anything superfluous or useless, often prefers to endow one thing with many effects."*

SETTING THE EARTH IN MOTION

But these advantages came at a price: Copernicus's system set the Earth in motion; it was now one of the planets, kin with Mercury, Venus, and the rest. "What appear to us as motions of the Sun arise not from its motion," he wrote, "but from the motion of the Earth and our sphere, with which we revolve about the Sun like any other planet." The model required one to think of the Earth as hurtling through space at enormous speeds—a shocking violation of common sense. No wonder Copernicus kept using the word "absurd" to describe his own theory.

Indeed, the notion of a speeding Earth had always been a key objection: Why don't we *feel* the Earth's motion? It is worth pausing to think about how our very conception of motion has evolved since Copernicus's day, as this may help us see why the idea of a moving Earth seemed so preposterous four hundred years ago. Today we get on a jetliner and experience (barring turbulence) a perfectly smooth ride; many of us have no trouble falling asleep at thirty thousand feet, as the plane whips along at six hundred miles an hour. We have a sense from movies like *Apollo 13*, and the *Star Trek* franchise, that traveling though space (in the absence of laser blasts and asteroid impacts) is even smoother—as indeed it would be. But traveling in Renaissance Europe was a good deal rougher. As John Gribbin reminds us:

> Remember that in the sixteenth century, motion meant riding on a galloping horse or in a carriage being pulled over rutted roads. The notion of smooth motion (even as smooth as a car on a motorway) must have been very difficult to grasp without any direct experience of it—as late as the nineteenth century there were serious concerns that travelling at the speed of a

---

* This idea is often associated with the medieval English monk William of Ockham (or Occam), and is known as "Ockham's razor." In his own words, "Entities are not to be multiplied beyond necessity." (quoted in I. Bernard Cohen, *Birth of a New Physics*, p. 127)

railway train, maybe as great as 15 miles per hour, might be damaging to human health.

A moving Earth seemed bizarre for many reasons. Think of a ball tossed into the air: If the Earth is rotating, shouldn't the ball land some distance from the spot from which it was thrown? Why aren't birds and clouds swept backward in the same manner? Indeed, Aristotle himself had ridiculed the idea of a moving Earth for just such reasons. (The same objections, incidentally, can be raised against the Earth's daily rotation, as well as its yearly journey around the sun.) Copernicus, however, had an answer: The Earth's atmosphere, and everything within it, must be carried along with it, so that no such motion is felt. (A complete explanation requires the concept of *inertia*, the tendency of moving objects to remain in motion and of stationary objects to remain at rest. Unfortunately, this idea was unknown at the time, making its appearance only with the work of Kepler and Galileo a half century later; it would eventually become a cornerstone of Newton's mechanics.)

And there were more objections: If the Earth really moved around the sun, the stars should appear to shift in position over the course of a year; in astronomical jargon, they should display *parallax*.* However, no such shift was seen in the positions of the stars. And if the Earth moves in a vast orbit, then it must be nearer to certain stars at certain times of the year; therefore, the stars should vary in brightness over the course of the seasons. The answer, Copernicus reasoned, is that the stars must be very far away compared with the size of our solar system.

This view of a large, possibly infinite universe was a radical departure from the prevailing medieval view. Suddenly, the stars were seen to lie at distances that defy the imagination. Yet for Copernicus, this new, larger universe—with its more coherent cosmological picture—was easier to swallow than Ptolemy's countless epicycles. "I think it is a lot easier to accept this [larger universe] than to drive ourselves to distraction multiplying spheres almost ad infinitum," he wrote, "as has been the compulsion of those who would detain the Earth in the center of the universe."

---

* You can see this effect without leaving your chair: With your arm outstretched and your thumb raised, close one eye. Now switch to the other eye. Note how the background, behind your thumb, seems to shift. That's parallax.

HYPOTHESIS OR HERESY?

As we have seen, science was not "at war" with religion at this time, and in fact many scientists, Copernicus among them, were members of the clergy. So what objection, if any, did religious thinkers have to this new arrangement of the heavens? To begin with, we must clear up one of the lingering misconceptions about Copernicus's system: that it somehow "dethroned" the Earth from a privileged, central position in the cosmos.* In fact, the Earth's position, according to the medieval view, was at the "lowest" position relative to the lofty spheres of the planets and the even loftier sphere of the fixed stars. At the center one would find hell—not a particularly privileged spot. Indeed, one could argue that Copernicus ennobled the Earth, lifting it up into the heavens with the other planets—closer to God, one might imagine. Having said that, the fact that Copernicus forced us to think of the Earth as a body in motion, and indeed one of many such bodies, must have been disconcerting. "Central" or not, the Earth had at least been unique, and static. As historian I. Bernard Cohen puts it, "The Aristotelian uniqueness of the earth, based on its supposedly fixed position, gave people a sense of pride that could hardly arise from being on a rather small planet (compared to Jupiter or Saturn) in a rather insignificant location (position 3 out of 7 successive planetary orbits)."

What Copernicanism threatened was Aristotelian physics rather than Christian dogma—but, as we've seen, these two ideologies had become intertwined by the sixteenth century. If the Roman Catholic Church had allied itself with some other way of thinking, history might have taken quite a different route. As Francis Johnson notes, "Had the Christian theologians not previously committed themselves to Aristotelian science, they would have found no great difficulty in reconciling passages in the Scriptures with the ideas of Copernicus." At first, the Church showed little sign of any discomfort with the heliocentric model. It was seen primarily as a mathematical tool, one that held the promise, perhaps, of improved astronomical predictions. When a Vatican council convened in 1515 to consider reforming the Julian calendar currently in use, they wrote to Copernicus to ask for his opinion. Indeed, Pope Clement VII's personal secretary lectured on the theory, with the pontiff and several cardinals in the audience.

---

* For a useful discussion, see Dennis Danielson's essay in *Galileo Goes to Jail and Other Myths about Science and Religion* (ed. Ronald L. Numbers, 2009).

And yet, the structure of the heavenly spheres—the *physical interpretation* of these complicated mathematical models—was another matter. This was something Church leaders were accustomed to dealing with, as they had spent centuries trying to reconcile Aristotelian physics with scripture, keeping only those views that were consistent with the Bible. A much-discussed biblical passage comes from the Book of Joshua, in which the Israelite leader commands the sun—not the Earth—to stand still in order to prolong the hours of daylight, allowing his army to prevail in battle:

> Then spake Joshua to the Lord in the day when the Lord delivered up the Amorites before the children of Israel, and he said in the sight of Israel, Sun, stand thou still upon Gibeon; and thou, Moon, in the valley of Ajalon.
>
> And the sun stood still, and the moon stayed, until the people had avenged themselves upon their enemies. . . . So the sun stood in the midst of heaven, and hasted not to go down about a whole day.
>
> (10:12–13)

It's not clear that the heliocentric model received a much warmer reception from Protestant thinkers; like their Catholic counterparts, they seemed more than willing to embrace the Copernican model as a mathematical theory, and to simply ignore whatever its claims to "reality" might be. The reformer Martin Luther is said to have offered a snide dismissal of the theory, chastising astronomers who wished to sound "clever" by proposing "to turn the whole of astronomy upside down." The truth, he declared, was set out clearly enough in the Bible: "I believe the Holy Scripture, for Joshua commanded the Sun to stand still and not the Earth." On another occasion Luther is said to have called Copernicus a "fool."*

When natural philosophy appeared to contradict scripture, there was little room for compromise. In 1270, the bishop of Paris issued a list of thirteen propositions, linked to the views of the more radical Aristotelians,

---

* These remarks—which may be little more than hearsay—are discussed in Dava Sobel's wonderful biography of Copernicus, *A More Perfect Heaven* (2011); see also Daniel Boorstin's *The Discoverers*, p. 302. Note that Luther (1483–1546) was an almost exact contemporary of Copernicus.

which were deemed false and heretical. Seven years later the list was expanded to 219 items. More than twenty of them refer to cosmology. Among them was a prohibition against claiming "that the world is eternal as to all the species contained in it; and that time is eternal, as are motion [and] matter. . . ."*

Perhaps, then, it is not so surprising that Copernicus, who had been quietly setting down a detailed exposition of his theory, was hesitant to publish. Only at Rheticus's urging did he finally allow his great work to see the light of day. Under Rheticus's supervision, the book was eventually printed in Nuremberg, as we've seen, with a copy of *De revolutionibus* reaching Copernicus as he lay on his deathbed.

"THE MARVELOUS SYMMETRY OF THE UNIVERSE"

Lengthy and highly mathematical, Copernicus's book was structured in a fashion parallel to Ptolemy's, with a series of sections ("books") outlining the main arguments. (On the title page, in Greek, one finds the Platonic motto: "Let no one who is ignorant of geometry enter here.") After that, however, came a surprise: Unknown to Copernicus, an anonymous preface had been added to the book. We now know it was penned by Andreas Osiander, a Lutheran clergyman who oversaw the final stages of the printing process after Rheticus was called away to an academic post at Leipzig. The preface served as a disclaimer, insisting that the heliocentric system was just a theoretical model rather than a true description of the cosmos; Osiander, it seems, was more nervous than Rheticus. The preface cautions the reader that "these hypotheses need not be true or even probable. On the contrary, if they provide a calculus consistent with the observations, that alone is enough." Indeed, Osiander says that astronomy isn't in the business of discovering "truth":

> So far as hypotheses are concerned, let no one expect anything certain from astronomy, which cannot furnish it, lest he accept as the truth ideas conceived for another purpose, and depart from this study a greater fool than when he entered it.

* However, historians caution against reading these condemnations as flowing from "the Church." They are more accurately seen as indicative of local conflicts; in this case, the condemnation was issued by a local bishop. (See Michael J. Shank's essay in *Galileo Goes to Jail.*)

Whatever the reader may have thought of this preface, they would soon come to Copernicus's own peace offering—a brief letter in which he dedicates the work to Pope Paul III. The letter begins:

> Holy Father, I can guess already that some people, as soon as they find out about this book I have written on the revolutions of the universal spheres, in which I ascribe a kind of motion to the earthly globe, will clamor to have me and my opinions shouted down.

Copernicus recognizes that some of these attacks may be religiously motivated. There will be "babblers," he fears, who "claim to be judges of astronomy although completely ignorant of the subject and, badly distorting some passage of Scripture to their purpose, will dare to find fault with my undertaking and censure it." He confidently states that such attacks are "unfounded."

Eventually he settles down to the business at hand. Having established the motion of the Earth, he summarizes this new heliocentric picture: "Truly indeed does the sun, as if seated upon a royal throne, govern his family of planets as they circle about him," he writes. "Thus we discover in this orderly arrangement the marvelous symmetry of the universe, and a firm harmonious connection between the motion and the size of the spheres. . . ." As for the sun, what better place for it than at the center of this wonderful arrangement? "For who would place this lamp of a very beautiful temple in another better place than this wherefrom it can illuminate everything at the same time?" Osiander's preface notwithstanding, Copernicus had been persuaded by the elegance and cohesive logic of the heliocentric model. The idea, if not the details, had convinced him that it was a true description of nature.

THE REVOLUTION THAT WASN'T?

In recent decades, scholars have taken a close look at the supposed "revolution" brought about by Copernicus's work, and found it to be less than revolutionary. To begin with, his model did not necessarily lead to more accurate predictions for the positions of the planets than Ptolemy's; indeed, from a purely mathematical perspective, they were virtually identical. (One may think of the Copernican system as roughly equivalent to the Ptolemaic system, but with the positions of the sun and Earth swapped.) In fact, the new model, in its simplest form, could not yield accurate planetary positions, because it contained a fatal error. Like Plato, Copernicus was committed to the idea that circles and spheres were the embodiment of perfection, and assumed that the planets moved with uniform speed along perfectly circular paths. It would be another seventy years before Johannes Kepler would correct the error, deducing their true shape (ellipses rather than circles).

Nor was the Copernican model "simpler" in any unambiguous way. In order to precisely match the observed motions of the sun, moon, and planets, the heliocentric system also needed eccentrics and epicycles, just as Ptolemy's did; it was very much an intricate system of circles on top of circles, just as its predecessor had been. (Though it did manage to do away with equant points, along with the largest of Ptolemy's epicycles, both of which are redundant in the Copernican model, and it managed to bring the total number of circles down from about eighty to thirty-four.) The complexity of the Copernican system may seem surprising, especially in light of the much-reproduced diagram that comes in book 1 of *De revolutionibus*—that famous drawing of neat and tidy concentric circles, seen on page 3. Historians believe that this was intended as a schematic, meant to ease the reader into the book; later on, Copernicus lays out the intricacies of his theory in all its mathematical and geometrical glory. The structural similarities between *De revolutionibus* and the *Almagest* make it hard to argue that the newer book is simpler than its predecessor; one might even view Copernicus's book more as a commentary on the earlier work (albeit a critical one) than as a "new" theory. "At the risk of verging upon the facetious," notes S. K. Heninger, "we might say that Copernicus did no more than a bit of tinkering with the existing world-view. . . . Copernicus himself thought in terms of simplifying the old, not in terms of introducing the new." And note that the title refers to the revolutions of "the heavenly spheres," not "the planets." As I. Bernard Cohen puts it, in *De revolutionibus* one finds "a kinship of geometrical methods and constructions" paralleling those of the *Almagest*, which "belies any simple claim that Copernicus's book is in any obvious sense a more modern or simpler work." Still, "simplicity" is not an easy word to define. One could argue that the Copernican model offers a more natural explanation of retrograde motion; while the epicycles are still there, they exist solely for the flexibility they provide.

Copernicus's book is an almost paradoxical mix of conservatism and innovation. Though he warned the pope of his "absurd" new theory, he also stressed, at every turn, its continuity with ancient thinking. Again and again, he references one or another ancient Greek philosopher; as Peter Dear puts it, *De revolutionibus* "was presented explicitly as a renovation of the ancient Greek astronomical tradition." And as Heninger notes, "only in retrospect does he assume the role of an intellectual radical."

Was the Copernican remodeling of the heavens a seminal event that changed the course of history—or an academic exercise that changed nothing? We can certainly view it *today* as an earth-shattering—or more literally, "earth-moving"—event. Three cheers for the power of hindsight. But how was it seen in Copernicus's time? How big a splash did his theory actually make? To I. Bernard Cohen, writing in the 1980s, *De revolutionibus* was so unrevolutionary as to be barely worthy of notice: "The idea that a Copernican revolution in science occurred goes counter to the evidence . . . and is an invention of later historians." He adds that the revolution that we usually associate with the name Copernicus is more properly attributed to the work of Kepler, more than half a century later. But philosopher Richard DeWitt, writing in the 2000s, assures us that, from the time of Copernicus's death to the end of the sixteenth century, "his theory was widely read, discussed, taught, and put to practical use." Perhaps both are right: Copernicus's book marked only the first phase in the revolution; but certain people definitely took notice. It was hardly, as Arthur Koestler once described it, "the book that nobody read."

A WORLD MADE FOR WHOM?

At least one aspect of the Copernican system was, in fact, revolutionary. His cosmos was truly vast: "So far as our senses can tell," he wrote, "the earth is related to the heavens as a point is to a body and something finite is to something infinite." That last word is tantalizing, though historians believe that Copernicus was unlikely to have imagined a literally infinite universe. Nonetheless, it was larger than anything the Western mind had had occasion to picture in the preceding centuries. One thirteenth-century astronomer had calculated that the most distant of the known planets, Saturn, lay some 73 million miles away—already "a staggering distance to the best-traveled medievals," as one historian puts it, with the fixed stars, presumably, lying just beyond. Interestingly, because he accepted the ancient Greek value for the distance to the sun, Copernicus's estimate of the distances to the planets actually shrank the solar system compared with the Ptolemaic model. The distances to the stars, however, increased tremendously. The absence of stellar parallax meant that the stars had to be *much* more distant than in the medieval view.* By one estimate, the Copernican universe was larger than its medieval predecessor by a factor of four hundred thousand. This was an age when it was taken for granted that the universe existed for our benefit. What, then, occupied all of this empty space between the planets and the stars? What could be its *purpose*?

Decades later, when Galileo was writing his treatise on the competing world systems, he was conscious of this (by now familiar) objection. His book was written in the form of a dialogue, and he has the character Simplicio, the defender of the traditional worldview, tackle the "empty space" problem:

> Now when we see this beautiful order among the planets, they being arranged around the earth at distances commensurate with their producing upon it their effects for our benefit, to what end would there then be interposed between the highest of their orbits (namely, Saturn's) and the stellar sphere, a vast space without anything in it, superfluous, and vain. For the use and convenience of whom?

Since the dawn of human thought, the universe was believed to have been made *for us*. The larger the universe became, the harder it became to sustain that belief. The wonders revealed via Galileo's telescope would compound the problem, but already in 1580—when Galileo and Shakespeare were in their teens—the French writer Montaigne had ridiculed the idea of a human-centered cosmos. He wondered how mankind had come to believe that the cosmos exists "for his convenience." Is it possible, he asked, "to imagine anything more laughable than that this pitiful, wretched creature—who is not even master of himself, but exposed to shocks on every side—should call himself Master and Emperor of a universe. . . ."

In Montaigne's skepticism, and in Galileo's hard-nosed reasoning, we see the dawn of a new way of thinking. We can also find this new perspective in the works of Shakespeare, as we will see. Removing our planet from the center of the universe, and setting it in motion, was the crucial first step. As Daniel Boorstin has put it, "Nothing could be more obvious than that the Earth is stable and unmoving, and that we are the

* Stellar parallax was eventually detected—but not until the 1830s.

center of the universe. Modern Western science takes its beginning from the denial of this commonsense axiom." Already in the closing decades of the sixteenth century, the writing was on the wall: Faith was not immediately threatened by the Copernican universe—but it would certainly have to adapt. As Paul Kocher writes, the stakes could not be higher:

> Was it still possible to believe that God made the world for man? Here lay the great question for Christianity as the geocentric gave place to the heliocentric universe. It was a highly complex question demanding no simple answer. Man's uniqueness, the quality of God's moral government of the universe for human good, the possibility of miracles, the authority of Scripture to teach truth about the physical world—these and many cognate issues all seemed at stake.

One of the defining characteristics of twentieth- and twenty-first-century science—one that would have been nearly unthinkable in early modern Europe—has been the idea that the universe was probably not made for our benefit after all; it simply *is*.* (Creationists, of course, reject such a view—but even secular liberals have trouble coming to terms with it.) One senses in Montaigne and Galileo—and, as we will see, in Shakespeare—the beginning of this profound change.

COPERNICUS'S UNIVERSE

Incidentally, the question of the universe's size, and of our planet's motion, are connected: The larger the former, the more plausible the latter. After all, why should the entire universe move about the Earth, if our planet is so minuscule? Or, as Copernicus phrased it: "How astonishing if, within the space of twenty-four hours, the vast universe should rotate rather than its least point!" The logic employed here goes back to ancient times, and involves an early version of "relativity"—not the Einsteinian sort, but rather the simple understanding that all motion is relative. Copernicus recalls Virgil's description of a ship at sea. He quotes from the *Aeneid*: "We sail forth from the harbor, and lands and cities draw

* "Nearly unthinkable," but not completely so: The idea of a purely naturalistic universe has ancient roots, notably in the writing of the Greek atomists and their followers, as we will see in Chapters 13 and 14.

backwards"—surely the more reasonable inference is that it is the ship which is in motion rather than the lands and the cities. "No wonder, then, that the movement of the earth makes us think the whole universe is turning round." Moreover, this larger universe hinted at the possibility of other worlds—or at least allowed for them. If our own sun harbored a family of planets, who could say how many planets might orbit other suns, at unfathomable distances from our own world?

In fact, the possibility of other worlds had been discussed often in the Middle Ages. Many medieval philosophers argued that an omnipotent God could have made as many worlds as he might have wished. Even so, the general consensus was that he only made one: our own blue-green world. (I've mentioned that list of heretical opinions issued by the bishop of Paris in 1277. Number 34 insisted, somewhat awkwardly, that the faithful concede that the Almighty *could have* created other worlds—but that no such worlds actually existed.) Nicole Oresme, a fourteenth-century French philosopher, considered the matter carefully, pondering the various mechanisms by which such worlds could be created, and where they might be located. In the end, however, he concluded, "But, of course, there has never been, nor will there be more than one corporeal world." Even if *De revolutionibus* wasn't met with panic in the streets—how many books are?—it indeed marked a turning point. As I. Bernard Cohen puts it, "the alteration of the frame of the universe proposed by Copernicus could not be accomplished without shaking the whole structure of science and of our thought about ourselves."

It would be a half century before an inquisitive Italian scientist would aim a new invention, the telescope, at the night sky.* When he did—as we will see in Chapter 9—the wild Copernican "hypothesis" suddenly became plausible physical fact. But even without a telescope, the evidence against the ancient cosmological system was mounting. A crucial event—one that I played with in the Prologue, and mentioned briefly in the Introduction—unfolded in the autumn of 1572. In November of that year, a bright "new star" appeared in the constellation Cassiopeia, lighting up the night sky for the remainder of the year and through the next. Today we would call it a *supernova*, the explosion that takes place when a massive star exhausts its nuclear fuel supply and sheds its outer

* As we will see, however, it is *possible* that English astronomers had something like a primitive telescope in the second half of the sixteenth century.

layers in a fiery burst of matter and radiation. The new star would come as an affront to the cosmology of Aristotle and Ptolemy. William Shakespeare was eight years old at the time—and a pompous Dane with a keen eye and a metal nose was twenty-six. After two thousand years, the ancient system of the world was beginning to show cracks in its very foundation.

3. "This majestical roof fretted with golden fire . . ."

TYCHO BRAHE AND THOMAS DIGGES

The story begins nine thousand years ago. Not in a galaxy far, far away, but in a relatively nearby section of our own Milky Way, located in the direction of the constellation Cassiopeia. Stars look pretty stable from night to night—even from century to century—but modern physics has revealed that stars actually take part in a continual tug-of-war between the forces of nature. Gravity strives to pull everything together; the energy produced by nuclear forces wants to blow everything apart. For most of a star's life, it shines by burning hydrogen through a series of nuclear reactions in its core, and the balance between the forces is maintained. But this particular star—today it goes by the less-than-imaginative name of 3C10—was an old one, and it had already exhausted its supply of hydrogen.* When that happened, gravity became the dominant force, and the star, now containing mostly carbon and oxygen, began to collapse. Once, it had been a red giant; now it had evolved into a "white dwarf." These dwarf stars are so dense that while they weigh as much as the sun, they are typically only the size of the Earth. (A lump of white-dwarf matter the size of a basketball would weigh about as much as an ocean liner.) Revolving in a mutual orbit with a larger companion star,

---

* Confusingly, the star has more than one name. 3C10 refers to its catalog designation from the 1950s, when remnants of the star were first identified using radio telescopes. Astronomers eventually concluded that this had indeed been the star that Tycho and others had observed in 1572. Astronomers sometimes call it "SN 1572" (SN for supernova; 1572 for the year it was first observed). It is often simply called "Tycho's star" in the popular literature, or, in reference to its current appearance, "Tycho's supernova remnant." (Also, there is more than one model for the physics behind this type of supernova; a merger of two white dwarfs might also produce such events.)

3C10 had been sucking hydrogen off of its neighbor, gaining mass in the process. This also caused the temperature in the core of the star to rise. Eventually, when its mass reached a bit less than one and a half times that of our sun, a new kind of reaction began: Carbon atoms were now fusing with each other, setting off an unstoppable nuclear chain reaction. A shock wave ripped through the star, radiating outward from the core, with a speed of more than ten thousand miles per second. The star exploded.

When it was a white dwarf, 3C10 had been much dimmer than the sun. Now it was a supernova, shining with the light of a billion suns. It would be—briefly—as bright as the rest of the galaxy combined. How many creatures, on how many planets, looked up and saw the spectacular death throes of this star? We don't know—but we do know *when* they would have seen it. Light travels at 186,000 miles per second. If there happened to be a civilization a thousand light-years from the star, they would have seen the explosion a thousand years later (that is, about eight thousand years ago). Because 3C10 happens to be located about nine thousand light-years from earth, light from the exploding star took nine thousand years to reach our planet.* Photons from that initial burst of light, having traversed nine thousand light-years of interstellar space, reached Earth in early November 1572. Before that moment, the remote nondescript star would have been invisible without a telescope (which had not yet been invented). Now, suddenly, it was as bright as the planet Venus.

Even before the light from 3C10 reached our planet, an inquisitive Danish nobleman named Tycho Brahe (1546–1601) had become hooked on astronomy. Tycho—like Galileo, he is remembered by his first name—was born in the province of Scania (today part of southern Sweden) three years after the publication of Copernicus's revolutionary book. He was born into a powerful noble family who assumed he would eventually serve the king as a soldier or as an administrator. Raised by an uncle, he enrolled as a law student at the University of Copenhagen, but soon became distracted by events that unfolded in the heavens.

In 1559 and 1560 Tycho observed first a lunar and then a solar eclipse.

---

* There's a significant margin of error in the distance to 3C10, and therefore also in the estimate of how long it took the light to reach Earth. If it were *exactly* nine thousand light-years distant, then we could say that it exploded more than 9,440 years ago (9,000 years for the light to reach Earth, plus the 440-plus years that have passed since Tycho observed it).

Still a teenager, Tycho was stunned to learn that astronomers could predict solar eclipses months and even years in advance. A few years later, while studying in Germany, he witnessed a close pairing of Jupiter and Saturn in the sky (astronomers call it a *conjunction*), an eye-catching celestial coupling that occurs about once every twenty years. But Tycho noticed that the published tables, whether based on Ptolemy's ancient system or the newer Copernican model, were inaccurate; the time given for the closest approach of the two planets was off by several days. Tycho became determined to improve on the existing tables; from that moment on, he would devote all of his energy to studying the night sky. Over the next few years he traveled widely within Europe, studying in various university towns and acquiring books and instruments as he went. In time, he became a master observer.

"A NEW AND UNUSUAL STAR"

But the most stunning celestial event—the one that cemented Tycho's passion for astronomy and would end up changing the course of Western thought—came in the autumn of 1572, when the new star exploded into view in the northern sky. His travels behind him, Tycho was living in Scania, where he had built a small observatory on grounds owned by a relative. Tycho first spotted the star on November 11. (A handful of other European observers, it turns out, had seen it a few days earlier.) His excited tone was still in evidence months later, when he set his thoughts to paper:

> Amazed, and as if astonished and stupefied, I stood with my eyes fixed intently upon it. When I satisfied myself that no star of that kind had ever shone forth before, I was led to such perplexity by the unbelievability of the thing that I began to doubt my own eyes.

Tycho studied the star's appearance over several weeks; he also compared notes with other observers across Europe. He hurriedly wrote and published a short book describing his account of the event, called *De Nova Stella* (*On the New Star*). (The full title was actually *De nova et nullius aevi memoria prius visa stella* [*Concerning the Star, New and Never Before Seen in the Life or Memory of Anyone*].) His tone was not exactly modest:

> I noticed that a new and unusual star, surpassing all the other stars in brilliancy, was shining almost directly above my head. And since I had almost from boyhood known all the stars of the heavens perfectly . . . it was quite evident to me that there had never before been any star at that place in the sky, even the smallest, to say nothing of a star so conspicuously bright as this.

The new star, appearing out of nowhere, was "the greatest wonder that has ever shown itself in the whole of nature since the beginning of the world."

Unfortunately, new stars *weren't supposed to happen.* It was an affront to Aristotelian and Ptolemaic cosmology, in which the heavens, by their very nature, were believed to be perfect and unchanging. How could one make sense of the appearance of a new star? One possible resolution was to imagine that it was actually a terrestrial phenomenon; perhaps it was located high in the Earth's atmosphere. (Think of the momentary difficulty one has today in distinguishing a star from an airplane.) If it were shown to be "sublunar," all would be well in Aristotle's world. Yet if the new star were a "local" phenomenon, it would display parallax—in this case, meaning that it would be seen at slightly different locations (relative to the background stars) by observers at different locations on the Earth—or, indeed, by a single observer watching over an interval of several hours, because the Earth, as it rotates, would carry the observer over a distance of several thousand miles.* The moon, which traditionally denoted the boundary between the corruptible terrestrial region and the perfect realm of the stars, was known to display a parallax of roughly one degree. But Tycho's new star showed no discernible parallax. In addition, it seemed to stay in the same location in the sky, relative to the other stars (displaying no "proper motion," as an astronomer would put it). "I conclude," Tycho wrote, "that this star is not some kind of comet or fiery meteor . . . but that it is a star shining in the firmament itself—one that has never previously been seen before our time, in any age since the beginning of the world."

Tycho was anxious to know what observers elsewhere in Europe thought about this strange and wonderful object. Thousands of people

---

* The technical term is *diurnal parallax.*

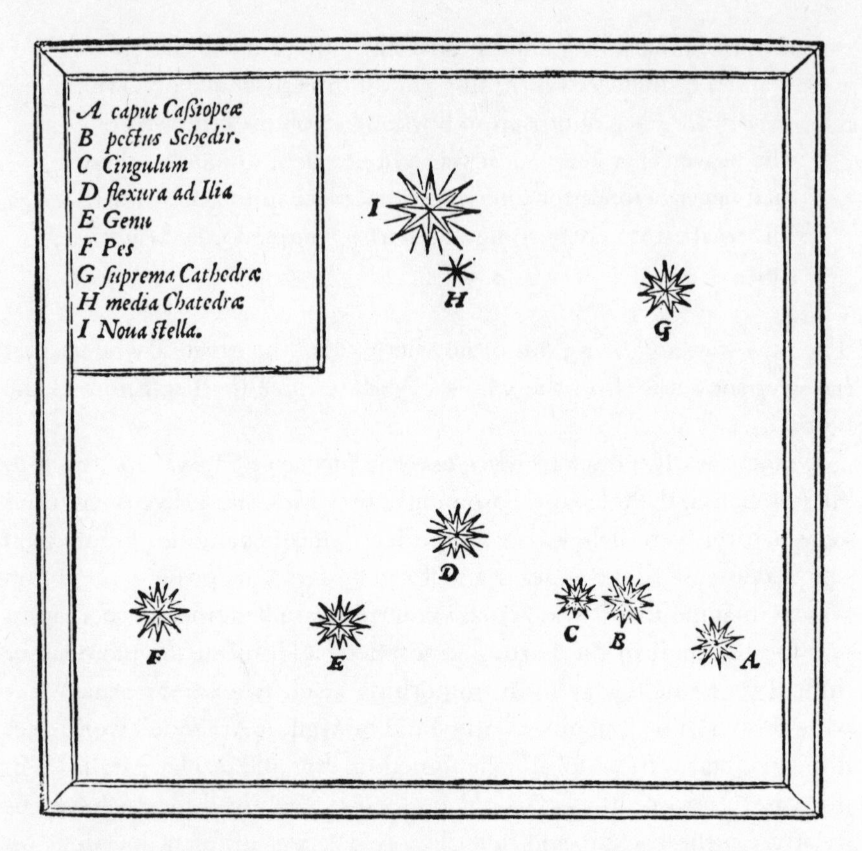

Fig. 3.1 Danish astronomer Tycho Brahe was among the first to see the "new star" of 1572 (depicted here above the "W" of Cassiopeia, and identified in Latin as "Nova Stella"). The star's appearance challenged the Aristotelian picture of the universe, in which the heavens were imagined to be perfect and unchanging. The Bridgeman Art Library, London

must have seen the new star, including dozens of professional astronomers, astrologers, and mathematicians. Many of them, like Tycho, rushed to get their observations into print. Among them was an English astronomer named Thomas Digges, one of the most important English thinkers of his day.

LEONARD AND THOMAS DIGGES

History has a funny way of allotting celebrity. Only the smallest handful of physicists and astronomers have come down to us as household names: Copernicus, Galileo, Newton, Einstein. From the next tier we have names like Tycho Brahe and Johannes Kepler—well known to those who have studied astronomy or taken a history of science course, but largely unknown to the general public. And then we have those who ought to be as well known, but, due to the vagaries of history, haven't received their due. It is from this group that we find Thomas Digges (ca. 1546–1595). Digges was a military engineer and a member of Parliament—but that is not why he is remembered. Instead, we know him as one of the first English Copernicans, a man who gave us a new vision of the cosmos.

Science seems to have run in the Digges family. Thomas's father, Leonard, was a distinguished mathematician who had studied at Oxford, where he developed an interest in problems of surveying—he invented the theodolite, for measuring angles—and in matters of defense. In 1555 he published an almanac, *A Prognostication of Right Good Effect*, which contained weather predictions as well as instructions for using various mathematical instruments for astronomy and navigation. It also included an outline of Ptolemy's description of the cosmos. Another of his books, on surveying, remained in use for the next 150 years, going through at least twenty editions. Striving to speak plainly on all manner of practical problems in surveying, he complained of the many books on geometry "locked up in strange tongues." He instructed his readers to go through the text once, and then a second time "with more judgement, and at the third reading wittily to practise" the various methods described, all in the aim of "profitable labour." Incidentally, all of these accomplishments might never have come to pass had history taken a slightly different turn: In 1553, Leonard got caught up in what would become known as Wyatt's Rebellion, a plot to overthrow Queen Mary. He was one of some five hundred alleged conspirators to be captured, tried, and convicted. But while the ringleaders were hanged—many were also drawn and quartered, for good measure—Leonard was among seventy-five men given a reprieve.

We know that Leonard Digges was intrigued by the night sky—but whether he gazed at it with the unaided eye, or had some kind of optical assistance, has been the subject of much debate. Did he invent a primitive telescope, and aim it at the heavens? If he did so, it would have been a full six decades before Galileo—but the evidence is scant. We have only Thomas Digges's account of his father's use of "perspective glasses" for long-distance viewing; supposedly he could read the lettering on coins that had been scattered in a field, and could see what people several miles away were up to. Digges goes on to discuss the "marvellous" things revealed by the use of "glasses concave and convex"; he describes aiming such a perspective glass at a village, from some distance away, and being able to discern "any particular house or room thereof . . . as plainly as if you were corporeally present. . . ."

The claims seem to ask to be taken seriously, but many historians are doubtful. Richard Panek, for example, assures us that such instruments did not yet exist. "Despite their seeming certitude, these writings (and many others) were speculations or embellishments." Richard Dunn describes the evidence as "uncertain," but concedes that by this time "many people were making investigations with lenses of sufficient quality for a working telescope to be a practical possibility." Thomas Digges, incidentally, says that he wrote a book on perspective glasses—but if he did, it has not survived, nor have any telescope-like devices from that period. (We will take a closer look at the plausibility of a "Tudor telescope" in Chapter 5.) Leonard Digges died when Thomas was just thirteen, at which point the younger Digges became a pupil of John Dee, one of the most influential philosopher-mystics of the age (and a figure we will look at in more depth in the next chapter). The two men would remain in close contact.

Thomas Digges was twenty-six when the new star of 1572 lit up the skies over England (the same star observed in Denmark by Tycho Brahe). Digges first noticed the star on November 17—six days after Tycho—and began to study the object carefully. The following year he wrote a short treatise on it. His opinion was in great demand; even the queen sought out his views on the new star. Astrological prognostications aside, Digges believed his observations could be used to test the Copernican theory. His own book about the star, *Alae seu scalae mathematicae* (*Mathematical Wings or Ladders*), appeared in 1573, before the star had faded from view. Like Tycho, Digges had managed to measure the parallax of the new star, or, more precisely, to determine that it displayed so little parallax that it must lie much farther away than the moon.

For Digges, measurements of parallax were more than just a clever substitute for a celestial yardstick. The data could also reveal which view of the cosmos—the Ptolemaic or the Copernican—was correct. That's because the distance between the Earth and the other planets (especially Mars) varied by a much larger amount in the sun-centered scheme than in the geocentric picture. With precise enough observations, one ought to be able to settle the matter. Digges was lucky to have access to the best instruments in England, including a ten-foot cross-staff, for measuring celestial angles, devised by a man named Richard Chancellor; when

Digges was a young man, he and Dee used the instrument together. We know that Digges was leaning toward the Copernican theory, and certainly admired Copernicus as the greatest observer of his age. Along with his own measurements of the new star, Digges included a data table with the positions of thirteen reference stars in the constellation Cassiopeia, "taken from Copernicus (with the printer's errors corrected)." He also confessed to working quickly in case "by order of the Most Powerful, it [the star] should recede again." The star did indeed fade in intensity over the next few months. By February 1574 it had finally disappeared from view. Yet interest in the strange apparition continued. Like Tycho, Digges was keen to read everything that astronomers throughout Europe had written about the wondrous new star.

"A LETTER SENT BY A GENTLEMAN OF ENGLAND"

Given that this was nearly four and a half centuries ago, one might imagine that fresh evidence concerning Digges and the new star would be rather unlikely to turn up. A few years ago, however, a British historian named Stephen Pumfrey stumbled across a published letter, apparently written by Digges, on the subject of this very star. The document, which does not give the author's name, is titled "A letter sent by a gentleman of England, to his frende, containing a confutacion of a French mans errors, in the report of the miraculous starre now shyning." Only one copy of the letter, published in 1573, seems to have survived. It had been dwelling, virtually forgotten, in the library in Lambeth Palace in London. In a journal article published in 2010, Pumfrey presents a compelling case for Digges being the author.

The letter was written in response to another letter, written by another anonymous individual—how much simpler life would be if people signed their names!—living in France.* The Frenchman is identified only by his (presumably Latin) initials, I.G.D.V. Pumfrey suspects this was the work of an astronomer named Jean Gosselin, who worked in the library of the French king, Charles IX. (As Pumfrey admits, the whole business is a mystery "fit for an Umberto Eco novel.") We know that the French publication dates from 1572. Obviously rushed into print, it has several pages of observations of the new star, as well as a diagram showing

---

* Today we usually think of letters as private—but in those days it was common to publish a letter if its contents were thought to be of interest to a larger audience beyond the named recipient. (A famous example from a half century later is Galileo's "Letter to the Grand Duchess Christina," which was widely circulated and clearly meant to be read by educated readers besides the duchess.)

its location. But it is deeply flawed: The author identifies the object as a comet, and states that it displayed a significant parallax.

The Frenchman's letter also provides one of many indications of the profound astrological significance associated with the new star. Although some saw it as a good omen, many feared it was a sign that something ominous was on the horizon—perhaps war, an ever-present danger. That was certainly true in France, a nation already torn by religious strife. Gosselin saw the new star as a sign that God would soon seek out and punish the sinful. He prayed that it would lead people to "mend their bad ways, and in the times to come to live in accordance with the holy law, in the Catholic faith." Both Tycho and Digges were similarly concerned with the star's astrological significance, which was widely debated in the royal courts of Europe. Were it truly a new star, it would (they believed) be the first since the birth of Christ, when a bright "star" was said to have led the Wise Men to Bethlehem; another such star, it was imagined, might be a sign of the coming Apocalypse. Such beliefs were widespread at the time, particularly by English Puritans, and were likely shared by Digges himself. As late as 1638—more than sixty-five years after the star had faded from view—an English poet named Francis Quarles was still describing the event from a biblical perspective. In one of his epigrams, he draws a parallel between Tycho's nova and the Star of Bethlehem:

> Both showed them light, and showed their blindness too.
> But why a star? When God doth mean to woo us,
> He useth means that are familiar to us.

Back to 1572, and Gosselin's letter: The English ambassador in Paris, Sir Francis Walsingham, obtained a copy and forwarded it to Sir Thomas Smith, one of the queen's senior advisors. Smith consulted with various Englishmen, including Digges, and wrote back to Walsingham, noting that "your astronomers and ours differ exceedingly" in the observation and interpretation of the new star. Because of the similarities between Smith's letter and the (presumed) Digges letter, Pumfrey concludes that Smith's letter was based on conversations with Digges.

Digges's response in the "Letter sent by a gentleman" is significant, Pumfrey argues, because it is the earliest document clearly revealing his support for the Copernican system, including the Earth's annual motion around the sun. However, Digges was not the first English scientist to take note of Copernicus's theory; as mentioned in the introduction, that honor belongs to Robert Recorde, and we'll take a closer look at his work in the next chapter. But Digges was the first Englishman to have been convinced of the validity of the Copernican model, a conviction that we now know he held as early as 1573. The "Letter sent by a gentleman" reveals something else: It highlights Digges's religious views, and once again illuminates the complex relationship between faith and "science." The letter also shows Digges's interest in the biblical "end times," believed by many to be close at hand. (For a true believer, of course, this was something to be welcomed rather than feared.) While criticizing the Frenchman's science, Digges agrees with the conclusion; the new star was "a forewarning of Gods inscrutable pleasure" and a "rare and supernaturall" sign.

The new star eventually receded from view; the apocalypse, if one was expected, failed to materialize. Life went on. But our picture of the universe would never be the same. No wonder the appearance of "Tycho's star" is often described as a pivotal moment in the history of astronomy. Breaking glass seems to be the metaphor of choice: Timothy Ferris says that the shock it dealt to the established worldview "could not have been greater had the stars bent down and whispered in the astronomers' ears"; Dava Sobel writes that "one could almost hear the tinkle of shattering crystal"; Dennis Danielson says that "we can almost hear the cracking of the foundations of medieval cosmology." Those who wanted to cling to the universe of Aristotle and Ptolemy may have dismissed the Copernican model of the solar system as a mathematical convenience, but there was no escape from the implications of Tycho's new star, observed by skywatchers across Europe and now proven to lie in the supposedly immutable heavens.

As the learned astronomers of Europe struggled to understand the new star, yet another cosmic surprise appeared, once again in November—a celestial sight now known as the Great Comet of 1577. (Tycho is said to have been fishing when he first caught sight of it.) The comet remained visible throughout the fall and into winter. Because most astronomers of the time were also astrologers, its appearance, like that of the new star, was seen as a portent. Indeed, comets had a long track record for disturbing the peace. They were often linked to disasters; the word "disaster" comes to us from the Latin *dis-astra*, "against the stars." The comet, like the new star, was too far away to be a terrestrial phenomenon.

Tycho's keen observations showed that it was at least as far away as the planet Venus.

Thanks to his work on the new star, Tycho had become famous—and was recognized by the king of Denmark, Frederick II, as a national treasure. The king approached Tycho with a remarkably generous offer: He could have his very own island from which he could conduct his observations. (He would also have the tenant farmers' rent as his income, adding to his already substantial personal fortune.) The island, Hven (now known as Ven), lies in the channel between present-day Denmark and Sweden. On Hven, Frederick assured Tycho, "you can live peacefully and carry out the studies that interest you, without anyone disturbing you. . . . I will sail over to the island from time to time and see your work in astronomy and chemistry, and gladly support your investigations." Tycho gratefully accepted.

THE LORD OF URANIBORG

Tycho would soon transform the three-mile-long island into Europe's foremost center of astronomical learning. Within a few months, Tycho and his assistants were observing the sun, moon, planets, comets, and stars. They invented new tools for astronomy and mapmaking, and used their own printing press to share their findings with the world. Learned young men from across the Continent descended on Tycho's laboratory, known as Uraniborg ("heavenly castle"); he referred to these young assistants as his *familia*. They were eager for a chance to work with the famous observer, even if they snickered behind his back at his deformity (as a twenty-year-old student, he had lost a good part of his nose in a duel; he wore a silver prosthetic, and was continually adjusting it and applying ointments to it). Among his visitors, incidentally, was King James VI of Scotland, who would eventually, on Elizabeth's death, assume the crown of England as James I.

Sadly, Tycho's castle no longer stands. But we know from numerous drawings that it was built in the grand style of an early Renaissance palace, with turrets, buttresses, and grand stone archways. As impressive as it must have been, it was not, in fact, very well suited for precision astronomy. Its platforms proved unsteady, and Tycho's instruments shook in high winds. And so he built a second observatory just down the road. It was known as Stjerneborg ("Castle of the Stars"), and its belowground foundations provided the steady support that his quadrants and large-scale sextants required. (Stjerneborg's foundations can still be seen today.) Tycho's most impressive instrument was his giant "mural quadrant." With a radius of more than six feet, it was oriented exactly north-south, and occupied an entire wall. Because of its size, it could not be turned—but as the Earth rotated, various celestial objects could be tracked. A star's altitude could be measured to a resolution of ten seconds of arc—that's about one-twentieth of the apparent size of the full moon. (The mural quadrant, like so many wooden instruments of the period, has not survived.)

Tycho's island was, in a sense, the world's first great scientific laboratory—a place where men could devote their lives to the study of nature, never wanting for expertise, equipment, or funds. The endeavor is said to have cost the king one ton of gold; historians estimate that the project absorbed between 1 and 1.5 percent of the Danish national budget. Yet it was more than just a place of scholarly learning. With its grand architecture, orchards, fish ponds, and aviary, it was also a place of sublime beauty. As historian John Robert Christianson describes it, "This was truly a microcosm, offering beauty, harmony, health, and delight to all the senses, for the abundance of the world was here: If ever it were possible to know the Creator through his Creation, this was the place."

Tycho spent more than twenty years on Hven. During this time, he and his assistants plotted the positions of 777 stars. For historian of astronomy Owen Gingerich, the sheer magnitude of Tycho's work became apparent during a visit to a Paris bookshop. An obscure volume, published in the mid-1600s, attempted to list all of the observations of the sun, moon, and planets recorded up to that time. The pre-Tychonic observations take up about ninety pages, and those carried out after Tycho occupy another fifty. Tycho's own observations fill the nine hundred pages in between. Without his contribution, the hefty book would be a mere novella. "The contribution, the sheer bulk of the observations that Tycho Brahe made, is staggering," Gingerich says. Tycho's detailed observations of the heavens would provide the most accurate stellar and planetary measurements carried out prior to the invention of the telescope.

One of the highlights of Tycho's tenure on Hven was the appearance of yet another comet, in 1585. Once again, it was too distant to be sublunar. How could these celestial bodies move to and fro in the heavens, seemingly passing right through Aristotle's crystalline spheres? As he ruminated over the structure of the heavens, he began to imagine that these spheres had simply never existed in the first place: Perhaps the planets simply moved through space unsupported and untethered.

A CELESTIAL COMPROMISE

One might imagine that, having pulled the rug out from underneath medieval cosmology, Tycho would have been eager to embrace the Copernican model. He had certainly read *De revolutionibus*, and greatly respected Copernicus. In 1574–75, he gave a series of lectures at the University of Copenhagen in which he expounded on the Copernican system, referring to its creator as "the second Ptolemy." But for Tycho it was too great a leap to imagine the Earth hurtling through space. Common sense showed that the Earth was stationary; he also took scriptural arguments against a moving Earth seriously. Plus, he was a committed Aristotelian; he believed that there was an unbridgeable divide between the earth and the heavens. And then there was the absence of stellar parallax, hinting, in the Copernican view, at a much larger cosmos—which Tycho was unwilling to accept.

And so Tycho developed a kind of compromise, a hybrid system that had some of the advantages of the Copernican system while keeping the Earth stationary at the center of the cosmos. In Tycho's model, the planets revolved around the sun, but the sun (along with the moon) revolved around the Earth (see figure 3.2). One might call it *geoheliocentric*—but that's a bit of a mouthful; "Tychonic" will have to do. As in the ancient Ptolemaic system, the sphere of the fixed stars defined the outer periphery of the universe; and, again in keeping with the ancient model, the stars revolved around the Earth every twenty-four hours. But Ptolemy's crystalline spheres were no more; the planets and the sun moved freely, through empty space. Mathematically, the Tychonic system is virtually identical to both the Ptolemaic and Copernican systems, but physically it is quite different.

Tycho had been considering this model for some time, but had worried about the paths that the planets appear to take. Those paths, as can be seen in the figure, intersect—which would be problematic so long as each planet was fixed to its own rigid, crystalline sphere. But his observations of comets had finally forced him to the conclusion that there were no such spheres after all. The planets were free to move through empty space, "divinely guided under a given law." The spheres, he wrote in 1588, "which authors have invented to save the appearances, exist only in the imagination in order that the motions of the planets in their courses may be understood by the mind."

The appeal of Tycho's system would gain wider currency with the invention of the telescope; many of the early observations carried out by Galileo clearly supported the idea that the planets revolved around the

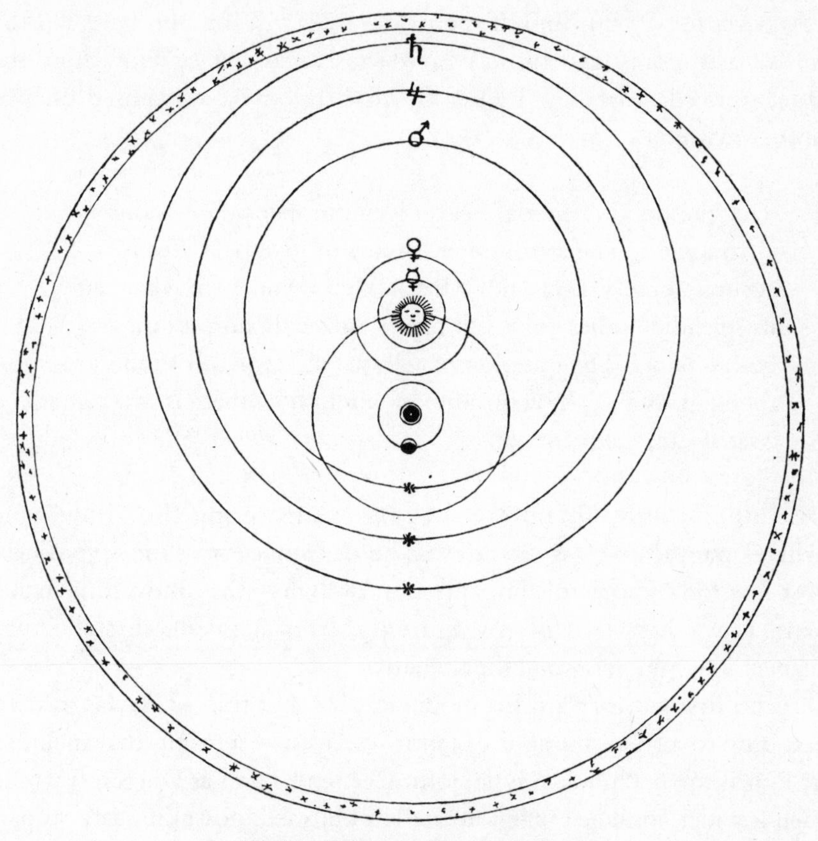

Fig. 3.2 Not quite Aristotelian, not quite Copernican: In this "hybrid" model of the cosmos, developed by Danish astronomer Tycho Brahe, the planets revolve around the sun, while the sun in turn revolves around the earth, which remains fixed at the center of the universe. The Granger Collection, New York

sun rather than the Earth. Yet one could still argue that the Earth itself remained stationary while the sun, with its cohort of planets, circled the Earth—as Tycho had suggested. And so, well into the early years of the seventeenth century, Tycho's system seemed to present a viable alternative to that of Copernicus.*

* One might imagine that the last adherents of Tycho's system would have died several centuries ago—but not so. A handful of biblically minded Creationists, primarily in the U.S., still argue for the Tychonic system over that of Copernicus; after all, it's consistent with most observations, and manages to keep the Earth at the center of the universe. (See DeWitt, p. 139.)

And yet for Tycho himself there was a clear distinction between the aims of astronomy and those of physics. He sought to show how the planets moved; *why* they did so, or what they were composed of, was another matter:

> The question of celestial matter is not properly a decision of astronomers. The astronomer labours to investigate from accurate observations, not what heaven is and from what cause its splendid bodies exist, but rather especially how all these bodies move. The question of celestial matter is left to the theologians and physicists among whom now there is still not a satisfactory explanation.

In other words, Tycho saw his job as answering the "how" and "where" questions: How did celestial bodies move across the sky? Where must one look for them? The question of "why" they moved in such a manner was best left to—intriguingly, from a twenty-first-century perspective—physicists *and* theologians.

Tycho first put forward his model in 1588, but this work was distributed only to a small number of fellow scientists—a group that included the Englishmen Thomas Savile, John Dee, and Thomas Digges. It would reach a wider audience when formally published, posthumously, as part of his *Astronomiae Instauratae Progymnasmata* (*Exercises for the Reform of Astronomy*) in 1602.

Life on Tycho's island began to sour when a new, budget-conscious king, Christian IV, came to power. In 1599, Tycho took a position in Prague under the patronage of Emperor Rudolf II. His career was winding down, though he was still keenly interested in matters of astronomy. He continued his astronomical observations in Prague, where he soon heard of a brilliant young German mathematician named Johannes Kepler. Tycho invited Kepler to join him, and the two men collaborated until Tycho's untimely death, in 1601. The story, which has passed into the folklore of science, is worth repeating: One day, Tycho was invited to a dinner party hosted by one of Prague's most important noblemen. Partway through the dinner, Tycho, who had been drinking "overgenerously," realized he had to go to the bathroom. Rather than excuse himself from the table and risk offending his host, he held it in, showing "less concern for the state of his health than for etiquette." (The description comes from Kepler, commenting on the final page of Tycho's astronomical logbook.) Eleven days later, the fifty-five-year-old Tycho, who was likely already suffering from prostate problems, was dead. His ornate tomb can be seen in the Church of Our Lady before Týn in Prague's Old Town Square.

"THIS SACRED CELESTIAL TEMPLE"

While Tycho had found it impossible to accept the Copernican view, Digges became one of its strongest supporters. A few years after the appearance of the new star, he seemed to have become even more fully immersed in the new astronomy. This time he was writing an appendix for a new edition of an almanac originally published by his father more than twenty years earlier; it would go by the title *A Prognostication Everlasting* (1576). The appendix was, in fact, a translation of part of *De revolutionibus*, focusing on its most crucial elements (including Copernicus's rebuttal to arguments against the possibility of the Earth's motion). Digges spoke of "that devine Copernicus of more than human talent," describing him as a "rare wit" who has recently proposed that "the Earth resteth not in the Center of the whole world, but . . . is carried yearly round about the sun, which like a king in the midst of all reigneth and giveth laws of motion to the rest, spherically dispersing his glorious beams of light through all this sacred celestial temple." At the same time, the Earth—described as a "little dark star"—is "turning every twenty-four hours round upon its own center, whereby the sun and great globe of fixed stars seem to sway about and turn, albeit indeed they remain fixed." Digges went on to explain that he included the excerpts from Copernicus in his almanac "so that Englishmen might not be deprived of so noble a theory."*

Digges does more than pay homage to Copernicus as a sophisticated natural philosopher. He insists that the author of *De revolutionibus* intended his description of the solar system to be taken as physical fact and—in spite of Osiander's disclaimer—not merely as a mathematical hypothesis. Copernicus meant for the heliocentric model to be employed not only "as Mathematicall principles" but to be recognized as "Philosophicall truly averred." Digges also goes out of his way to counter some of the arguments most commonly put forward against the idea of a

---

* Owen Gingerich eventually tracked down Digges's own copy of *De revolutionibus*, which is now in a library in Geneva. There aren't too many annotations, but, tellingly, Digges wrote on the title page, "*Vulgi opinio Error*" ("the common opinion errs"). Gingerich also found two copies that had been owned by Dee (Gingerich, *The Book Nobody Read*, pp. 119, 242).

moving Earth. As we've seen, it had long been argued that on a rotating Earth, an object dropped from a tall tower—or, say, from the mast of a tall ship—would land some distance away from the base. Not so, says Digges, who may well have conducted such experiments himself; it would land at the base of the mast, just as if the ship were at rest. Finally, using a trick that Copernicus and others had often employed, he tried to make it sound like his vision of the cosmos was both novel and, at the same time, rooted in the most noble thinking of antiquity. The crucial chapter on the Copernican system is titled "A Perfect Description of the Celestial Orbs According to the Most Ancient Doctrine of the Pythagoreans, Lately Revived by Copernicus and by Geometrical Demonstrations Approved." Thanks to the work of Digges, references to Copernicus in English books, both scholarly and popular, become much more frequent from the mid-1570s. By that time, as Johnson notes, "nearly every writer on astronomy felt it necessary to pay some attention to the heliocentric theory, if only to try to refute it by the conventional Aristotelian arguments."

Digges even speculates on the nature of gravity. For Aristotle, gravity was a force that drew objects—no matter where they were in the cosmos—to the center of the universe. The Earth, being heavy, occupied that central position, so gravity pulled other objects toward the Earth. But if the Earth was one of the planets, then what? In such a system there was clearly more than one "center." Digges reasons that "it may be doubted whether the center of this earthly gravity be also the centre of the world. For gravity is nothing else but a certain proclivity or natural coveting of parts to be coupled with the whole. . . ." Isaac Newton would develop these ideas further in the next century.

Digges was fluent in Latin, but chose to write in the vernacular. His reasoning was practical: He wanted to put the knowledge in the hands of men who had not attended university, but could nonetheless profit from such learning. Dee had expressed a similar motivation, and the trend would continue: Robert Norman and William Borough would write in English about the workings of the magnetic compass; John Blagrave wrote about astronomical instruments. Science was becoming a pursuit not just for scholars, but for ordinary literate citizens.

TO INFINITY AND BEYOND

Perhaps more important than Digges's words, however, was a fold-out diagram he included with the text (see figure 3.3). The central part of the diagram reproduced Copernicus's new vision of the solar system, with the Earth as one of the planets revolving around the sun. Beyond it, however, the stars could be seen extending outward in all directions, perhaps, one might imagine, to infinity. A few lines of text embedded in the diagram reinforce what it already plainly shows: This "orbe of stares" extends "infinitely up . . . in altitude," and is "therefore immovable." The "primum mobile" is no longer necessary. The heavens, which had since ancient times been seen as moving around the Earth, had been brought to a standstill.

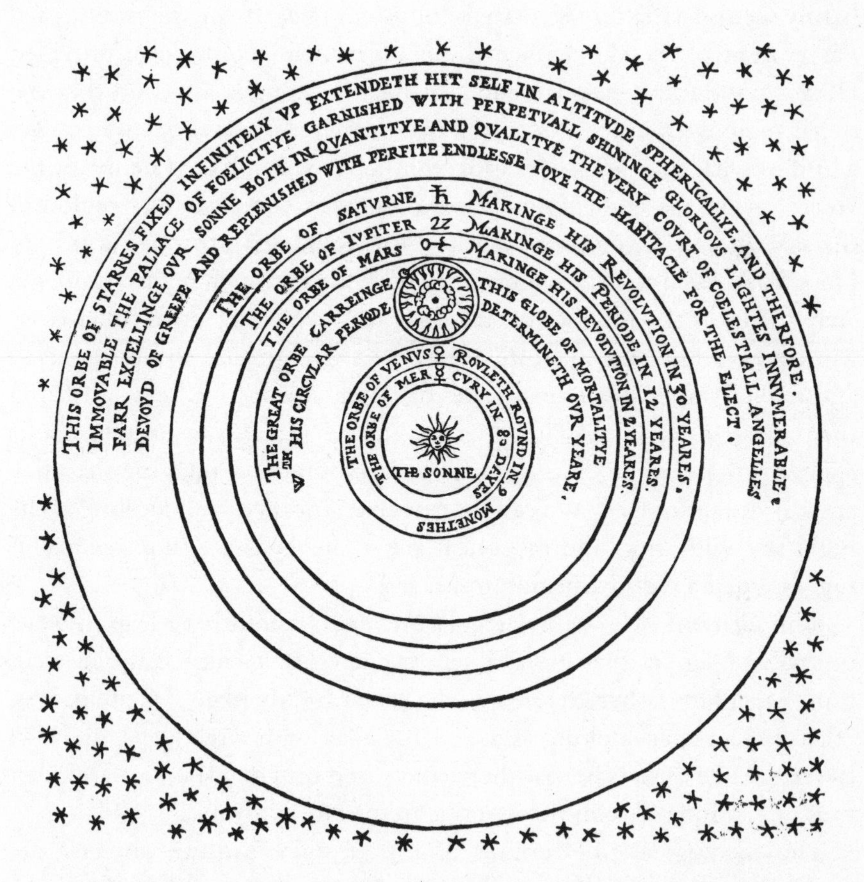

Fig. 3.3 In 1576, the English astronomer Thomas Digges published a new edition of an almanac written by his father, Leonard. Thomas added an enthusiastic synopsis of the Copernican theory, and—perhaps even more profoundly—a diagram of the universe in which the stars are seen to extend outward without limit. The Granger Collection, New York

---

The diagram, reprinted in each subsequent edition of the *Prognostication Everlasting*, was seen and pondered widely, says Francis Johnson; it "was the representation of the new Copernican system most familiar to the average Englishman of the Renaissance." Moreover, because of the book's popularity, many English thinkers came to think of the infinite universe as an integral part of the Copernican theory.

What was the inspiration for this remarkable and daring expansion of the celestial canvas? One theory is that his father had indeed invented something like a telescope, and that the younger Digges had an opportunity to look through it. Indeed, whether it was actually a telescope in the modern sense, or just a lucky combination of lenses or mirrors, hardly seems to matter. As Francis Johnson notes, "Even the most casual observation of the sky with such an instrument would have provided Digges with ample experimental justification for his assertion that the stellar regions should be conceived of as infinite." Even today we can get a hint of what such an observation would have been like. The next time you're out in the countryside, have a look at some particular region of the sky on a clear, moonless night. Now look again through a pair of binoculars. Even the cheapest, most poorly crafted such device immediately increases the number of stars that are visible dozens of times over. Moreover, while this magnified view reveals a greater *number* of stars, and makes the bright ones appear brighter, it does not seem to bring them any closer. Even with a very good telescope, the brightest stars appear as mere points. It is not a great leap of faith to imagine that they are *very* distant indeed. We cannot perceive "infinity," but looking at the night sky with any kind of optical aid, even the most rudimentary, is suggestive of a world without limits.

John Gribbin, who calls Digges's move an "astonishing leap into the unknown," agrees that it was likely triggered by gazing at the heavens using one of his father's telescopes. It "seems highly likely," Gribbin says, "that he had been looking at the Milky Way with a telescope, and that the multitude of stars he saw there convinced him that the stars are other suns spread in profusion throughout an infinite Universe."

The theological implications of this possibly infinite universe are worth considering. If God and the angels were supposed to dwell beyond the sphere of the fixed stars, then where, exactly, would they reside in this new, larger universe, in which there is no "last star"? As Owen Gingerich explains, Digges came up with a rather ingenious solution: They could dwell *among* the stars. In yet another line of text embedded in the diagram, Digges explained that the realm of the stars constitute ". . . the very court of celestial angels, devoid of griefe and replenished with perfite endless joye, the habitacle of the elect." What better place for God than in an infinite and unmoving heaven?

Thomas Digges's astronomical research came to an end around 1580, when, because of his skills as a military engineer, he was called on to assist in the fortifications at Dover and later to support English forces in the Netherlands. But his influence continued to be felt. His emphasis on observation and experiment marks him as an early proponent of what would eventually become known as the scientific method. His revised edition of the *Prognostication* would go through at least seven editions before 1605, with an eighth published in 1626. By Owen Gingerich's estimate, some ten thousand copies must have been printed by that point. Sadly, fewer than forty copies are known to exist today. I had the privilege of viewing one of them in Gingerich's private collection, at his office at the Harvard Observatory. Here one finds a set of books, maps, and manuscripts that would put many smaller museums to shame, and a few larger ones, too. The more valuable items, of course, are kept in a safe—and this was where Gingerich reached to pull out his copy of the *Prognostication*, gently unfolding its famous diagram for me to inspect. He held up the centuries-old picture of the heavens, and I was at once filled with awe. I was also reminded just how difficult it is to put ourselves in the mind of someone who walked and ate and slept and pondered the night sky four and a half centuries ago. How did the most brilliant natural philosopher in Shakespeare's England picture the universe? This diagram is a large part of the answer. Here was the first detailed description of the Copernican system written in English—and with it comes a vision of a boundless cosmos, a universe even more expansive than that imagined by Copernicus.

4. *"These earthly godfathers of heaven's lights . . ."*

THE SHADOW OF COPERNICUS AND THE DAWN OF SCIENCE

In London, history is often on full display; at other times it lurks in dark corners, hidden from view. Everyone knows of the great museums of London's South Kensington neighborhood—the Natural History Museum, the Science Museum, and the Victoria and Albert Museum. Together, they draw more than nine million visitors each year to their grand exhibit halls. Few, however, know about the museum "annex" in West Kensington. Occupying a stately turn-of-the-twentieth-century building known as Blythe House, the annex serves as a storehouse for millions of "overflow" items from the three museums. My guide there is a young, sharply dressed man named Boris Jardine, a historian and curator for the Science Museum. We walk past row after row of metal shelves containing thousands of items, ranging from the familiar—I recognize the astrolabes and sextants and telescopes—to the obscure. The endless rows of tightly packed metal-frame shelves are straight out of the final scene of *Raiders of the Lost Ark*, in which the Ark of the Covenant is ignobly and anonymously placed to rest in a vast military warehouse. There are no holy relics in Blythe House that I know of, but there are indeed treasures, and they span some twenty-five centuries. None are normally on display to the public.

We stop to look at a small armillary sphere—a kind of 3-D model of the Earth and the heavens—measuring five or six inches across. Later armillary spheres represent a sun-centered system, but this one is firmly Ptolemaic: it depicts the geocentric model of the heavens, with our home planet at the center. "These were very popular astronomical teaching tools," Jardine says, as he slips on a pair of white latex gloves and lifts the object carefully from the shelf. A brass plate indicates that it was made in 1542—one year before the publication of *De revolutionibus*—and, like many of the finest instruments of that era, it was an import from continental Europe. (In a few decades, however, English craftsmanship would come close to matching that found on the Continent.) "At least through most of the sixteenth century, they have the Earth at the center," he says, "because the Ptolemaic view of the universe was the one that was widely accepted at that time."

TALKING TO ANGELS

Jardine leads me down to the basement, and into a room that looks, at first glance, like a storeroom for Hollywood horror-movie props. I can't help noticing the array of iron torture instruments and related objects: chains with spikes; a prisoner's mask; a mask worn by an executioner. Jardine tells me they span the seventeenth to the nineteenth centuries. There's also a disturbingly lifelike wax head, apparently from the early twentieth century. One of the items in the room, however, has a remarkable connection to the period we've been looking at. Jardine opens a drawer to reveal a tiny display box—it fits easily in his hand—containing a small, convex glass disk, rather like a semi-transparent lens, measuring about an inch across. It belonged to Queen Elizabeth's court astrologer.

"This is actually John Dee's pendant, which he used in order to communicate with angels," Jardine says, taking the crystal gently from its case. With a bit of imagination, one can picture the inquisitive Dr. Dee holding the crystal in front of a candle or fireplace, peering at the shimmering image it would have presented. "It's effectively a lens, and it distorts whatever is behind it," Jardine says. "So clearly, in the right conditions, you could use it to conjure up some rather extraordinary visions."

Stored alongside the crystal is a short manuscript, in Latin, written on vellum parchment. It was penned by a man named Nicholas Culpeper, a seventeenth-century doctor famous for his herbal recipes and treatments. Jardine reads from an English translation of Culpeper's testimony: In November 1582, it says, "the angel Uriel appeared in the window of the museum of Dr. Dee, situated to the west, and gave him this transparent stone or crystal." The note, written in 1640, goes on to explain how the crystal came into Culpeper's possession: Dee's son, Arthur, gave it to him as a reward "for having cured a liver complaint of his with the greatest rapidity." Culpeper went on to make use of the crystal in his own practice: "I have used this crystal in many ways, and have thus cured illness. But with its use, a very great weakness always set in and a lethargy of the body."

In other words, the crystal was believed—by Culpeper, at least—to have been a gift from God, and to possess magical properties. "That is apparently its origin," Jardine says. "It was a divine gift to Dr. Dee, the famous astrologer. To his enemies, he was a conjurer; to his friends, he was a mathematician and great teacher of Continental learning." Did Dee believe in his own supposedly divine gifts? Did he really think the crystal came from God? "I don't know whether that's *his* claim or not," says Jardine. "It's certainly his claim that using this stone, this transparent crystal, along with an obsidian mirror and some other artifacts, that he actually communicated with angels, together with some of his friends and acquaintances. The question of whether they really believed what they claimed to be seeing is an incredibly difficult one to answer."

As the "magic crystal" suggests, John Dee (1527–1608) was a remarkable figure whose way of thinking embraced both science and magic—which is less of a surprise when we remember that he lived at a time when a distinction between the two was only just emerging. In his only known portrait, he sports a long white beard and a black skullcap. Wherever a magician-like character turns up in Elizabethan literature, scholars wonder if it is a reflection of the enigmatic John Dee. He has long been seen as a potential inspiration for Shakespeare's Prospero—and perhaps for the title characters in Ben Jonson's *The Alchemist* and Christopher Marlowe's *Doctor Faustus*. More recently he is said to have inspired the character of Albus Dumbledore in the Harry Potter series. (His portrait "would only need a crystal ball, a cat and a pointy hat with astrological symbols to complete the picture," as Nigel Jones noted recently.) The son of a wealthy textile trader, Dee was a skilled mathematician, astronomer, and navigator, as well as an astrologer and alchemist. He also had a sizable ego, believing that he understood God's will better than the biblical prophets did. Though he was fascinated by the occult and magic of all kinds, in some ways he nonetheless had a very modern perspective on what science is and how it should be conducted. He seems to have embraced something close to what we would now call the "experimental method"—the combination of careful observation and recording of data, the use of mathematics to analyze the data, the development of hypotheses, and the devising of new experiments to check those hypotheses.

Dee wasn't only a man of science. As a student, he had a passion for the theater. While an undergraduate at Cambridge, Dee helped to construct the props for a student production—in particular, a mechanical flying beetle. The play, an early Greek comedy, calls for a dung beetle that flies toward the sun. Dee would later claim that his beetle was so realistic that audiences ran from the theater in terror. He explained that his contraption was harmless; it was just a machine—albeit a sophisticated one—that could be understood through mechanics and mathematical laws. Many didn't believe him; he was called a "conjurer," a practitioner of dark magic—a label that would haunt him throughout his career.

Dee was a skilled mathematician, and, when still in his early twenties, lectured widely on the Continent. He took an interest in astronomy, and was ready to embrace the Copernican theory—up to a point. In 1557 he wrote a preface for a set of ephemeris tables* penned by the astronomer John Field. In the preface, Dee lamented the shortcomings of the older, Ptolemaic tables. The figures in the new tables were worked out using the new model, which Dee recognized as yielding improved accuracy. He praised Copernicus "for the more than Herculean labours which he endured in giving a new impetus to the study of the heavens and confirming it most strongly by his calculations," though he cautioned that "this is not the place to discuss his hypothesis." (Field himself seems to have been more confident; in his own foreword to the ephemeris, he spoke of the new theory having been "established and based upon true, certain, and genuine demonstrations.") Dee was clearly prepared to utilize the Copernican model when it proved advantageous; but did he accept it as the true description of the cosmos? We can't be sure—but at the very least, he was clearly aware of the problems raised by the conflict between the two systems of the universe.

We've already seen how theology and science were intertwined at this time, and it's no surprise that Dee referred to Psalm 19 with approval when he described the structure of the universe:

> The whole frame of Gods Creatures, (which is the whole world,) is to us, a bright glasse: from which, by reflexion,

* Ephemeris tables list the positions (coordinates) of celestial objects over a given time period.

> rebboundeth to our knowledge and perceiverance, Beames, and Radiations: representing the Image of his Infinite goodness, Omnipotency, and wisedome. And we therby, are taught and persuaded to Glorifie our Creator, as God: and be thankfull therefore.

The passage suggests a universe that was created by God but that could be understood through science—an approach hinted at in Dee's optical metaphor, with the universe compared to a glass, and light representing divine knowledge.

MORTLAKE'S MAN OF SCIENCE

Dee was England's foremost man of science in the second half of the sixteenth century. Working from his home at Mortlake, near Richmond (now fully engulfed by Greater London), he constructed a multipurpose laboratory and collected books and astronomical instruments. His vast personal library—probably the largest private library in all of England—included two copies of *De revolutionibus*, along with Ptolemy's *Almagest*. Dee was an influential teacher, instructing many (perhaps most) of the great English mathematicians and astronomers of his day, and maintaining a correspondence with scientists from across Europe, including Tycho Brahe. "Dee knew everyone who was anyone," as Nigel Jones notes; and as Lesley Cormack puts it, "Anyone who was anyone in Elizabethan natural philosophy knew Dee."

Could "anyone" have included William Shakespeare? Perhaps. As Peter Ackroyd notes, the playwright's company, the King's Men, stayed in Mortlake briefly in 1603, during an outbreak of the plague. "It is possible that the actors encountered the notorious Doctor Dee during their residence in Mortlake," Ackroyd writes—though of course we cannot be sure. What we can be somewhat more certain of is that whoever did visit the Dee residence would have been exposed to the ideas of Copernicus. Historians believe that the Copernican theory was freely discussed within Dee's circle of friends—more freely, certainly, at his home than would have been possible at Oxford or Cambridge at that time—and that Dee himself played an important role in spreading word of the Copernican system. His house at Mortlake, as Antonia McLean puts it, "became the focal point of all scientific advance in mathematical and allied sciences in the first half of the reign of Elizabeth." Dee was also a familiar presence at court, and served as an advisor to Queen Elizabeth, consulting on matters of astrology and alchemy.

Another preface written by Dee hints at his desire to foster what we would now call scientific literacy. This time it was an English edition of Euclid's *Elements*, published by a merchant named Henry Billingsley. Their goal, it seems, was to convince their countrymen of the value of mathematics; to show how mathematical proficiency would benefit the entire nation. Dee sought to make Euclid's sprawling text as comprehensible and accessible as possible. The more one understood of numbers and their manipulations, he argued, the more one could make sense of the natural world. Billingsley, meanwhile, wrote that understanding mathematics required immersion in "the principles, grounds, and elements of geometry." And it required dedication. The student must be prepared for "diligent study and reading of old, ancient authors." But there was more: Dee linked the study of mathematics to that of music, painting, and even medicine. It was, in effect, a summary of Elizabethan knowledge—and a call to arms to work at growing that knowledge base. Even after nearly 450 years, it remains, as McLean puts it, "one of the most comprehensive and important statements on learning ever written by an Englishman."

Dee also mentions the "perspective glass," presumably referring to a device similar to that mentioned by Thomas Digges. Dee writes that a military commander "may wonderfully helpe him selfe, by Perspective Glasses, in which (I trust) our posterity will prove more skillfull and expert, and to greater purposes, than in these days. . . ." He appreciated that such novel devices could fill the user not just with awe but with fear. Was the effect real, or was it magic? Dee showed that the telescope's effects were simply the result of the laws of perspective and optics. He even urged skeptical readers to visit the London home of a former student, Sir William Pickering, where they could partake of a demonstration.

I've mentioned the possibility that Shakespeare may have encountered Dee at Mortlake—but there is another (admittedly quite speculative) link between the two men. It has occasionally been claimed that the design of the Globe Theatre, where so many of Shakespeare's plays were performed, was based, in part, on Dee's writings. In his preface to Euclid, Dee wrote of the harmony of geometric forms, and this may be reflected in the Globe's design, which consists of a square stage bounded by a circular floor, bounded in turn by a hexagonal outer structure. The design "bears a direct and important relation to the Preface," writes McLean, "simply because it puts the theories contained [in Dee's preface] into practise. The Globe therefore becomes the first example of a *classical* Renaissance building in London."

Above all, Dee stressed the value of mathematics for the national interest. Armed with such knowledge, Englishmen were poised to discover "new works, strange engines, and instruments for sundry purposes in the commonwealth." In the twenty-first century, it is commonplace to hear science-policy makers stressing the value of math and science for the national good (especially in the context of a competitive global marketplace); Dee, writing more than four centuries ago, was perhaps the first to make the case.

We have now met Thomas Digges, one of the first English thinkers to embrace the Copernican theory, and John Dee, who also lent it significant support, and was prepared to weigh it against the traditional view. As we've seen, however, both of these thinkers were preceded by a nearly forgotten Welshman named Robert Recorde (c. 1510–1558), who referred to the new theory less than a decade after the publication of *De revolutionibus*.

SETTING THE RECORDE STRAIGHT

In spite of his relative obscurity, Robert Recorde was probably the most influential English scientist of his day. A man of diverse talents, Recorde studied at Oxford and later earned a medical degree at Cambridge; he taught at Oxford and in London, and helped to train England's first generation of navigators. Recorde was something of a polymath: He was an expert on languages, metallurgy, and mathematics, served as a physician in the court of Edward VI, and in his final years held the title of General Surveyor of Mines, as well as being appointed comptroller of the Bristol Mint. He was also what we would now call a popularizer of science; though he had mastered Greek and Latin, he taught in English, and chose to write for a lay audience in clear and elegant English prose.

In 1542 Recorde published the first English textbook on arithmetic, called *The Grounde of the Artes*, which remained in print to the end of the seventeenth century and introduced the + , − , and = signs for addition, subtraction, and equality. Nine years later he published *The Pathway to Knowledge*, the first book in English on geometry. In 1556, he tackled the subject of astronomy, with a book called *The Castle of Knowledge*, the first comprehensive astronomy text written in English. The book is presented in the form of a dialogue between a master and a young scholar, and the crucial passage runs as follows:

> MASTER: How is it that Copernicus, a man of great learning, of much experience, and of wonderful diligence in observation, hath renewed the opinion of Aristarchus of Samos, affirming that the earth not only moves circularly about his own centre, but also may be, yea and is, continually out of the precise centre of the world eight and thirty hundred thousand miles: but because the understanding of that controversy dependeth on profounder knowledge than in this Introduction may be uttered conveniently, I will let it pass till some other time.
>
> SCHOLAR: Nay sir, in good faith, I desire not to hear such vain fantasies, so far against the common reason, and repugnant to the content of all the learned multitude of Writers, and therefore let it pass for ever, and a day longer.
>
> MASTER: You are too young to be a good judge in so great a matter: it passeth far your learning, and theirs also that are much better learned than you, to improve his supposition by good arguments, and therefore you were best to condemn nothing that you do not well understand.

Recorde acknowledges the counterintuitive nature of the Copernican model; the young scholar balks at the proposal, calling it a "vain fantasy" that stands opposed to "the common reason." But the master, older and wiser, cautions that first impressions can be deceiving; that the truth sometimes requires us to abandon our preconceptions. Along with his favorable mention of the Copernican system, we should also note Recorde's skepticism toward the veneration of ancient texts. He cautions the reader to "be not abused by their authority, but ever more attend to their reasons, and examine them well, ever regarding more what is said, and how it is proved, than who said it: for authority often times deceives many men." *The Castle of Knowledge* went through a second and third printing before the century was over, and was one of the most popular mathematical works in England at the time.

Better-known than Robert Recorde is William Gilbert (1540–1603), another "early adopter" of Copernicanism, best known for his work on magnetism. Naturally occurring magnets, known as lodestones, had been known since ancient times; we now recognize these brownish-black stones as fragments of a mineral called magnetite. It was everyday knowledge that lodestones could attract needles or flakes of iron—but *why* they did so was a mystery. It is easy to see why, in Elizabethan England, magnetism was associated with magic—indeed, a magnet was the very prototype of a "magical" body. And yet, it was magic that could be put to profitable use. The magnetic compass is a medieval invention, and philosophers were aware that the Earth itself has magnetic properties. Even so, there was disagreement as to why a compass needle pointed (roughly) north: Was Polaris, the pole star, endowed with magnetic properties? Might there be a magnetic island to the north of Scotland or Scandinavia? A turning point came in 1600, when Gilbert published his treatise *De magnete*. (The full title translates as *On the Magnet and Magnetic Bodies, and on the Great Magnet the Earth*.) As the title suggests, Gilbert's breakthrough was the realization that the Earth itself could be regarded as a giant magnet.

MAGNET MAN

Gilbert was born in Colchester and graduated from St. John's College in Cambridge; he then moved to London, where he worked as a doctor, later serving as personal physician to Queen Elizabeth (and briefly to King James). He was also an astronomer, and in fact is credited with sketching the first known map of the moon. Frankly, it's not a very good map—but, as Stephen Pumfrey notes, it was probably motivated by very specific considerations that had little to do with accuracy or detail. It was, Pumfrey suggests, part of an ongoing effort to deduce whether the moon rotates relative to the Earth, or remains—as it would appear—to keep one side facing permanently Earthward. In the decades that followed, astronomers would come to realize that the moon indeed "wobbles," turning just enough to show a thin sliver of its far side as it does so; and then, weeks later, another thin sliver from its opposite edge. These wobbles are called "librations," and their discovery has traditionally been credited to Galileo, who noticed them telescopically; but Pumfrey believes Gilbert was the first to discern them, more than twenty years earlier.

Gilbert went further than Copernicus by proposing a mechanism for the motion of the planets. (Even so, he was noncommittal regarding *the*

Earth's annual revolution about the sun.) To begin with, Gilbert imagined that his theory of magnetism could explain the Earth's daily rotation. The Aristotelians had said that a large body like the Earth would naturally be at rest; after all, what would cause such a massive body to move? Gilbert believed that magnetism provided the answer: He believed (wrongly, it turns out) that magnetic fields from a spherical lodestone would set it spinning, and that similar fields caused both the Earth and the moon to rotate. He also believed that magnetic forces emanating from the sun, together with the sun's rotation, caused the planets to move in their heliocentric orbits. The entire sixth "book" of *De magnete* is taken up with this effort to link magnetism to this one facet of Copernicus's theory.* Remarkably, Gilbert predicted (correctly) that all the bodies in the solar system would tug on each other, and that these irregularities ought to have observable consequences; one result would be that planetary orbits ought not to be perfectly circular—a bold notion that contradicted traditional thought. The lunar librations were one such piece of evidence, offering, as Pumfrey puts it, "visible proof of his radical cosmology."

Gilbert was certain that the ancient view of the heavens was wrongheaded. He eagerly embraced Thomas Digges's view that the stars were infinite in number, and that they were located at various distances from the Earth, extending, quite possibly, without limit. The idea that the vast universe, filled with countless stars, revolved around Earth was simply untenable. Just as the planets lie at unequal distances from the Earth, he wrote,

> so are those vast and multitudinous lights separated from the earth by varying and very remote altitudes; they are not set in any sphaerick frame or firmament (as is feigned), nor in any vaulted body. . . . How immeasurable then must be the space which stretches to those remotest of the fixed stars!

Gilbert was one of the foremost scientists of the Elizabethan age, every bit as influential as John Dee. And yet, should we attempt to label Gilbert as a "modern" figure we run into difficulties. Newton, the greatest

* Note that a theory of "gravity," as we think of it today, would remain elusive until the work of Newton more than sixty years later.

mind of the next century, was comfortable discussing the motion of the planets in terms of inanimate mechanical forces—but Gilbert saw such motion in almost psychological terms. (In describing his book, he used the phrase "physiologia nova"—he was bringing a "new physiology" into the science of cosmology.) The planets were animate; they possessed souls. (One of his chapter titles was "The Magnetic Force is Animate, or Imitates a Soul; In many Respects it surpasses the human soul while that is united to an organic body.") As John Russell puts it, the idea of mechanical force plays only a minor role in Gilbert's theory; instead, the universe "is more like a community of souls stimulating each other to activity." (We might note that Kepler, who did more than anyone else to develop Copernicus's original theory, also believed for many years that the Earth had a soul.) At the same time, we see a decidedly modern emphasis on experiment and observation. Gilbert has no patience for those who blindly regurgitate the theories of the ancients—the "probable guesses and opinions of ordinary professors of philosophy"—without bothering to engage their own senses. His own theories, in contrast, are "demonstrated by many arguments and experiments." The end result is that "causes are made known of things which, either through the ignorance of the ancients or the neglect of the moderns, have remained unrecognized or overlooked." As I. Bernard Cohen puts it, *De magnete* "contains the seeds of revolution."

Gilbert's book was tremendously influential, and drew praise from both Kepler and Galileo. In fact, magnetism itself became a topic of general interest in the decades following *De magnete*, and the effects could be seen on the London stage. Ben Jonson's final comedy, first performed in 1632, was called *The Magnetic Lady*. The main character is a wealthy woman named Lady Loadstone, who tries to marry off her niece, Placentia Steel, while taking advice from one Master Compass. (For good measure, the Lady is assisted by a steward whose best friend is named Captain Ironside.) Shakespeare, in contrast, does not seem to have used the words "magnet" or "magnetism" in his works—but magnetism is referenced briefly in *Troilus and Cressida*. When Troilus pledges his love, he vows to stay with Cressida "As iron to adamant" (3.2.174)—with footnotes in today's scholarly editions helpfully explaining that adamant is another word for lodestone.

As we examine the work of early scientific thinkers like Dee and Gilbert, we are confronted with a peculiar (to us) mix of medieval and modern.

In the case of John Dee, we sense a man who was certainly part magician—and yet, with hindsight, we can also see something of a modern scientist in his attitude and his work, something we perceive in Gilbert as well. An even more curious case is the figure of Giordano Bruno (1548–1600), the Italian philosopher and mystic who would pass through England in the 1580s. Even the statue of Bruno that now stands in Rome's Campo de' Fiori—the "field of flowers"—has a certain somber darkness to it. The statue, erected in 1889, was supposed to face south, but at the last minute its orientation was changed so that it faced northward, toward the Vatican. (Facing away was seen as disrespectful.) As a result, the hooded figure, depicted with arms crossed and clutching a hefty book, is nearly always in shadow. Nor is it likely that the grand statue is much of a likeness of the doomed philosopher; it towers over the square, even though Bruno himself was a diminutive man. Appropriate, perhaps, for a man who, at least in his own mind, was a larger-than-life figure—a firebrand whose every thought seemed to challenge the established order.

Fig. 4.1 Philosopher, mystic, heretic: Giordano Bruno, condemned by the Roman Catholic Church and burned at the stake in 1600, is now honored by a statue in Rome, at the site of his execution. Author photo

Bruno was trained as a Dominican monk in his native Naples, but even as a young man his unorthodox views turned many of his friends into enemies. Among other things, he imagined a broader kind of Christian faith, larger than that embraced either by the Catholics of his home country or by the Protestants north of the Alps; under his guidance, he imagined, religious fighting would become a thing of the past, and a new golden age could be ushered in. Although Bruno succeeded in attaining the priesthood, he was also suspected of heresy and was forced to flee the land of his birth.

THE WANDERER

Bruno would spend more than twenty years wandering across Europe—years that he appears to have spent teaching, writing, and, when money was tight, proofreading. And perhaps above all, arguing—and making even more enemies. (One gets a sense of this even from his scribblings in other people's books. In the margin of one text, he wrote, "Remarkable that this ass professes himself a doctor.") He was also known for his feats of memory, and taught mnemonic techniques for improving memory (even demonstrating his methods to the pope, Pius V, before fleeing Italy).

Bruno, like Tycho and Digges, lived at a unique moment in the history of our understanding of the cosmos, a time when ancient wisdom was confronted with new doubts. The poet Dante Alighieri, Bruno's countryman, had described the structure of the heavens in his fourteenth-century epic poem *Divina Commedia* (the *Divine Comedy*), built on a strictly Ptolemaic view of the universe—but such descriptions could no longer be taken at face value. As Bruno's biographer Ingrid Rowland writes:

> Dante's certainties about natural philosophy were no longer certain. Astronomy had begun to split away from astrology, substituting a mechanical system of stars and planets for a system tied to the gods of Olympus. Mathematicians like Copernicus could track their movements with the help of complex equations, an activity as absorbing, in the end, as tracking their personalities had been in times past. In a whole series of applications, from cosmology to mechanics to geometry, algebra promised more exciting discoveries than numerology.

Bruno had read the works of Aristotle and the other influential thinkers of antiquity, and he respected their opinions. But he didn't grant them any special authority just because of their age or their perceived wisdom. He believed that much of what had been learned in the intervening millennia might be equally useful. Moreover, not lacking in ego, he imagined his own intellect as very much on par with that of the ancients: he could build on what they had begun.

Bruno's wanderings included more than two years in England, from the spring of 1583 to the autumn of 1585. When crossing the English Channel for Dover, he apparently used the voyage as a chance to test his ideas concerning the curvature of the Earth. Certainly he had already begun to embrace the Copernican model by this time. He had recently been in Paris and Wittenberg, both of which were home to at least a handful of Copernicans. (In Paris, he likely encountered a French translation of book 1 of *De revolutionibus*, published in 1552.) He was also keenly interested in a philosophical tradition known as Hermeticism, a cult based on obscure sacred texts, believed (incorrectly) by its practitioners to be of Egyptian origin. Followers of Hermeticism believed that nature's secrets could be discerned through careful study, and that this knowledge could, in turn, be used to influence natural forces. In his embrace of Hermeticism, Bruno would have had the sympathy of John Dee, who had similar occult leanings.

While in England, Bruno published two books, both of which contain a vigorous defense of the Copernican theory: *La cena de le ceneri* (*The Ash Wednesday Supper*) and *De l'infinito universo e mondi* (*On the Infinite Universe and Worlds*), both dating from 1584. These were part of a series of six cosmological "dialogues," penned in the Italian vernacular. (Why not Latin? As Giovanni Aquilecchia has noted, it could be partly because of the appreciation of all things Italian in the Elizabethan court—but also, perhaps, because thinkers such as Robert Recorde and Thomas Digges were publishing in their own vernacular.)

In England, Bruno was naturally drawn to Oxford, already a renowned center of learning. There he delivered a series of lectures on cosmology and philosophy. Outspoken and argumentative, Bruno's reputation would have preceded him; indeed, the French ambassador had written to the queen's advisor and "spymaster," Francis Walsingham, to warn of Bruno's imminent arrival in England. (Just for good measure, Walsingham also had a spy working in the French embassy.) And so we can imagine a packed hall each time Bruno approached the lectern. While the details of the lectures are lost, we know that Bruno was met with at least some measure of ridicule. As Rowland points out, this was probably as much the result of how he spoke as what he said: His Latin would have been Italian-sounding, very different from the Latin of the Oxford dons. They may well have mocked him for his diminutive size and his awkward manner. Still, if they listened attentively, they must have at least caught the drift of his arguments. And when he returned to Oxford in August of the same year, we know that he spoke in support of Copernican astronomy. George Abbot, who would later become the archbishop of Canterbury, noted that Bruno "undertooke among very many other matters to set on foote the opinion of Copernicus, that the earth did goe round, and the heavens did stand still; whereas in truth it was his owne head which rather did run round."

WORLDS WITHOUT END

Bruno was likely familiar with Tycho's alternative model of the solar system, and this, too, may have been mentioned in his Oxford lectures, though in the end he clearly preferred the vision of Copernicus:

> For he [Copernicus] had a profound, subtle, keen and mature mind. He was a man not inferior to any of the astronomers who preceded him . . . his natural judgment was far superior to that of Ptolemy, Hipparchus, Eudoxus, and all the others who followed them, and this allowed him to free himself from many false axioms of the common philosophy, which—although I hesitate to say so—had made us blind.

Indeed, Copernicus was "ordained by the gods to be the dawn which must precede the rising of the sun of the ancient and true philosophy, for so many centuries entombed in the dark caverns of blind, spiteful, arrogant, and envious ignorance."

Bruno, however, would go further than Copernicus, embracing the idea of an infinite cosmos and an infinity of worlds (and arguing the case in both of the books he published while in England). Indeed, Bruno was more explicit on the subject of infinite space than even Digges had been:

> There is a single general space, a single vast immensity which we may freely call Void: in it are innumerable globes like this on which we live and grow: this space we declare to be infinite, since neither reason, convenience, sense-perception nor nature assign to it a limit.

Not only is the Earth one of the planets, Bruno argued, but our planet is "merely one of an infinite number of particular worlds similar to this, and that all the planets and other stars are infinite worlds without number composing an infinite universe, so that there is double infinitude, that of the greatness of the universe, and that of the multitude of worlds." Bruno was not the first to entertain such thoughts; indeed, the problem of the "edge of the world" had been debated from antiquity. The ancient

Roman philosopher Lucretius (whose influence we will examine in Chapter 13) pondered the matter in his remarkable poem *De rerum natura* (*On the Nature of Things*). He writes:

> Suppose now that all space were created finite, if one were to run on to the end, to its furthest coasts, and throw a flying dart, would you have it that that dart, hurled with might and main, goes on whither it is sped and flies afar, or do you think that something can check and bar its way?

With infinite space, one might imagine, it is a short step to the possibility of the stars extending off to infinity as well. As we've seen, Thomas Digges was perhaps the first to widely publicize the idea—but a century earlier, a German cardinal, Nicholas of Cusa, had presented a similar argument. Each star, Nicholas imagined, might well appear sun-like, if only we could view it from a close enough distance. Bruno, however, took the idea further, arguing not only for infinite space, but infinite time as well. While most of his contemporaries imagined the universe to be about six thousand years old, Bruno—inspired by the writings of Plato as well as more obscure ancient texts—was willing to grant it an eternal past.

Could Bruno have been influenced by the work of Digges, perhaps encountering his ideas during his stay in England? Historians suspect that even if Bruno never met Digges, he would have at least known of the Englishman's work. As Hilary Gatti notes, Digges's mentor, John Dee, taught mathematics to the courtier and poet Sir Philip Sidney—to whom Bruno dedicated two of his dialogues; and Dee once hosted Sidney and his entourage, after the group had visited Oxford (a meeting at which Bruno is known to have been present).

Digges's almanac—the one with the diagram showing the stars receding to infinity—had been published in 1576, and had gone through at least two editions before Bruno's arrival in England. As Gatti points out, Bruno was not fluent in English and would have needed help with Digges's text—but if he saw Digges's diagram, showing the stars extending outward without limit, he would have understood its meaning instantly.

Did Bruno have any sense of what might be in store for him? He surely knew what his English hosts thought of his sharp tongue, and must have understood the risks of speaking openly on such dangerous subjects—especially if he were to leave the relative safety of England. Mocked in Oxford, Bruno was eventually accused of plagiarism, and it became clear he was no longer welcome there. He was hardly more welcome in London, where, in his own view, he was seen as a dangerous revolutionary threatening to subvert "a whole city, a whole province, a whole kingdom." If the English distrusted Bruno, he was equally repulsed by their manners and their unbridled xenophobia. He writes:

> England can brag of having a populace that is second to none that the earth nurtures in her bosom for being disrespectful, uncivil, rough, rustic, savage, and badly brought up. . . . When they see a foreigner, they look, by God, like so many wolves, so many bears who have that expression on their faces that a pig has when its meal is being taken away. . . .

Bruno was forced to take refuge at the French embassy in London, where he served as a secretary to the ambassador. Although he scorned the supposed intellectuals at Oxford, he nonetheless held the English queen in the highest regard. Elizabeth was, in his words,

> superior to all the kings of this world, for she is second to none of the sceptered princes for her judgement, her wisdom, her advice and her government. As for her knowledge of the arts, her notions of science, her intelligence and expertise in the use of those European languages which are spoken by the erudite and the ignorant, there is no doubt that she compares favourably with all the other princes of our time.

One can sense why the very mention of Bruno's name would have been met with disdain back in his home country. Nearly every word he uttered was an affront to Christianity, whether Catholic or Protestant; that he bestowed praise on a foreign head of state only made matters worse. And his arrogance was beyond the pale. Here was a man who insisted that millennia of religious teachings had been wrong—that only *he* had the wisdom needed to set the record straight. God didn't create the world, Bruno argued; he couldn't have, because God is *a part of nature*. Bruno was an atomist: Echoing the sentiments of Lucretius, he believed that the universe is made up of a myriad of atoms, each imbued with a divine essence; from these atoms were composed not only our own world but an infinity of worlds, all of them teeming with life. Infinity holds the key, "for from infinity is born an ever fresh abundance of matter." Nothing dies; living beings simply take part in what we might call an eternal cosmic recycling program. Heaven and hell are mere fantasies.

An infinite cosmos was problematic on many levels: If the universe was created for man's benefit, why must it be so large? Why all of that extra space? In this infinite cosmology, there is no longer any essential difference between the substance of the Earth and the substance of the heavens, no distinction between the terrestrial elements and the "quintessence" of space. This allowed Bruno to take the next logical step: to imagine that these infinite worlds were populated, perhaps with creatures not so very different from man. But what kind of faith would such creatures have? How would they receive Christ's message? How would they be saved? Christians believe that God sent his only son, Jesus Christ, to offer salvation to human beings. If there were people on other worlds, how would they know of God's word? Would there have to be multiple messiahs and multiple Crucifixions? The very idea was blasphemous. For Bruno, the notion of a singular savior, in the form of Jesus Christ, was no longer tenable. This was too narrow a view of the divine. He believed that "God was present in everything, everywhere, always," as Rowland puts it. Unfortunately, one man's expanded view of God is another man's heresy. The very idea of a multitude of inhabited worlds threatened the established view of man's central—and unique—place in the cosmos, and struck at the very heart of the Christian faith.*

ON THE ROAD AGAIN

Eventually, the French ambassador fell out of favor with his English hosts, and Bruno, his tenant, took to the road once again. He lived briefly in France before moving on to Germany and then to Prague, where the emperor, Rudolf II, had a penchant for filling his court with

* Times, however, have changed, and the Vatican has recently made peace with the idea of alien civilizations. In 2008, the director of the Vatican Observatory, José Gabriel Funes, told the Vatican's newspaper that the existence of intelligent aliens "doesn't contradict our faith" because such beings would still be God's creatures." In the interview, headlined "The Extraterrestrial Is My Brother," Funes also acknowledged that the universe was billions of years old and likely began with the big bang—although it was designed by God and is "not the result of chance." (Ariel David, "Heavens big enough for both God and aliens, says Vatican astronomer." *The Globe and Mail*, May 14, 2008, p. A3.)

all manner of alchemists, astrologers, and magicians. Though London was now a thousand miles behind him, Bruno had at least one more encounter with a noteworthy Englishman: In 1587, none other than John Dee had arrived in town, traveling with a colleague, a shadowy figure named Edward Kelley. (Convinced that others could wield the magic of various crystals more effectively than he could, Dee had employed a number of assistants, Kelley being the latest.) By the time of Bruno's arrival, the emperor had grown tired of the Englishmen's antics, and Dee and Kelley had been banished to the countryside. (Things went downhill further when Kelley tried to convince Dee that the angels wanted the two men to share their wives.) His fortunes in decline, Dee had no choice but to return to England, where, long out of favor at the royal court, he struggled to evade his creditors and those who accused him of witchcraft. He lived out his final years in poverty, his daughter selling off his books one by one to buy food.

For Bruno, however, there was work to be done, and, still residing in Prague, the Italian set out to describe his vision of the cosmos in more detail. The result would be his final poem, *De innumerabilibus, immenso, et infigurabili* (*On the Innumerable, Immeasurable, and Unfigurable*), published in 1591. Latin verse was now his instrument, in homage to Virgil and Horace—and especially to Lucretius, whose *On the Nature of Things* had influenced him deeply. Bruno readily embraced Lucretius's theory of atoms and the void. The atoms were many, and existed in a state of endless flux, while God was everywhere and unchanging. For Bruno, God was "an all-pervading world-soul" that he sometimes compared to the ocean. And the stars? Of course, they appeared to spin around the Earth every twenty-four hours, but Bruno understood that this was an illusion. Wouldn't it be "the mother of all follies," he writes, to imagine

> That this infinite space, with no observable limits,
> Laid out in numberless worlds, (stars is how we define
> them) . . .
> Should be creating but one single continuous orbit
> Around this point, rotating in such measureless circles
> In such a short time?

As with Dee, we have in Bruno a figure whose thinking encompasses science and theology, the rational and the magical. And perhaps the political: As Rowland puts it, Bruno's universe "is a republic of stars, not a monarchy, in which all stars are created equal, all circled by equal 'earths' . . ." It was not a popular view.

"AN IMPERTINENT, PERTINACIOUS, AND OBSTINATE HERETIC"

Bruno eventually returned to Italy—foolishly, we might imagine, unless he *wanted* a martyr's fate. (Perhaps he did.) He first settled in Padua, and then Venice—where one of his patrons turned against him and reported him to Church authorities. He was arrested by the Holy Office of the Inquisition, questioned, and eventually charged with a litany of heresies, including the belief in a multitude of inhabited worlds and an infinite cosmos. (There is no evidence that Copernicanism per se played a role in Bruno's trial—but the theory seems to have become tainted by association. This fact is particularly salient when we think of what lay in store for his countryman, Galileo, three decades later.) Bruno's trial, beginning in Venice and ending in Rome, dragged on for eight years, with Bruno clinging to his beliefs right to the end. He was finally sentenced on February 9, 1600. Bruno, the Inquisition concluded, was "an impertinent, pertinacious, and obstinate heretic." Even with his fate sealed, priests and churchmen continued to try to sway his opinions—to at least save his soul—but to no avail. Eight days later, the sentence was carried out: The prisoner was stripped naked, his tongue was clamped, and, by tradition, he was set on a mule that would carry him to the Campo de' Fiori, where he was tied to a stake and burned alive.

In the curious figures of John Dee and Giordano Bruno we see a peculiar mixture of science and mysticism. Peculiar to our sensibilities, that is; as we have seen, such mixtures were simply part of the intellectual landscape in early modern Europe. We can also view Dee and Bruno as Renaissance polymaths, men for whom all fields of knowledge were inexorably linked. In hindsight, we can look at the gradual acceptance of the Copernican theory—with people like Dee and Bruno as early adherents—as one of the indications that a new worldview was taking shape, one that bore at least some resemblance to the one that defines our world today. This, of course, is partly a reflection of our modern biases, and the tendency to read historical figures as engaging in a struggle to be "like us" is, as mentioned, an obvious and ever-present danger. Still, the gradual decline and fall of the Ptolemaic system surely signaled that change *of some kind* was in the air. Of course, paradigms do not shift easily, and many decades would pass before the "new astronomy" gained widespread acceptance.

Nonetheless, historians have come to see these years as a critical period in Europe's intellectual history. Mordechai Feingold calls these years an "incubatory period" in which great minds grappled with an array of "rival theories, old and new cosmologies, rational and irrational elements of science." If rival theories were at war in late-sixteenth-century England, there were three main battlefields: the university cities of Oxford and Cambridge, and the commercial and cultural heart of the nation, found in London. Academia and commerce both had use for science, but took radically different approaches. The latter was, naturally, far more concerned with practical matters; but, as we will see, even the most down-to-earth merchants and tradespeople saw the value in looking up.

5. *". . . Sorrow's eye, glazed with blinding tears . . ."*

THE RISE OF ENGLISH SCIENCE AND THE QUESTION OF THE TUDOR TELESCOPE

It has been said that necessity is the mother of invention—and fear probably helps, too. In the second half of the sixteenth century, a handful of science-minded Englishmen had been calling for mathematical lectures to be held regularly in London, to aid in the training of sailors and navigators. It didn't come to pass, however, until England had seen its ships doing battle with Spanish forces just off the coast. The armada was defeated, of course—and the importance of a strong navy (and maritime prowess in general) was hammered home. This, in turn, depended on scientific knowledge: Navigators needed to understand mapping systems and the shape of the globe, and they had to calculate and plot routes, skills that were rooted in arithmetic and geometry. Astronomy played a key role as well, since the stars were the primary guideposts for the seaborne traveler. In November 1588—only months after the defeat of the armada—a lectureship covering these subjects was established, along with a detailed plan of action in case a hostile fleet, Spanish or otherwise, were to make its way up the Thames in the future. The annual lecture was delivered in the Staplers' Chapel in Leadenhall, in the commercial heart of London, and was open to the general public. The text of the first of these lectures, delivered by an instructor named Thomas Hood, has survived, and, together with his published books, gives us some insight into his teaching strategy: One must begin with arithmetic, geometry, and basic astronomy; when the student has mastered these, he can move on to the use of maps and various instruments, and tackle practical problems in surveying and navigation. Because theoretical aspects of astronomy were not a primary concern, we don't know what Hood thought of the Copernican theory—but we know that he routinely used Copernicus's figures and calculations. His students, as Francis Johnson notes, "had little scholarly training but were inflamed with a passionate desire for useful knowledge."

THE CLEARINGHOUSE

What started as a lecture series eventually became a college. In the final years of the sixteenth century, London had, for the first time, a school dedicated almost exclusively to what we would now call science. Founded in 1597, Gresham College was named for its founder, a prominent merchant named Sir Thomas Gresham. Students at Gresham learned practical skills associated with navigation, commerce, and medicine. Astronomy and geometry figured prominently in the curriculum; one could also study divinity, music, and rhetoric. In the first half of the seventeenth century, the college would become, as Johnson puts it, "a general clearinghouse for information concerning the latest scientific discoveries."

The learned men who made up the Gresham faculty had known one another for years, in some cases decades, before the founding of the college. They also had close ties to the city's craftsmen and instrument makers. (I suspect that the college, had it been founded in the twentieth century rather than the end of the sixteenth, would have called itself a "polytechnic institute.") The college would go on to play a crucial role in the founding of one of the world's first scientific societies: Many of those who met and taught at Gresham would help to establish the Royal Society of London in 1660.

While Hood may have been hesitant to fully support Copernicanism, other members of the faculty at Gresham seem to have been largely committed to the new theory. Allan Chapman notes that "pretty well all the Gresham Astronomy Professors after 1597 were Copernicans: Henry Briggs, Henry Gellibrand, John Greaves, Sir Chrisopher Wren, Robert Hooke, and so on." Chapman's list gets us a little ahead of ourselves—Christopher Wren lived well into the eighteenth century—but even so, it is intriguing to think of London's aspiring young seamen, in the final years of Elizabeth's reign, pondering whether the sun moved around the Earth or the Earth moved around the sun.

* * *

The spirit of inquiry embodied by Gresham College spilled out beyond its walls; indeed, it seemed to reflect the mind-set of an ever-increasing number of ordinary Londoners. This was enabled, in part, by a sharp rise in literacy: More than a hundred grammar schools had been built in England in the previous half century, and more people could read and write than ever before. The publishing business was booming, with dozens of printers and booksellers at work in the capital (many of them plying their trade from the churchyard beside St. Paul's Cathedral). Those who couldn't buy could borrow: Libraries were becoming common throughout England in the early seventeenth century, though members of the public were able to borrow books from London's Guildhall from the early fifteenth century. As Johnson notes, at least one out of every ten books published in England between 1475 and 1640 dealt with the natural sciences. Some of these volumes were written by the scientists themselves, others by what we would now call science popularizers; and then, as today, they varied widely in quality. The influence of these books, Johnson says, "was not confined to scholars, or to those who had studied at the universities, but extended throughout all literate classes." (We will take a closer look at the grammar schools and also the world of books and publishing in the next chapter.) Along with books, customers could pick up newssheets, treatises on medicine and surgery, and mathematical instruments that came with booklets explaining their use. Almanacs were in particularly high demand, and one gets a sense of their popularity from a scene in *A Midsummer Night's Dream*: The "mechanicals" (craftsmen) are planning a performance of the play *Pyramus and Thisbe*; because the lovers meet by moonlight, they want to know if the moon will be shining on the night of their performance. Bottom demands, "A calendar, a calendar! Look in the almanac; find out moonshine, find out moonshine!" (3.1.49–50).

INSTRUMENTS OF KNOWLEDGE

Even though, as mentioned, the word "science" had not yet acquired its current meaning, many Londoners were in fact earning their living from scientific pursuits. There were mathematicians and doctors, botanists and apothecaries, builders and inventors. They worked in hospitals, laboratories, and family-run workshops. Craftsmen worked with metal, wood, and ivory; they built, among other devices, theodolites for surveyors; rangefinders and gunsights for artillery officers; astrolabes, quadrants, cross-staffs, and backstaffs for navigators; and drawing instruments for a multitude of professions.

This burst of activity was enabled, in part, by the capital's unique position: Though located in the Continent's northwest corner, London was effectively at its crossroads. To live in the English capital at the close of the sixteenth century was to bear witness to an endless parade of new peoples, new inventions, and, perhaps most significantly, new ideas. As Deborah Harkness puts it in her wonderful book *The Jewel House* (2007): "Every ship that put in at a London dock might contain new materials that had to be classified and understood, each new book rolling off the presses at St. Paul's could contain a radical idea about the natural world, and the experiments undertaken in London had, at any moment, the potential to bring long-held beliefs into question." Novelty was everywhere in the bustling city; indeed, there was an *appetite for the new.* On every corner, one might stumble across some peculiar wonder from a far-off land: an ostrich egg, money from China, a canoe from Lappland, a stuffed, two-headed snake. In the quest for knowledge, one might still turn to the words of ancient philosophers—but their limitations were becoming increasingly obvious. Londoners by this time were creating *new* knowledge—and indeed much of what they were finding was either at odds with what the ancient writers had described, or was simply too novel to have been known to them. It is hardly surprising that the very word "news" dates from this period.

Not only literacy but also numeracy was on the rise. Private tutors taught mathematics to the city's merchants and their apprentices; some instructors offered room and board to their pupils. An educator named Humphrey Baker boasted that he taught "after a more plain manner than has heretofore been usually taught by any man within this City." One of his favorite tools was the now ubiquitous "math word-problem." In his book *The Well Spryng of Sciences* (1562), Baker posed this problem to his readers:

> Three merchants have formed a company. The first invested I know not how much, the second put in 20 pieces of cloth, and the third had invested £500. So at the end of their business, their gains amounted to £1000, whereof the first man ought to have £350, and the second must have £400. Now I demand: how much did the first man invest, and how much were the 20 pieces of cloth [worth]?

A merchant could easily face a variation of a problem such as this, and lessons from teachers like Baker gave some measure of preparation. (We also have here the forerunner of all those vexing exam questions that cause so much grief for students to this day—"a train leaves Chicago heading east at eighty miles per hour, while at the same time a train leaves New York heading west. . . .") Baker also saw to it that his pupils could use instruments such as the quadrant, square, staff, and astrolabe—devices crucial in surveying, navigation, and astronomy. The shops of instrument makers—English born, and also French and Flemish—lined the busy shopping streets above the Strand and Fleet Street in the heart of London. While many of these devices were imported from the Continent, a growing number of local instrument makers were crafting equally fine products. "By the end of Elizabeth's reign," writes Harkness, "Londoners had fully embraced mathematical instruments, and many of her citizens had achieved levels of mathematical literacy that previous generations would not have dreamed possible."

At least some of these instrument makers knew about Copernicanism and took the new theory into account when developing their instruments. One of these local craftsmen, John Blagrave, designed a new astrolabe in 1596 in accordance with the Copernican model and, in a book describing its operation, made it clear that he accepted the theory as more than a mathematical convenience. In fact, his support for the heliocentric model was set right on the book's title page, which noted an approach in which ". . . agreeable to the hypothesis of Nicolaus Copernicus, the starry firmament is appointed perpetually fixed, and the earth and his horizons continually moving from west toward the east once about every 24 hours. . . ." In traditional astrolabes, the horizon was represented by a fixed metal plate, while the brightest stars were inscribed (actually pierced as small holes) on a movable plate that rotated relative to the main plate. In Blagrave's astrolabe, however, it is the Earth—that is, the observer's horizon—that moves, while the stars remain fixed. As Johnson notes, aside from his countryman Thomas Digges, Blagrave "probably did more than any other sixteenth-century Englishman to disseminate an intelligent knowledge of the Copernican theory." The new cosmos, once just an abstract idea, could now be held in one's hands. And it wasn't just Blagrave. Nicolas Hill, a natural philosopher active at the same time, was convinced of the Earth's rotation and of the Copernican theory, and developed a version of the atomic theory. Mark Ridley, a doctor, was similarly attracted to the Copernican view, and published a treatise on magnetism. (Mind you, scientists on the Continent continued to lead the way: In 1551, Gerard Mercator had built the first celestial globe based on Copernicus's data. And it is worth noting that John Dee had worked with Mercator at the time he was working on his globe.)

Shopkeepers, carpenters, clockmakers, surveyors, sailors—all were motivated to learn and master basic mathematics. One needed mathematical knowledge to sort out different measures of weight, dimension, and currency; more generally, it was said to sharpen the mind. In his popular book on arithmetic, *The Ground of the Artes*, Robert Recorde described mathematics as "the ground of all men's affairs." Without mathematical literacy, he noted, "no tale can be long continued, no bargaining without it can be duly ended, nor no business that man has justly completed." It's hard to know exactly how many mathematical books were published in London in the sixteenth century, since many have been lost; but historians put it at about five per year in the early years of Elizabeth's reign, with texts on navigation and surveying proving the most popular; and the number was certainly higher by the century's close.

OXFORD AND CAMBRIDGE

With all of this activity—all of this learning—happening in London, one may reasonably inquire what was going on up the road at Oxford and at Cambridge, where professors had been professing and students had been studying for half a millennium. It is sobering to remember that the university at Oxford, the second-oldest surviving university in the world (after the University of Bologna in Italy), was already five centuries old by Shakespeare's time. In fact, the actual date when the University of Oxford was founded is not known, though the record of teaching there goes back to 1096, and it grew rapidly after 1167, when English students were prohibited from studying at the University of Paris. (We might note that even Oxford's "New College" dates from 1379.) The university at Cambridge was founded in 1209 when some disgruntled Oxford students fell into a dispute with townsfolk and fled eastward.

The students were younger back then: Officially, one had to be fifteen, but the sons of noblemen were often admitted earlier. (Robert Devereux, the Second Earl of Essex and one of Elizabeth's favorites, was admitted at age ten, though he didn't matriculate until two years later.) A bachelor of arts normally took four years, a master of arts an additional three. If money helped one get in, it could also help you get out: For those who had not yet met the BA requirements by the end of the fourth year of study at Oxford, ten shillings would get you your degree. While the student was enrolled, discipline was strict. At Oxford, students found lurking about inns and taverns, or even tobacco shops, could be flogged; the same penalty awaited those with the gall to play football on university grounds.* (Cambridge may have been slightly more laid-back: There, football was deemed a "legitimate" sport, along with archery and quoits, a game similar to horseshoes.) Not everyone was cut out for the scholarly life. In *Twelfth Night*, the buffoonish Sir Andrew Aguecheek says he always regretted that he didn't attend university: "I would I had bestowed that time in the tongues [i.e. learning languages] that I have in fencing, dancing, and bear-baiting. O, had I but followed the arts!" (1.3.90–93). The "arts" he refers to are the so-called liberal arts that had formed the backbone of Western higher education since ancient times—and which Prospero, in *The Tempest*, says he studied (". . . and for the liberal arts / Without a parallel; those being all my study . . .") (1.2.73–74). The standard program of study consisted of the "trivium" of grammar, dialectics, and rhetoric, along with the "quadrivium" of arithmetic, geometry, astronomy, and music—a program of study which, as J. A. Sharpe puts it, was "basically the education appropriate to the free man of a Greek city-state in the fourth century B.C." The natural philosophy taught at the two august institutions had been Aristotelian through and through—and so it remained, by and large, in the sixteenth century. But a student's experience depended critically on who his teachers were. As Francis Johnson puts it, education for this select group of young men would have been "superficial and elementary"—unless the student was lucky enough to be taught by one of the handful of brilliant young mathematicians "during the brief period of their active connection with the university." If a student did receive a first-class education, it would most likely have been an accident, "entirely dependent upon the enthusiasm and enterprise of individual scholars on the faculties." The keenest learners would have shunned the dry Latin texts offered by the schools in favor of the more accessible—and more up-to-date—popular works.

Given that the universities at this time were not exactly hotbeds of innovation, one might expect that any astronomy that was taught would

---

* While tennis was enjoyed by the nobility, football ("soccer" to North Americans) was seen as a vulgar sport suitable only for the lower classes. In *King Lear*, Kent contemptuously calls Oswald a "base football player" (1.4.74).

have been strictly Ptolemaic. And yet, perhaps surprisingly, Copernicus's theory did come up for debate, on occasion, in the second half of the sixteenth century. Certainly the universities on the Continent were ahead of the game: By 1545, Erasmus Reinhold's *Commentary*—published a few years earlier, and containing a favorable reference to the Copernican theory—had become the standard astronomy textbook at the University of Wittenberg (where Hamlet is said to have studied). By 1560, as Paula Findlen notes, students at Wittenberg were learning astronomy from Copernicus's own book; at Salamanca, by the 1590s, *De revolutionibus* was required reading. This doesn't mean that Copernicanism had triumphed; rather, it was being taught as an alternative. It "enriched the calculating skills of astronomers," Findlen writes, without necessarily making them confront problems in cosmology or physics. "It was possible to read Copernicus as a manual rather than a manifesto."

And so it was at Oxford and Cambridge. At Oxford, records from 1576—the same year that Digges published his treatise on the Copernican system—show that one of the questions assigned for disputation for prospective MA students was "*An terra quiescat in medio mundi?*" ("Is the entire earth at rest in the middle of the world?") We don't know which side was argued by the student and which by the proctor, but, as Johnson puts it, "we may be sure that the theories of Copernicus and the opposing doctrines of Aristotle were the chief subjects of debate."

Even so, historians are divided: John Russell writes that there is "no good evidence that Copernican ideas had made any serious impact [at Oxford] at this time." And we might recall the mockery that Bruno faced when he lectured on Copernicanism at Oxford in the 1580s. But Mordechai Feingold notes that the heliocentric model was taught, "and sometimes taught well," even if those who taught it didn't necessarily accept it fully. Typically, the instructor left it up to the students to weigh the arguments and reach their own conclusions. They were welcome to read further on their own, and the more gifted among them undoubtedly did just that.

A student named Edmund Lee, for example, kept a notebook with commentaries on numerous scientific ideas spanning the fields of physics, astronomy, and mathematics; among his dense notes, he comments favorably on the Copernican system. The mathematician Sir Henry Savile, meanwhile, was appointed as the first professor of astronomy and geometry at Oxford, and taught there in the 1560s and 1570s. His notebooks, now in the Bodleian Library at Oxford, show that he taught from

Ptolemy's *Almagest*—but with references to Copernicus, including a chapter-by-chapter comparison between the *Almagest* and *De revolutionibus*, as well as other medieval and contemporary thinkers.* His own leanings seemed to favor the new cosmology, and we catch a glimpse of his enthusiasm from a remark jotted in his notebook, "*Copernicus Mathematicoru Modernoru Priceps*" ("Copernicus, the prince of modern mathematicians"). William Camden, an Oxford student and friend of Savile, may have owned his own copy of *De revolutionibus*, and, like Thomas Digges, made careful observations of the nova of 1572. At Cambridge, a student named John Mansell (later Master of Queens' College) responded to a question on the structure of the solar system, in 1601, by defending Copernicus and describing the heliocentric model in detail. The works of Copernicus and Digges could be found at the library in Cambridge by 1580, along with astrolabes, quadrants, globes, and elaborate sundials. A couple of decades later, an Oxford tutor named Richard Crakanthorpe again used Aristotle and Ptolemy as his starting point, but covered all the latest astronomical observations and ideas: the new star of 1572; the great comets of 1577 and 1580; the telescopic discoveries of Galileo. We know that he consulted the works of Kepler and Digges, as well as Galileo's *Siderius Nuncius* (*The Starry Messenger*). Sir William Boswell, a Cambridge mathematician, corresponded with Galileo himself, and played a vital role in making the Italian scientist's work known in England.

However much theoretical science was being disseminated at the two universities, they were almost certainly lagging behind the capital in terms of practical learning. Although written a few decades later, a letter penned by mathematician John Wallis is telling. When he moved from Cambridge to London in the 1640s, he found a greater number of people interested in his craft, "[for] the study of *Mathematicks* was at that time more cultivated in London than in the Universities." We have also seen that England was hardly a scientific backwater; the notion that it lagged far behind continental Europe is simply unfounded. One sign is the relative openness to the Copernican theory: While Tycho Brahe in Denmark and Christopher Clavius in Rome had been vocal opponents of the

---

* Full confession: I can't actually read Latin, beyond picking out a few key words; however, it would have been a shame to visit Oxford without taking the time to view some of these remarkable documents firsthand. This, of course, meant reciting the Bodleian Library's centuries-old oath, which includes a promise ". . . not to bring into the Library, or kindle therein, any fire or flame. . . ." If only those in charge of the ancient library at Alexandria had been as vigilant. . . .

heliocentric model, we can add Blagrave, Hill, and Ridley to our ever-growing list of English thinkers—with Recorde, Dee, Thomas Digges, and perhaps a handful of others—who embraced the "new astronomy."* By the latter years of the sixteenth century, England, as we've seen, was far from being an intellectual hinterland. As Francis Johnson writes, "[In] England, perhaps more than in any other country, an intelligent knowledge of the Copernican theory was spread among all classes of practical scientific workers before 1600, and nowhere was there a keener interest in the implications of the new astronomy, or a more earnest search for a satisfactory physical explanation of various features of the Copernican hypothesis."

Intriguingly, some of the people whom one might imagine to have eagerly embraced Copernicanism in fact rejected it. The philosopher and statesman Francis Bacon was just such a thinker: It's not that the Copernican model was too radical for his tastes; it's that there was not, at this early stage, enough direct observational evidence in support of the new theory. It is easy to imagine, in hindsight, that the Copernican theory was "obvious"—but of course it was anything *but* obvious, and one might legitimately harbor doubts about the heliocentric model, at least until the telescopic work of Galileo.

THE PHILOSOPHER-STATESMAN

As we noted in the introduction, the standard view is that Shakespeare lived "too early" to have been a witness to the Scientific Revolution. But we should remember that Bacon, one of the key figures of modern science, was an almost exact contemporary of the playwright, entering the world three years ahead of him, and departing ten years after him. Bacon's first important scientific work, *The Advancement of Learning*, was published in 1605—around the time that Shakespeare was finishing *King Lear* (and in fact Bacon had written about the nature of science two years earlier, in *Valerius Terminus: On the Interpretation of Nature*, although the work is fragmentary and was never published).

Francis Bacon (1561–1626) was a well-connected man: His aunt was married to Queen Elizabeth's key advisor, William Cecil. Bacon studied

* Another such figure is Robert Fludd (1574–1637), an English philosopher, medical doctor, and alchemist. Fludd was a vocal supporter of the Copernican model—which he tied to "a mystical explanation for the formation of the universe involving an angelic hierarchy and the mutually influential macrocosm and microcosm," as Lesley Cormack describes it (Cormack, "Science and Technology," p. 517).

law, served as a member of Parliament, and eventually held the titles of Attorney General and Lord Chancellor. But we remember Bacon not for his statesmanship but for his philosophy. He is considered the father of empiricism, the idea that knowledge ultimately rests on what we can observe and study via the senses. For Bacon, science was about much more than esoteric knowledge, and he refused to accept ancient wisdom just because it's ancient. He had grand aims for science, whose "explanations take the mystery out of things." It wasn't foolproof—the experiment may be flawed; the observer may make a mistake—but the process is self-correcting. Yes, the senses may sometimes deceive; "but then at the same time they supply the means of discovering their own errors."

Bacon wasn't himself a scientist; in fact, he made no significant discoveries on his own. The one experiment that he did (allegedly) attempt—at least, the only one we have a detailed record of—also appears to have killed him: He is said to have caught a chill while trying to determine if one could preserve a chicken by stuffing it with snow on a freezing March day; a few days later he was dead, from either bronchitis or pneumonia.* Nonetheless, Bacon thought a great deal about *what science ought to be*. In *The Advancement of Learning*, he sets out to partition science into various branches, including physics, metaphysics, mathematics, astronomy, engineering, and medicine—although we should note that he included theology, poetry, and drama among the sciences, and considered them to be equally deserving of study. (God, he says, made the world—but not in order for us to be mystified by it. As Philip Ball puts it, Bacon sees the world as "an intricate puzzle," one that God hopes mankind will rise to the challenge of solving.) Bacon argues passionately for the importance of scientific learning, asserting that natural philosophy, and the improved technologies it would lead to, would better the lot of mankind. He declares that "heaven and earth do conspire and contribute to the use and benefit of man." In *The New Atlantis* (1627), Bacon describes something like the ultimate scientific laboratory—he calls it Solomon's House—in which many of these new technologies are presented in detail. Here we find, as John Cartwright notes, dozens of ideas "that anticipate the technology that science did deliver in the centuries that followed." These include "the genetic engineering of plants

* If not a chicken, an edible bird of some kind. The account from John Aubrey refers to it as a "fowl."

and animals, zoological gardens, robots, telephones, refrigerators, weather observation towers, and all sorts of flying machines."

SCIENCE FIT FOR A QUEEN (AND A KING)

At least some measure of what we would today call "science literacy" filtered through all levels of society, including the very top. Queen Elizabeth herself was something of a science buff. We have already noted her consultations with Thomas Digges, as well as her close association with John Dee, who advised her on both astronomy and astrology. According to Bacon, Elizabeth read extensively about the philosophical and scientific ideas of the day: "This lady was endued with learning in her sex singular, and rare even amongst masculine princes," he wrote, "whether we speak of learning, of language, or of science, modern and ancient, divinity or humanity: and unto the very last year of her life she accustomed to appoint set hours for reading, scarcely any young student in a university more daily or more duly."

We also know that Elizabeth showed an interest in the latest gadgets and gizmos, at one point commissioning the construction of a complex set of musical chimes. She even wore a tiny "alarm watch" on her finger; at the appointed hour, a small prong would extend and give a gentle poke. Of course, timepieces weren't just any gadgets; they were, in fact, among the most sophisticated mechanical devices of the age. The first mechanical clocks had appeared in the late thirteenth century, with the first pocket watches dating from the early 1500s. (I explore the history of clocks and timekeeping in some detail in Chapter 5 of my earlier book *In Search of Time*.) But clockwork wasn't just for clocks: A German visitor, strolling along Whitehall in 1598, described a sophisticated "jet-d'eau" which splashed passers-by: It employed "a quantity of water, forced by a wheel, which the gardener turns at a distance, through a number of little pipes" and "plentifully sprinkles those that are standing around."

Elizabeth's successor, James I, was even more passionate about science and technology. He admired the work of Kepler and Tycho—as mentioned, he once visited Tycho at his island observatory—and even composed a brief verse in praise of the Danish astronomer.* Elaborate

---

* Perhaps curiosity ran in the Stuart family. His great-grandfather, James IV of Scotland, wondered what language a child would speak if raised in isolation and without exposure to any specific "mother tongue." A chronicler named Robert Lindsay reports the king's cruel but at least quasi-scientific experiment: He "caused two children to be marooned with a deaf-and-dumb nurse on the island of Inchkeith . . . [and] furnished them with al Necessaries . . . desiring to understand

mechanical devices held him in awe. Word of his interest reached the Continent, and in 1609 Rudolf II, ruler of the Holy Roman Empire, presented James with a clock and a celestial globe. James's son, Prince Henry, seems to have shared his father's scientific leanings. As Scott Maisano has pointed out, the young Henry relished visits from French and Italian engineers and inventors, and acquired a substantial library; in 1610 he even asked an Italian contact for "the latest book by Galileo." (Henry would have inherited the throne had he not died of typhoid fever at the age of eighteen; the crown would instead pass to his younger brother, Charles.)

James's relationship with the Dutch scientist Cornelis Drebbel is especially noteworthy. Drebbel moved to London in 1604, at James's request, and presented the king with a (purported) perpetual motion machine—apparently a sophisticated clockwork mechanism that displayed the time, date, and season. The Dutchman is also remembered for building the first working submarines. Using plans drawn up by mathematician William Bourne, Drebbel built and demonstrated a series of underwater craft during his time in London. The craft were built from wood and covered with grease-soaked leather, and employed pigskin bladders that could be filled with water or emptied in order to dive and to ascend, while oars provided forward propulsion. The largest of these vessels is said to have carried up to sixteen passengers and crew, and could remain underwater for three hours—long enough to travel from Westminster to Greenwich and back, at a depth of about fifteen feet—with those on board breathing through a hollow tube that reached the surface.* When James himself was offered a ride, he became the first king in history to travel underwater.

The submarine was of course a novelty—too novel even for the navy to fund further research at this early stage—but just about everything else involving the seas and navigation was regarded as a national priority.

---

the Language the Bairns [children] could speak when they came of lawful age." Unfortunately, we have no record of the result. Lindsay provides only the rumored outcome: "Some say they spak guid Hebrew. But as to myself, I know not" (quoted in Guthrie, p. 1193).

* Even with the breathing tube, carbon dioxide would have built up within the vessel, probably to dangerous levels. It's not certain how Drebbel solved this problem. One guess is that he heated "nitre"—either potassium nitrate or sodium nitrate—in a metal pan, causing it to release oxygen; the residue would also act to absorb carbon dioxide. (A diarist noted that Drebbel used a "cheymicall liquor" that worked to "restore the troubled air.") (http://www.nmmc.co.uk/index.php?/collections/featured_boats/the_drebbel_submarine)

This push for maritime supremacy was inexorably linked to the English drive to explore and exploit the New World, which had begun under Elizabeth and continued under James, and to defend trade routes, particularly against competing Spanish, French, and Dutch forces. The attempted English settlement on Roanoke Island (in present-day North Carolina) failed, but the later effort at Jamestown, Virginia, succeeded, giving England a permanent presence in America from 1607. A second foothold to the north—at Plymouth, on Massachusetts Bay—would be established in 1620. But the colonies were a small operation in a vast and distant land, and in Elizabeth's time, the New World was almost completely unexplored. (That is, unexplored from the European point of view; the natives who had been living there for millennia of course knew parts of the Continent quite well.) The various Englishmen who helped to explore the American continent are well known, the figure of Sir Walter Raleigh being almost a household name (and deservedly so; he was a poet, linguist, philosopher, astronomer, and all-around man of action). Less well known is the scientist who sailed to the New World under his patronage, a man of multiple talents named Thomas Harriot (ca. 1560–1621).

EXPLORING A NEWFOUND LAND

Harriot was probably born in or near Oxford. He enrolled at the university at the age of seventeen, taking his entrance oath at the Church of St. Mary the Virgin. He lived and studied nearby, at St. Mary's Hall. (The church is still there, and still looks much as it did in Harriot's day. St. Mary's Hall—not actually a building but rather a kind of collective—has since been absorbed into Oriel College.) Graduating in 1580, Harriot moved to London, and soon after began working for Raleigh, teaching mathematics and navigation to Raleigh's sailors. He set out across the Atlantic with Raleigh's men in 1585, acting as chief scientist and surveyor for the new colony of Virginia. His observations of the new land and its people appeared in a book called *A Briefe and True Report of the New Found Land of Virginia*, published in 1588—the first book about the New World to be printed in English.

Harriot turned out to be an acute observer: He described in detail the vegetation, animal life, and natural resources of the land, and took an intense interest in the indigenous people he encountered. He studied native customs and religious practices, and learned the Algonquin language, even developing a system for transliterating Algonquin words into English so they could be written down. He also may have taken a telescope-like device with him on the voyage. Harriot notes that on the tenth day at sea—a short way into the eleven-week voyage—he observed a solar eclipse from the deck of his ship. Whether he used his optical device to aid in observing the eclipse, we simply don't know; but he definitely showed off some kind of optical instrument to the native people that he encountered in America. He mentions a number of objects which he says fascinated the local inhabitants, including "a perspective glasse whereby was shewd manie strange sightes." The natives were also much taken with the guns, books, and clocks that he showed them, all of which "were so strange to them, and so far exceedeth their capacities to comprehend . . . that they thought they were rather the works of Gods rather than of men. . . ." Harriot was accompanied on the voyage by an artist named John White, whose detailed sketches would appear alongside the scientist's text in *A Briefe and True Report.* The book was partly propaganda; one of its aims was to encourage settlement, and to portray the new lands as bountiful and rich. Yet Harriot himself was more engaged in the new land and its people than many other early European visitors were.

After spending just over a year in Virginia, Harriot returned to England. He continued to work for Raleigh, living briefly in Ireland, where he administered one of Raleigh's estates. He then returned to London, where, beginning in 1595, he found employment with a new patron, Henry Percy, the Ninth Earl of Northumberland, known as "the Wizard Earl" because of his passion for science.

ENGLAND'S GALILEO (ALMOST)

Over the next decade or so, Harriot seems to have developed a keen interest in astronomy. In 1607 he made naked-eye observations of the comet now known as Halley's Comet (long before Edmond Halley's birth). Soon—probably sometime in the summer of 1609—he began to use a new invention from Holland, the telescope, to study the night sky. Clearly this device was different from the perspective glasses that he and other English scientists had been using in earlier decades, and that he had taken to Virginia nearly twenty-five years earlier. Whether it was an improvement on a familiar device or an entirely new design isn't clear—but it definitely opened up new vistas in a way that the earlier instruments had not. Using a device that magnified celestial objects by a factor of six, Harriot was able to sketch the surface of the moon, observe sunspots, and determine the sun's rotation speed. He also observed the moons of Jupiter, calculating their orbits, and observed the phases of Venus.

This was around the same time that Galileo was making his groundbreaking observations in Italy—the observations that would end up in *The Starry Messenger*, published in 1610—and it is only natural to ask who was first. While some of Harriot's observations postdate those of Galileo, his study of the moon, at least, seems to come first. One of his lunar drawings is dated July 26, 1609, while Galileo's first observations likely date from November or December of that year. Allan Chapman writes, "As far as we can tell from the historical record . . . it was Thomas Harriot who became the first person to look at an astronomical body through a telescope, on or before 1609 July 26, when he came to realize that the image of the moon produced by it was very different from what was seen by the naked eye, although he did not publish his discovery." The big difference, of course, is that while Galileo shouted his discoveries from the mountaintops, so to speak, Harriot kept his findings under wraps, perhaps telling a few trusted associates but no one else. (Indeed, more than 150 years would pass before Harriot was recognized for his astronomical work. His manuscripts are now in the British Museum, and at Pentworth House in Wiltshire.)

However, he had enough of a reputation in his day that if someone in England wanted a telescope, Harriot was known as the man to ask. Correspondence between Harriot and Sir William Lower, an astronomer and member of Parliament, has survived, and shows that by February 1610—still a month before the publication of Galileo's book—Lower was using telescopes supplied by Harriot to observe the moon. The letters also reveal that Harriot had received a copy of *The Starry Messenger* soon after it was written; indeed, within three months of the book's publication, he had summarized its contents to Lower, and Lower had written back. Clearly, the publication of Galileo's book marked a turning point (which we will examine more closely in Chapter 9).

Although Harriot had been making telescopic observations of the night sky before reading the Italian scientist's book, his encounter with *The Starry Messenger* seems to have sparked a renewed interest in all things astronomical, and he appears to have begun a regular observing program at this time. We know that from 1610 to 1613 Harriot made numerous astronomical observations, most of them carried out from the grounds of Syon House, Northumberland's grand estate near Richmond,

Fig. 5.1 The grounds of Syon House, in west London, part of the Earl of Northumberland's estate. In the late sixteenth century, it was home to the ninth earl, Henry Percy, patron of astronomer Thomas Harriot. It was from here that Harriot observed the night sky with a telescope—beginning, it seems, a few months ahead of Galileo. Author photo

---

on the outskirts of London (figure 5.1). He made detailed maps of the lunar surface, produced nearly a hundred drawings of Jupiter with its four bright moons, and made several dozen drawings of the surface of the sun. We might note that his drawing of the full moon, likely dating from the summer of 1610, is, arguably, *better* than Galileo's (compare figure 5.2 with figure 9.1). Though he doesn't have Galileo's knack for realistic, three-dimensional topography, Harriot does a better job of showing various prominent craters and the lunar "seas" (now known to be plains of hardened lava) in their true positions. (Perhaps, having seen Galileo's engravings by this point, he had become more confident of his own observing skills.)

Harriot kept up a diligent correspondence with other like-minded scientists of the day—including Kepler (at that time living in Prague), with whom he discussed optical theory and techniques for building telescopes. Within England, he had a diverse array of protégés who watched the sky and reported back to him. By the time of his death, Harriot had accumulated a sizable collection of telescopes, which he bequeathed to

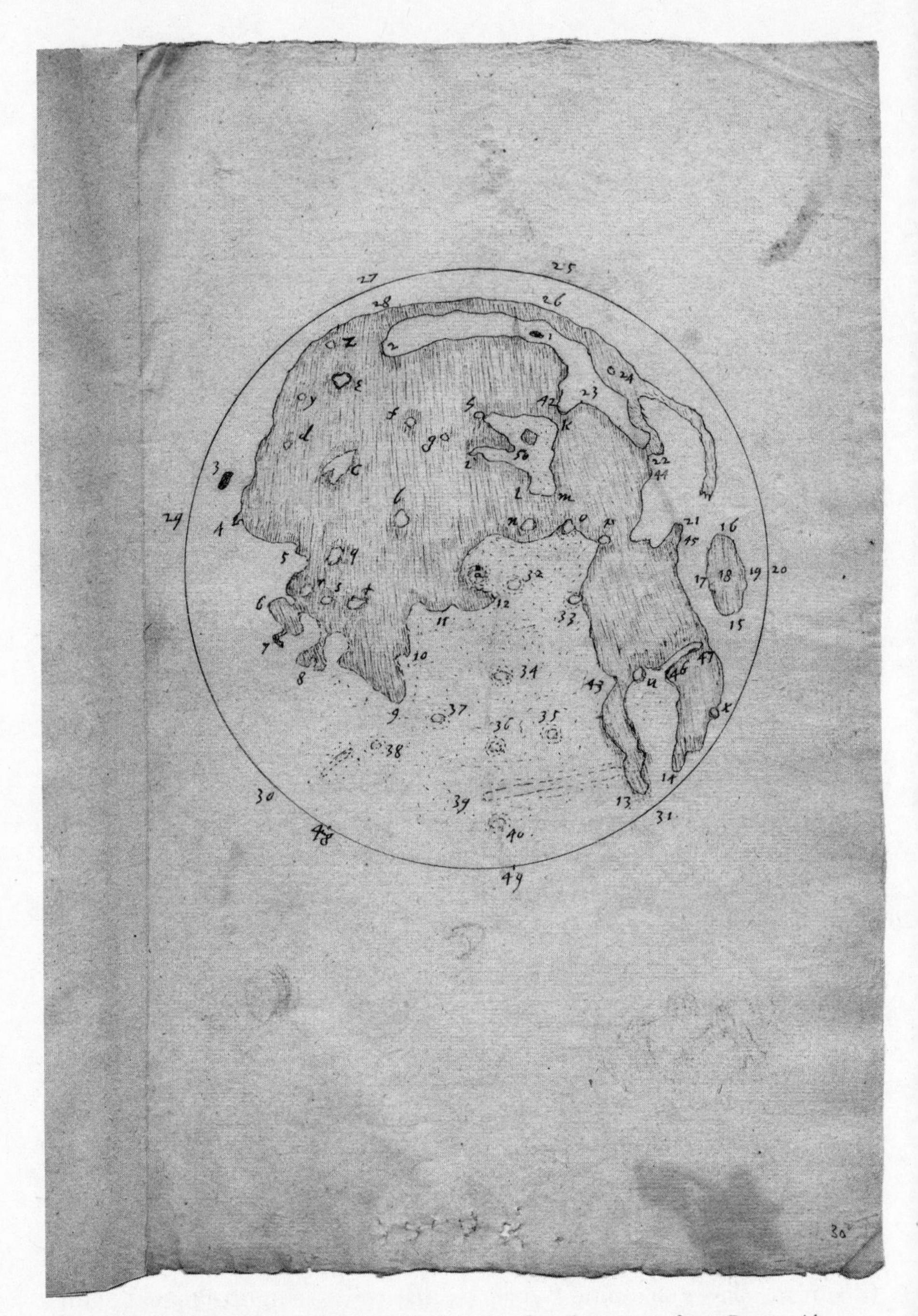

Fig. 5.2 Thomas Harriot's drawing of the full moon, from the summer of 1610. Drawn with telescopic aid, the sketch shows the relative sizes and locations of various lunar features (if not their topography) more accurately than those of Galileo (compare with Fig. 9.1). Lord Egremont

his friends and patrons. His will states that he left to the Earl of Northumberland his "two perspective trunks wherewith I use to see Venus horned like the Moone and spotts on the Sonne."

Harriot seems to have been every bit as well connected as John Dee. Indeed, historians suspect that all of the notable English philosopher-scientists we've looked at so far would have likely known each other. Harriot certainly knew Dee; in his diary, Dee mentions two visits by Harriot in the early 1590s. And since Dee was Digges's teacher, it is likely that Digges knew Harriot as well. Harriot likely also knew Gilbert, and may even have met with Bruno at some point during the Italian's stay in the south of England. (We know that Bruno's books were in Northumberland's library.) "There can be no question," writes Johnson, "that Harriot and his group of astronomers in England were not only fully abreast of all the latest developments in their science, but were also carrying on independent researches of their own along similar lines." We might also note that Northumberland's younger brother, Sir Charles Percy, was a patron of the arts and a Shakespeare fan; he once commissioned a production of *Richard II* from Shakespeare's acting company. There is no evidence that Shakespeare ever met Harriot himself, although it is certainly possible, and it seems likely that the playwright would have at least heard of his work. When I spoke with Stephen Greenblatt, he weighed in on the question of who knew whom in this tightly knit world: Greenblatt notes that Raleigh and Harriot were both said to have had connections with Shakespeare's colleague Christopher Marlowe; and this group "could well have intersected with Shakespeare himself." (We will look at Marlowe in more detail in Chapter 14.) Greenblatt has also identified a link (albeit a tenuous one) connecting Shakespeare and Bruno: In *Will in the World*, he notes that Shakespeare's friend, the printer Richard Field, had served as an apprentice to Thomas Vautrollier, who had published Bruno's books. During our interview, he acknowledged that "there is a possibility that Shakespeare could have encountered Bruno," although it is somewhat of a long shot.

Harriot, like Bruno, seems to have been a Copernican, and his telescopic observations would certainly have led him in that direction even if he had not been previously inclined toward the new astronomy. And like Bruno, he endorsed the plurality of worlds as well as the atomic theory of matter first proposed by the Greek philosophers. Although both theories could be considered dangerous, the atomic theory was, at the time, certainly the more repugnant of the two, having long been associated with irreligion and atheism. As with Dee, accusations of atheism haunted Harriot, especially in his later life. Perhaps in hesitating to publicize his work he was simply thinking of his own safety (or at least his peace of mind). As Allan Chapman puts it, Harriot "was quite happy to accumulate 35 years' worth of work, and publish nothing. . . . One suspects that, like Copernicus, he found controversy to be distasteful."

Harriot's work in mathematics had more of an impact than his astronomical work. He wrote a groundbreaking book on algebra, known as the *Artis analyticae praxis* (*The Application of the Analytic Art*), or the *Praxis* for short, published a decade after his death. In the *Praxis*, Harriot explained how to solve polynomial equations, conceived of negative numbers, and introduced the "greater than" and "less than" algebraic symbols (> and <), as well as the modern symbol for "square root" ($\sqrt{\quad}$). He studied the parabolic paths of projectiles and determined the densities of numerous materials. He investigated optics, and understood the principles behind the rainbow. And he independently discovered what we now call Snell's Law governing the refraction of light (doing so a full twenty years before the Dutch scientist for whom it is named).

Matters of state held little interest for Harriot, but he nonetheless got swept up in the political turmoil of the early seventeenth century. His second patron, Northumberland, was linked to the Gunpowder Plot of 1605. A distant relative of Northumberland's, Thomas Percy, had been one of the five conspirators in the failed plot to blow up Parliament; Northumberland, considered guilty by association, spent seventeen years in prison. Harriot, too, was jailed, but was released after a few weeks. (His first patron, Raleigh, was not so lucky. Imprisoned several times in the Tower of London for a variety of alleged crimes, he was finally beheaded for treason in 1618.) When not gazing skyward from the grounds of Syon House, Harriot could be found at another of Northumberland's homes, located on Threadneedle Street in the heart of London. It was here that he died in 1621.

Even though he was a keen observer of the heavens and one of the most important mathematicians and experimental scientists of his day, Thomas Harriot remains an obscure figure. Part of the reason is that Harriot, unlike his better-known contemporaries, published very little; only the *Briefe and True Report* appeared in his lifetime. (The vast majority of his scientific work—including his astronomical observations—exists only in manuscript form.) A plaque honoring Harriot's scientific achievements was finally unveiled on the grounds of Syon House in

2009, during the International Year of Astronomy (honoring the four hundredth anniversary of the modern telescope). An older memorial plaque had been placed near Harriot's grave in the parish church of St. Christopher le Stocks in London, not far from the home on Threadneedle Street; when the church was demolished to make way for an expansion of the Bank of England in 1781, the inscription was copied onto a new plaque, set on a wall within the bank, where it can be found today—or rather, where it *can't easily* be found; in quintessential British fashion, the plaque is mounted in a hallway that is not normally open to the public.

Why was Harriot so reluctant to publish? Surely he recognized the novelty of the sights he was seeing through his telescope—and yet he seemed perfectly content to keep them to himself, or to share them with only a handful of colleagues. Allan Chapman notes that, although there was nothing particularly dangerous about the study or practice of astronomy in Elizabethan and Jacobean England, Harriot may simply have sought to avoid unnecessary risks. His secrecy may reflect "a reluctance to put his head above the parapet in dangerous times." Besides, there was nothing to gain by going public with his discoveries. Unlike Galileo, Chapman notes, Harriot did not desire fame; thanks to his well-heeled patrons, he already had everything he needed. "A peaceful and intellectually productive life," Chapman suggests, "was best enjoyed as a private gentleman, not as a public figure."

WHITHER THE "TUDOR TELESCOPE"?

Just how good were telescopes in England in the sixteenth century, and the early years of the seventeenth? As we've seen, they almost certainly underwent a significant evolution over this time period, from the simple perspective tube or trunk purportedly used by Leonard Digges in the 1550s, to the device taken by Harriot to Virginia in the 1580s, to the more sophisticated device that Harriot was able to use to observe the moon and other celestial objects beginning in 1609. The latter development may have come just as word was arriving from Holland of a new-and-improved telescope, and of course it was just such a device, further refined by Galileo, that would herald the dawn of modern astronomy in the months that followed.

What, if anything, did Shakespeare have to say about these early optical devices? We may find a clue in an intriguing passage from *Richard II* (as it happens, the play once commissioned by the brother of Harriot's

Fig. 5.3 A matter of perspective: Viewed from a sharp angle, a hidden image—of a human skull—appears at the bottom of Hans Holbein the Younger's painting, *The Ambassadors*. A reference to "perspectives" in Shakespeare's *Richard II* may refer to such illusions. The Bridgeman Art Library, London

patron). The play was likely written in 1594–95, and was first published in 1597. In act 2, one of the king's supporter's, Sir John Bushy, speaks metaphorically of the task of reading a person's emotions (in this case, the Queen's); he says it's like gazing at an object by means of a "perspective":

> For sorrow's eye, glazed with blinding tears,
> Divides one thing entire to many objects,
> Like perspectives which, rightly gaz'd upon,
> Show nothing but confusion; eyed awry,
> Distinguish form.
>
> (2.2.16–20)

This may plausibly allude to a "perspective glass" of some kind, as David Levy argues in his book *The Sky in Early Modern English Literature* (2011). But most Shakespeare editors are skeptical. Anthony Dawson and Paul Yachnin, in the Oxford edition, say that Bushy is playing on two different meanings of the word "perspective": As an optical device, it can mean "a glass instrument whose multi-prism lenses show the viewer multiple images of an object"; but it can also refer to a particular kind of painting or drawing in which a hidden, secondary image appears only when the work is viewed from an oblique angle. The most famous of these, as Dawson and Yachnin note, is Hans Holbein the Younger's *The Ambassadors*, dating from 1533 (see figure 5.3): At first glance, it is a portrait of two accomplished young men; when viewed from a sharp angle, however, a hidden image appears—the technique is called "anamorphosis"—and we see a skull at the bottom of the canvas, "a stealthy reminder of mortality." (We might note that the German-born Holbein was living in England at the time, where he also painted iconic portraits of Henry VIII and members of his court.) Peter Ure, in the Arden edition, notes both of these possible allusions but believes that Bushy is more likely referencing the kind of hidden image typified by Holbein's painting. He also offers an additional word of caution, noting that neither of these two meanings "must be confused with the *perspective* (or "prospective") *glass*, a kind of magic crystal which could be used to look into the distance or the future." Things are not looking so good for the Tudor telescope *or* the Tudor time machine!

The case for this passage from *Richard II* being an allusion to a special kind of painting may be bolstered by a similar passage from *Antony and Cleopatra*, which seems more clear-cut: The queen is describing her lover's character; his nature, it seems, depends on one's perspective: "Though he be painted one way like a Gorgon, / The other way's a Mars" (2.5.116–17). The footnote in the Arden edition explains: "The allusion is to a 'perspective' picture of a kind which was popular in Shakespeare's time. They were painted on a furrowed surface in such a way that if looked at from the left they showed one portrait and from the right they showed another. Viewed straight from the front they appeared confused. . . . Cleopatra says that, seen from one point of view, Antony appears like the Gorgon Medusa (whose head was crowned with snakes and whose gaze turned men to stone), but seen from the other he looks like the god of war." Telescopes aside, Shakespeare seems to have been well versed in the methods of contemporary painters.

There are further references to a "glass"—for example, in *Macbeth*. In act 4, scene 1, the witches present Macbeth with "a show of eight kings"; as the stage direction indicates, the last of them enters "with a glass in his hand." In the New Cambridge edition, A. R. Braunmuller interprets "glass" as a "magic crystal permitting visions of the future . . . not a looking-glass or a mirror." The glass's magical properties are the key; there is, once again, little evidence that Shakespeare was referring to a telescope-like device. One finds a similar reference to a "perspective" in act 3, scene 4 of Ben Jonson's *The Alchemist*, and the scholarly verdict is much the same: As Douglas Brown sees it, the reference is to "a specially devised optical instrument, or, perhaps primarily, a design constructed to produce remarkable effects." That seems to cover nearly all the bases: These Elizabethan writers may have been referring to just about anything one might be able to achieve with a specially crafted glass, or to anything that mimics such a change in perspective, *except for* producing a magnified view of distant objects.

So why weren't inquisitive Tudor tinkerers building and using telescopes? Such devices would be only as good as their lenses; but this by itself may not have been a limiting factor, since high-quality spectacles—that is, reading glasses—had been common since the fifteenth century (as one can surmise from portraits dating from that period), as had simple magnifying glasses. Surely an inquisitive young spectacle maker must have played with various arrangements of lenses; indeed, it's hard to imagine a bored optician *not* fiddling with a variety of lenses in such a manner. Richard Dunn notes that "by the late sixteenth century many people were experimenting with lenses and mirrors and thinking about their potential." The idea of a telescope seems to have been "in the air," so to speak, and perhaps it was inevitable that someone would invent a device similar in principle to a modern telescope before too long—and yet there seems to have been only halting progress toward such a device over a period of many decades.

As we've seen, the strongest claim comes from Thomas Digges, who says that his father, Leonard, had used a sophisticated telescope-like instrument in the 1550s, one that could reveal minute details of people and objects at a great distance from the observer. For those who take Thomas's story at face value, it offers the possibility, as we have seen, that it was a look through one of his father's instruments that set Digges on the path toward envisioning an infinite cosmos. But this is speculation—and recall that Leonard died when Thomas was thirteen. His account may well be based on fading recollections of his father's stories, or the stories of others who knew him, and may be little more than embellished family folklore. Intriguingly, however, we do have an additional secondhand account, which happens to come from Bourne, the man behind the submarine designs. Sometime after Thomas Digges published his description of his father's experiments, Bourne wrote a letter to William Cecil, Elizabeth's chief advisor, assuring him that Digges's optical device, and its reported capabilities, were real: A certain kind of "looking glasse," he wrote, could be used to see "things of a marvellous largeness in a manner incredible to bee believed of the common people." He added that "those things that Mr Thomas Digges hathe written that his father hathe done, may be accomplished very well, withowte any dowte of the matter. . . ." And John Dee, as we've seen, had written of the military value of such a device.

It's a classic conundrum for historians: A smattering of enthusiastic written accounts for an early telescope-like device—but no physical evidence, and no evidence that others adopted Digges's design. (If it was so good, why wasn't it widely copied?) Among twentieth-century historians, Francis Johnson was probably the strongest proponent of the Diggeses, suggesting that it "seems entirely probable" that Thomas Digges had access to one of his father's early telescopes and "may have used them for examining the heavens." But more recent voices are more cautious. David Levy, an astronomer who has also written extensively on the history of the field, says that we "cannot conclude . . . that telescopes were invented in England," even if various people had tinkered with an early form of such a device. Allan Chapman, who specializes in the history of English astronomy, is even more doubtful about the existence of a Tudor telescope decades before the telescope's "official" invention, circa 1609. He notes, for example, that Harriot showed the Native Americans "strange sights" with his optical devices, but makes no mention of achieving a *magnified* view. "I simply do not believe that the historical evidence for a Tudor telescope stands up to detailed scrutiny," he concludes. (We will return to this issue in Chapter 8, when we consider a rather bold claim regarding Shakespeare and Elizabethan telescopy.)

None of the scientific developments that we've been looking at in this chapter were revolutionary on their own. Acceptance of the Copernican system, as we've seen, came slowly. And while Bacon's first writings came during this period, his major works still lay a few years off.

William Harvey's treatise on the circulation of the blood—a crucial breakthrough—would come only in 1628. And yet these first stirrings of science in England, and in London in particular, laid the groundwork for what was to come. "There would have been no Scientific Revolution in England without the intellectual vitality present in Elizabethan London," writes Deborah Harkness, "for she provided later scientists with its foundation: the skilled labor, tools, techniques, and empirical insights that were necessary to shift the study of nature out of the library and into the laboratory." We find, if not science in the modern sense, at least "the seeds of modern scientific thought," as Freyja Cox Jensen puts it. Mordechai Feingold calls the period between 1560 and 1640—which happens to encompass Shakespeare's life—the "prologue of modern science"; the disciplines that were beginning to take shape during those years would, by the second half of the seventeenth century, evolve into highly specialized fields of study. "Prologue," "seeds," "foundation": Whichever word we apply to this incubation period, we are witnessing the beginning of a profound and far-reaching change.

We have taken a broad look at the growth of English science, with a focus on astronomy in particular; and along the way we have met some of the great minds at work in Oxford, Cambridge, and London in the second half of the sixteenth century. Now let us turn our attention to a provincial town in Warwickshire, and a very different kind of genius.

6. *"Who is it that can tell me who I am?"*

A BRIEF HISTORY OF WILLIAM SHAKESPEARE

"Right, shall we start walking then?"

The guide is Barbara, a middle-aged woman with gold hair and boundless energy. She corrals the dozen or so tourists who have signed up for the town's most popular Shakespeare walk, and sets off from the meeting place—the Swan Fountain by the River Avon, a hundred yards or so from the Royal Shakespeare Theatre. Either Barbara or one of her colleagues conducts the walk every day, as they have for the past eleven years. I follow the group as we explore the church where the playwright was baptized, and where his bones now rest; the famous Birthplace on Henley Street; the grammar school; and the handful of other locations linked in some way to the town's most famous resident. From time to time we interrupt Barbara to ask questions. The man with the German accent seems to be the most inquisitive, or at least the most persistent. Barbara handles the queries deftly. Sometimes we clog up the sidewalk; we have to be reminded that there are regular Stradfordians here, ordinary men and women, including a large number of pensioners, who are just trying to get their shopping done. They must have mixed feelings about sharing their streets with so many bardolators.

It is 2012, but in our minds it is the late 1500s. Stratford at that time was a provincial market town of some two hundred houses, known for its bustling fairs and for its strategic location on a crossing of the River Avon. The town's name, as Barbara points out, tells of its origins: "Stratford" comes from the Old English word for street, combined with ford, the name for a river crossing; in other words, this was the site where an ancient Roman road crossed the Avon. In Shakespeare's day, London would have been two days away by horseback; at least four days by foot. Oxford, with its famous university, was about half that distance.

It was around 1550 that a man named John Shakespeare moved to Stratford from the nearby village of Snitterfield. John was a glove maker and also traded in wool and meat. He married a woman named Mary Arden, who came from a wealthy family in Wilmcote, a few miles away. John Shakespeare apparently did quite well for himself. He served on the town council and later became an alderman and, eventually, a bailiff (a position similar to that of mayor). He ended up buying two large properties in the town, including the one on Henley Street, now a mecca for Shakespeare fans from around the world. The two-story wattle-and-daub structure is built around a wooden frame; the sturdy oak beams would have come from the nearby Forest of Arden. The house is not remarkable architecturally, but was large enough to accommodate living quarters as well as a workshop, and would have marked the owner as a man of some means.

Fig. 6.1 Birthplace of genius: The house on Henley Street, in Stratford-upon-Avon, is now a tourist attraction; in the 1550s, it provided John and Mary Shakespeare with space to raise their children, and for a workshop. Author photo

---

John and Mary Shakespeare had eight children, including two daughters who died in infancy. Childbirth took place in the home, and was inherently risky for both mother and child. Infant mortality was high—perhaps twelve times higher than today. One out of five children died before their first birthday. Only three-quarters would live to the age of ten.* William, their third child and first son, was one of the lucky ones. We don't know the exact date of his birth, but we do have the record of his baptism, from Stratford's Holy Trinity Church. The parish records for April 26, 1564, indicate the baptism of "Gulielmus filius Johannes Shakspere" ("William, son of John Shakespeare"). Traditionally, a child was baptized three days after its birth, and so we celebrate Shakespeare's birthday on April 23—which, by pleasant coincidence, is also the feast day of Saint George, England's patron saint.

ONE LAND, TWO FAITHS

The England of William's youth was very different from that of his father's childhood. The old Catholic religion had been swept away—in theory, at least—with the reforms begun by Elizabeth's father, Henry VIII, and a new Church of England installed in its place. Shakespeare's father was caught up in this transformation. Stratford's Guild Chapel was already three hundred years old when John Shakespeare arrived in town; its colorful stained-glass windows displayed Catholic saints, and its walls were covered in the iconography of the old faith. John Shakespeare was in charge of renovating the chapel to conform to the new order. Under his supervision, on a midsummer's day in 1571, the windows were smashed and replaced with plain white glass. An earlier effort at "correcting" the chapel's imagery had taken place a few years earlier, when the ornate frescoes on the walls were covered in whitewash. The Bible, too, had been replaced. A scholarly English translation of the scriptures, known as the Geneva Bible, had been published in England in the 1570s; this, along with the Bishop's Bible of 1568 and the Book of Common Prayer, first published in 1549, were the primary religious works that Shakespeare would have encountered in his youth. Echoes of each of these books can be found throughout Shakespeare's plays. (By

---

* Knowing that each child may not be long for this world, were parents reluctant to bond with their offspring? Historians have found no evidence to support such a notion. Jeffrey Forgeng notes the following words written by a nobleman, addressed to his newborn son: "I love thee, boy, well. I have no more, but God bless you, my sweet child, in this world forever, as I in this world find myself happy in my children. From Ludlow Castle this 28th of October, 1578." The note is signed, "Your very loving father." (Forgeng, *Daily Life in Elizabethan England,* p. 47)

the end of the playwright's career there was yet another version of the Bible—the King James edition of 1611.) The scriptures, in one form or another, were ubiquitous. By law, the parish minister was required to give religious instruction to local boys, age six and up, on alternate Sundays and on holy days—though basic religious instruction would have already begun in the home.

What kind of faith would John and Mary Shakespeare have instilled in their children? We will probably never know, but biographers aren't shy about guessing. Much has been written on the question of John Shakespeare's possible Catholic sympathies.* Perhaps, muses Stephen Greenblatt, the elder Shakespeare may have been of two minds on the issue of faith. He may have wanted "to keep both his options open" so as to cover all the spiritual bases, so to speak. "He had not so much a double life as a double consciousness." For James Shapiro, the walls of the guild chapel—with the images of the old faith just visible beneath the fresh layers of paint—is an appropriate symbol of the religious confusion that hung over the times:

> To argue that the Shakespeares were secretly Catholic or, alternatively, mainstream Protestants misses the point that except for a small minority at one doctrinal extreme or other, those labels failed to capture the layered nature of what Elizabethans, from the queen on down, actually believed. The whitewashed chapel walls, on which, perhaps, an image or two were still faintly visible, are as good an emblem of Shakespeare's faith as we are likely to find.

Faith, once a cornerstone of English life, had become uncertain, a thing that could be tampered with or overhauled by powerful men and crafty politicians. Some people surely practiced a combination of the two faiths—adopting Protestant practices in public, while quietly keeping the Catholic faith at home. Religious tension, as Norman Jones notes, "was a fact of daily life." How this tension affected William Shakespeare's life—and his career—has been endlessly debated. As Jonathan Bate puts it, "His mind and world were poised between Catholicism and Protes-

* And we shouldn't make too much of the fact that Shakespeare's oldest sister was baptized a Catholic, as she was born when Mary was still on the throne. William was born six years after Elizabeth's accession.

tantism, old feudal ways and new bourgeois ambitions, rational thinking and visceral instinct, faith and scepticism."

Something else was in the air: For the first time, the people of England may have felt some small measure of personal freedom. One could choose—to some extent—one's destiny. As Norman Jones notes, however, this could be a mixed blessing: "One had to make the right choices. God still ran the world, demanding obedience. But obedience to which theology? Which Church? Which economic order? Which master? Where could a person turn for intellectual certainty in a world of choice and confusion?" But perhaps there was an upside. Uncertainty may have nurtured creativity. "Confusion," Jones notes, "made Shakespeare's age one of the most culturally productive in English history." It is sobering to imagine that Shakespeare, had he lived in a less turbulent time, would perhaps have been content to take up his father's glove business.

The country itself, meanwhile, was growing. The population stood at about four million, a figure that was increasing by roughly one percent every year during Elizabeth's reign. Because of the rapid population growth, the demographics skewed young: About one-third of the population would have been under fifteen; half were under twenty-five. Average life expectancy at birth was forty-eight—but most of the danger was in the first few years. Those who lived to thirty would likely make it to sixty. Death, of course, was not a stranger. Disease was a perpetual threat, the bubonic plague the most feared of all. There were at least five outbreaks during Shakespeare's lifetime. The plague hit the crowded cities harder than rural communities, but no corner of the country was safe. The plague-free years must have been a welcome respite; but even so, there were dangers. The combination of a rising population and uncertain harvests meant that demand often outstripped supply; when harvests failed, people died. Food prices rose, wages fell, and the gap between rich and poor widened. Lack of food could trigger riots, as depicted in Shakespeare's *Coriolanus*, and the "Poor Laws" were established to alleviate the suffering of the least well-off—and to reduce the chances of more severe rioting. The "working poor" seem to have been tolerated; those classified as "idle," "rogues," or "vagabonds" were demonized as disease carriers and threats to "good order"; they could be flogged and driven from town.

For those who did have money, there were a variety of ways to show off one's good fortune, with expensive clothing being the first choice. (As one clergyman noted, the land was full of "fickle-headed tailors" who were only too happy to satisfy demand for whatever the latest trend called for, adding that "nothing is more constant in England than the inconstancy of attire. Oh, how much cost is bestowed nowadays upon our bodies, and how little upon our souls.") Many fashion trends originated on the Continent, and wine, silk, and lace were typically imported; but many items once brought from abroad were increasingly made in England, including felt hats, playing cards, soap, and fine cloth. For the poor, of course, such luxuries could only be dreamed of.

It helped if you were a man. Women had limited social and legal standing, and only marginally better economic status. If a woman was married, it was assumed that she would look after the home and the children, leaving her husband free to serve as breadwinner. (Widows and unmarried women had slightly more freedom, such as the right to own property and to sign contracts.) In language that is difficult to stomach today, a seventeenth-century political theorist declared, "Women [were made] to keep home and nourish their family and children, and not to meddle with matters abroad, nor to bear office in a city or commonwealth no more than children or infants." A Dutch visitor, however, noted that English women "are not kept so strictly as they are in Spain or elsewhere":

> They go to market to buy what they like best to eat. They are well dressed, fond of taking it easy, and commonly leave the care of household matters and drudgery to their servants. . . . All the rest of their time they employ in walking or riding, in playing at cards or otherwise, in visiting their friends and keeping company, conversing with their equals (whom they term gossips) and their neighbours, and making merry with them at childbirths, christenings, churchings, and funerals; and all this with the permission and knowledge of their husbands, so such is the custom.

ENTER ROSALIND, READING A PAPER

Schooling, beyond the most basic home instruction, was almost exclusively reserved for boys and young men. However, girls from well-heeled families could attend an elementary school, and, on rare occasions, were allowed to matriculate at a grammar school. Even then, they were allowed to stay for only the first few years, and were not taught Latin. (The universities were strictly off limits.) Nonetheless, literacy rates among women were on the rise—partly as a result of the Protestant desire to give the largest number of people the ability to read the Bible. While only a handful of works were penned by English women before 1500, the number rose steadily over the next half century—during which time more than a hundred works were composed or translated by English women, including religious works, poetry, essays, advice books, diaries, and letters. (As Stephen Greenblatt has noted, it is remarkable how many of Shakespeare's women are depicted reading.)

Schooling began at home, and the young Shakespeare would have been taught to read beginning at age four or five. If we are to try to picture William learning his letters, we should bear in mind that, at that age, boys and girls alike would have worn long gowns or dresses; it was only at about age six that a boy would be "breeched"—fitted with the breeches of the style worn by adult men. (In *The Winter's Tale*, Leontes looks at his young son, and imagines himself at that age: ". . . I did recoil / Twenty-three years, and saw myself unbreeched . . ." [1.2.153–54].) At age seven or eight William would have begun his studies at the local grammar school, adjacent to the guild chapel, an institution reestablished by Edward VI in the 1550s as the King's New School. (We don't have any actual record of William's attendance, but as the son of a prominent town official, it's a fairly safe assumption that he was in fact schooled there.) As noted in the previous chapter, the reign of Elizabeth would see an explosion of these grammar schools; by the end of the sixteenth century, England had about 160 such institutions, about one for every twelve thousand inhabitants (a much higher proportion than one finds even in Victorian times). As a result, basic literacy is thought to have reached 30 percent for men and perhaps 10 percent for women—higher, of course, for privileged classes and townsfolk than for the rural poor.

"THE WHINING SCHOOLBOY, WITH HIS SATCHEL"

The school day was long, running from six in the morning (seven in winter) to five or six in the afternoon. If the family had means, the boy would carry a lantern to light the way in the dark of winter. There was only a short break for recess, and another for lunch—for which William presumably headed home; the house on Henley Street was just a few blocks away. Students would recite the alphabet from a "hornbook" and read from the Bible. Writing was done with a goose-quill pen and inkhorn. The flavor of Elizabethan schooling is captured in Jacques's famous speech on the seven ages of man, in *As You Like It*, in which the second age sees "the whining schoolboy, with his satchel / And shining face, creeping like snail / Unwillingly to school" (2.7.145–47). Later, from age eleven or so, William's studies would have continued with grammar, logic, and rhetoric. A barrage of Latin grammar was unavoidable; Latin maxims were to be learned by heart. The older boys would learn the poetry of Ovid, Virgil, Cicero, and—as hinted at in the prologue—Horace. (Shakespeare's plays echo the themes employed by all of these classical writers.) Older children were meant to speak exclusively in Latin, and could be punished for reverting to English.

Every bit as important as the book and the pen was the birch rod, the chief implement for enforcing discipline, as depicted in numerous woodcuts from the era. As one schoolmaster explained, corporal punishment was simply a part of God's plan, a practice that "God hath sanctified . . . to cure the evils of [students'] conditions, to drive out that folly which is bound up in their hearts, to save their souls from hell, to give them wisdom: so it [the rod] is to be used as God's instrument to these purposes." For the Elizabethans, a child, as one scholar has put it, was, "just a diminutive and exceptionally troublesome adult."

What did young William think of his schoolmasters? We can get a sense, perhaps, from the mocking tone in which the teacher Holofernes is portrayed in *Love's Labour's Lost*, and similar scenes in *As You Like It* and *The Merry Wives of Windsor*. In the latter play, a lad named William Page gives the master, a Welshman named Sir Hugh Evans, a hard time:

> EVANS:
>
> What is "lapis," William?
>
> WILLIAM:
>
> A stone.
>
> EVANS:
>
> And what is "a stone," William?

WILLIAM:

A pebble.

EVANS:

No, it is "lapis," I pray you remember in your prain.*

(4.1.27–31)

If the plays of the ancient Romans whetted the young Shakespeare's appetite for acting and dramatic writing, he would have another taste whenever traveling "players" (theater troupes) passed through town. (And owing to its location in the heart of England, Stratford would have witnessed more such shows than most towns of a similar size.) We know that the Earl of Leicester's players dropped by in 1573 and 1576; Lord Strange's Men in 1579; those of the Earl of Essex in 1584, and the Queen's Men in 1587. It is not hard to picture William sitting wide-eyed in the audience for these touring performances, listening to each word and observing each gesture.

The players were not the only special visitors to pass through. In the summer of 1575 Queen Elizabeth herself visited Warwickshire, during one of her many ceremonial tours of the countryside, known as "progresses." She stayed at Kenilworth Castle, not far from Stratford, as a guest of the Earl of Leicester. Country folk came from miles around to see their queen, with the festivities—music, plays, fireworks—stretching for three full weeks. Shakespeare would certainly encounter Elizabeth years later, when his own theater company performed before the Court; but perhaps the eleven-year-old William caught a glimpse of the middle-aged Elizabeth on this tour. We know that she liked to "work the crowd," so to speak. A few years earlier, the Spanish ambassador observed that the queen "ordered her carriage sometimes to be taken where the crowd was thickest and stood up and thanked the people." Elizabeth was educated, witty, and erudite, and could switch effortlessly between modern and ancient languages (apparently she was fluent in Latin and Greek by

* "Prain"? What's a prain? None of the major editions have a footnote to assist the reader, but, as Scott Maisano explained it to me, it's simply "brain" pronounced in a way that reflects Evans's Welsh accent. (The joke is that the teacher is chiding the student over his slow progress with Latin, while he himself hasn't yet mastered English.) An online search of the canon seems to back up this interpretation: Shakespeare uses "prain" a handful of times—but it is only spoken by Evans and by another Welshman, Captain Fluellen from *Henry V.* In each case the context does seem to suggest "brain."

age twelve). She excelled at dancing and riding, and could fire an arrow as well as any huntsman. One can sense why her courtiers referred to her as "Gloriana." Another eyewitness account of her demeanor, from twenty years later, focuses on her physical appearance. The queen, in her later years, was

> very majestic; her face oblong, fair, but wrinkled; her eyes small, yet black and pleasant; her nose a little hooked; her lips narrow, and her teeth black (a defect the English seem subject to, from their too great use of sugar). She had in her ears two pearls, with very rich drops. Her hair was of an auburn colour, but false; upon her head she had a small crown.

It would have been the younger, sprightlier version of the monarch that William would have seen in Warwickshire; a woman who had yet to face down a foreign armada; a woman who had yet to declare, "I know I have but the body of a weak and feeble woman, but I have the heart and stomach of a king."

Whatever he thought of Her Majesty, William spent more time thinking about a woman named Anne Hathaway, eight years his senior. (The audio guide on one of the bus tours naughtily refers to William as her "boy toy.") They married in 1582, when he was eighteen and she was twenty-six—and pregnant. (Contrary to what we might imagine based on the youth of the lovers in *Romeo and Juliet*, the average age for marriage at that time was twenty-seven for men; twenty-four for women.) Their first child, Susanna, was born the following year, and twenty months later they had twins—a boy named Hamnet and a girl called Judith.

THE GENIUS FROM WARWICKSHIRE

Unlike some of the other successful playwrights of the time, Shakespeare did not attend university. A common refrain from the "anti-Stratfordians"—those who believe that someone other than the actor from Stratford wrote the works of Shakespeare—is that someone from such a lowly background, with such a modest education, could hardly have written about affairs of state, the courtly intrigues of kings and princes, military struggles, sea voyages, and the ways and customs of foreign lands. How could a country bumpkin have become the greatest writer in the English language? What the doubters seem to forget is that there is more to learning than mere schooling. As one biographer puts it:

> It may, in fact, have been a positive advantage for Shakespeare to have had no experience of a university. Many of his contemporaries who prided themselves on their learning have since been criticized for artificiality, whereas Shakespeare had enough education to profit from it, but not so much that it spoiled him.

Or, as another similar work notes:

> This was an age in which a good secondary school pupil endowed with a strong sense of curiosity could become thoroughly self-educated. The fact that Shakespeare lacked higher education and social advantage proved no obstacle. His love of language and innate mastery of the art of the theater—combined with a tremendous capacity for work and fertile powers of invention—were enough to enable him to produce work of astonishing range.*

Moreover, Shakespeare's humble origins were hardly unique: Jonson had worked as a bricklayer, and Marlowe was the son of a cobbler. Such arguments probably do little to satisfy the anti-Stratfordians—but then, people who buy into conspiracy theories aren't usually interested in what the experts have to say. (After all, if you're enamored with the conspiracy, the "experts" are part of the problem.) An exploration of the psychology behind conspiracy theories is beyond the scope of this work, but the motivation behind the so-called authorship question seems at least partly rooted in the simplistic disbelief that underlies a wide range of such theories. A country lad "couldn't possibly" have written *King Lear*, just like ancient Egyptians "couldn't possibly" have built the pyramids, and the NASA of the 1960s, equipped with chemical-powered rockets

* It may be a useful intellectual exercise to reach back 110 years, to the time of Albert Einstein's youth, rather than the nearly 450 years in the case of Shakespeare. Einstein, still working on his PhD, was employed as a patent clerk when he came up with the insight that led him to the first part of his theory of relativity. A miracle? No—just the outcome of an extraordinarily agile mind, a nurturing network of friends, and nerve-racking hours, days, and months of hard work.

and computers with two kilobytes of memory "couldn't possibly" have flown to the moon—except, of course, that they did. In the case of Shakespeare, class prejudice is probably also a factor. As James Shapiro notes, those "who believe that a genius of Shakespeare's order had to be from a higher social station, or have a university degree, and so on, reveal more about their prejudices than they do about the nature of genius."

We don't know what drove Shakespeare to leave his hometown, nor do we know exactly when he left. Perhaps he already had an eye on becoming an actor, and sensed that any real career advancement could come only in the capital.* One seventeenth-century account—seen as somewhat dubious today—says that he was caught poaching deer on the estate of a wealthy landowner in Charlecote, just across the river from Stratford; if it's true, perhaps he simply felt that some distance between himself and his native Warwickshire was in order. Whatever the reason, sometime in the mid-1580s, William Shakespeare said good-bye to his wife and children, packed his bags, and set off for London.

LONDON CALLING

In Elizabethan London, the term "urban sprawl" was unknown—but the concept would not have been a foreign one. In the Middle Ages, the population had managed to constrain itself more or less to the space within the city walls, built on the ruins of the ancient wall established by the Romans more than a thousand years earlier. By Shakespeare's time, this was no longer possible. Under Henry VIII, London's population had been just fifty thousand, but by midway through Elizabeth's reign, that figure had swelled to some two hundred thousand—about one-twentieth of its current size, and yet infinitely bigger than the town the young actor had left behind. London was literally bursting at its medieval seams.

When I walk along the narrow streets of "the City"—London's ancient center—I like to play a game that, for lack of a better name, we might call "time machine." St. Paul's Cathedral is as good a place as any to begin. If I was in a time machine of the sort that H. G. Wells imag-

---

* We will never know if Shakespeare agreed with Francis Bacon's (horribly sexist) sentiment that having a wife and children spelled the end of one's creativity: "He that have wife and children hath given hostage to fortune; for they are impediments to great enterprise, either of virtue or mischief. Certainly the best works, and of greatest merit for the public, have proceeded from unmarried or childless men, which both in affection and means have married and endowed the public." (quoted in Pritchard, pp. 28–29)

ined in his novel, I would simply dial up some numbers on a console, turn a lever, and the scene would begin to evolve backward through time. As Wells describes it, the day would turn to night, and then to day and night again, and so on; as one pushed the lever further, the days and nights would eventually blur together into a kind of nondescript gray. Eventually, as the years pass in reverse, the glass-and-steel office towers deconstruct, the Starbucks and Pret a Manger outlets transform into the small independent shops that they replaced, and, eventually, the roads turn from pavement to cobblestones; the cars and buses to horse-drawn carts and carriages. An astute observer would notice the colors of the buildings change: Granite and limestone facades that gleam white today—including St. Paul's—would give way to their blackened, soot-coated appearance of earlier decades. Structurally, however, the cathedral would show almost no perceptible change for more than seventy years, until we reach the time of the Blitz, when fires raged all around it; as we speed backward through time, the smoke quickly clears and the cathedral is once again a grand, imposing building. As we continue our backward journey, we reach the year 1709, the year of the present cathedral's completion. Now the dome begins to dissolve, top to bottom, to be replaced with the debris from the Great Fire of 1666. As the years whizz backward, the fire itself flashes in, and then out, of view: black smoke, orange flames, chaos. Then the cathedral suddenly pops into view again—this time in its gothic, medieval form. If we're aiming for the London of Shakespeare's day, we'll know we've gone too far when the cathedral's great spire snaps into place. The soaring spire, measuring almost five hundred feet from base to tip, was destroyed by lightning in 1561. (It is indicative of the religious mood of the time that both Catholics and Protestants saw the destruction of the spire as a sign of God's displeasure with the other group.) And so we turn the lever in the other direction and travel forward until we reach the late 1580s—the approximate time of Shakespeare's arrival in London.

Stepping out from our time machine, we find ourselves in a smaller London—but also a noisier, dirtier, smellier London. A few streets, like Cheapside, are broad thoroughfares, but most are narrow and crowded, and filled with the sound of vendors' cries and rattling of wagon wheels. Of course, the iconic red phone booths, mailboxes, and double-decker buses are gone—and yet there will be familiar sights. The Tower of London, dating from the time of William the Conqueror, is already six centuries old. (Shakespeare imagined it was a thousand years older—it's

Fig. 6.2 Heart of a bustling city: The Dutch artist Claes Visscher created a remarkable panorama of London as it appeared around 1600—the year of Shakespeare's *Hamlet.* In the detail seen here, St. Paul's Cathedral towers over the city; the Bear Gardens and the Globe Theatre can be seen on the south bank of the Thames. The Bridgeman Art Library, London

referred to as "Julius Caesar's ill-erected tower" in *Richard II*, reflecting the popular mythology of the day.) The medieval Guildhall is there too; so is the Priory Church of St. Bartholomew the Great, and a handful of other places of worship. Remarkably, one will also find the Staple Inn on High Holborn, dating from 1585—the only half-timbered structure to survive the Great Fire (and presumably, in Shakespeare's day, absent the Vodafone shop). Another landmark would be familiar by name but not by sight: The Royal Exchange had been founded in 1571, though it has been destroyed and rebuilt twice since then.

And of course there is St. Paul's itself; though located on the site of the present-day cathedral, the building we find in Shakespeare's day bears little resemblance to Christopher Wren's masterpiece. In fact, the gothic structure was bigger than today's cathedral, stretching some 585 feet in length; it was the largest church in all of Europe. And it was not just a place of worship, but a place of business, too: Everyone from lawyers to barbers plied their trade within its walls, which would have been plastered with advertising. There were booksellers and other vendors, an endless parade of deliverymen (many of them simply using the cathedral as a shortcut), and of course beggars and pickpockets. Just like today, visitors could pay a fee to climb the stairs and enjoy the view from a balcony near the top of the tower; in Shakespeare's time, the admission was one penny.

Should we stroll down to the river, we would find it spanned by only one bridge: the remarkable London Bridge, a wonder of medieval engineering. Built in the late twelfth century, it stands on a series of nineteen stone arches and is lined with multi-storied homes and shops (not to mention the heads of traitors). Beneath it flows the mighty Thames—a far more polluted waterway than its twenty-first-century counterpart. The river carries everything from royal barges to tiny skiffs and rowboats; farther downstream, great sailing ships bring goods from across Europe, the Mediterranean, and beyond. Looming over the river's southern bank we would also recognize Southwark Cathedral, at this time known as the Church of St. Mary Overy, begun in the thirteenth century and largely unharmed by the passing centuries.

The city's crowds, noise, and filth must have had an immediate impact on the young Shakespeare. He would have encountered, first of all, *people*, of every shape, size, and social class, and speaking a multitude of tongues. As Thomas Dekker described the scene:

> For at one time, in one and the same rank, yea, foot by foot and elbow by elbow, shall you see walking, the knight, the gull, the gallant, the upstart, the gentleman, the clown, the captain, the apple-squire [pimp], the lawyer, the usurer, the citizen, the bankrupt, the scholar, the beggar, the doctor, the idiot, the ruffian, the cheater, the puritan, the cut-throat, the high man, the low man, the true man, and the thief. . . .

They came from all over England, seeking work in the bustling capital. Increasingly, they also came from the Continent, including growing numbers of French and Italians, as well as a wave of Dutch immigrants fleeing religious persecution. There were handfuls of Africans and Turks and Jewish *conversos*—Jews from Spain or Portugal who had converted to

Christianity.* London was becoming cosmopolitan, and would remain so. But life was difficult for newcomers, no matter where they were arriving from. As one official observed, the city was home to "great multitudes of people" forced to live "in small rooms, whereof a great part are seen very poor, yea, such as must live by begging, or by worse means . . . heaped up together, and in a sort smothered with many families of children and servants in one house or small tenement." (Sewage systems, as we know them today, did not exist, and household waste was emptied directly onto the street.) "This city of London is not only brimful of curiosities," a Swiss visitor noted in 1599, "but so popular also that one simply cannot walk along the streets for the crowd."

Westminster, the site of Court and Parliament, was a separate city, although the road that connected it to London was rapidly being developed. Other neighborhoods that today are considered part of central London were semirural, and sometimes their names tell a story: One went to Notting Hill to gather nuts, while sheep grazed at Shepherd's Bush; hogs could be found at Hoxton. While some of London's nooks and crannies must have been repugnant to the eye and nose, it was also, undeniably, a city seething with activity and abuzz with creative energy. There were painters and musicians, poets and playwrights. There was high life and low life and everything in between. One could attend a play one afternoon and watch the spectacle of bear baiting the next—often in the same venue.† Cockfights and dogfights were popular; so were the brothels. Indeed, as Jonathan Bate notes, "the link between the theatre industry and the sex trade was symbiotic." The prostitutes plied their trade at the playhouses, and George Wilkins—Shakespeare's coauthor for *Pericles*—owned a chain of brothels. It was, needless to say, a cutpurse's paradise.

---

* The Jews had been a presence in England (and especially London) since the time of William the Conqueror, but had been expelled by Edward I in 1290. While some may have remained and practiced their faith in secret, there was no active Jewish community from the expulsion until the time of Oliver Cromwell in the mid-seventeenth century (and we know from court records that a number of Jews were discovered and deported during this period).

† The bear was tied to a post, while hungry dogs were set loose in the enclosure. While unspeakably cruel to twenty-first-century eyes, it was clearly not seen that way at the time. One contemporary observer described bear baiting as "a sport very pleasant," as the bear tried to fight off the dogs, "with biting, with clawing, with roaring, tossing and tumbling . . ." (quoted in Ridley, p. 269). Henry VIII enjoyed bear baiting; his daughter Elizabeth even more so. Shakespeare alludes to the spectacle metaphorically in *Macbeth*: As the protagonist's world closes in around him, he vows that "bear-like I must fight the course" (5.7.2). Bear baiting was finally banned in 1835.

"VOLUMES THAT I PRIZE ABOVE MY DUKEDOM"

Returning to the neighborhood immediately surrounding St. Paul's, we find what we might call the publishing district, where printers had their workshops and booksellers hawked their wares. As many as twenty booksellers plied their trade in the churchyard itself, and more in Paternoster Row behind the cathedral. (The area remained the center of London's book trade right up to the Second World War.) Shakespeare surely gleaned many of his ideas as he perused the stalls and leafed through the latest offerings. And what a deluge of offerings it was: Between 1558 and 1579, some 2,760 books were published in London. Between 1580 and 1603, the number rose to 4,370. As Frank Kermode points out, the percentage of the population that could read was small, but there were still enough readers that a popular work could quickly sell out—as was the case with Shakespeare's *Venus and Adonis*, which went through nine editions during the poet's lifetime. A curious browser could find religious works, poems, plays, romances, and etiquette guides; as we've seen, there were also numerous science-themed offerings, including texts on botany, medicine, astronomy, and astrology, along with almanacs and atlases. For those on a budget, there were the penny and halfpenny "ballads," the forerunners of today's newspapers. These illustrated, single-sheet publications contained bits of news, stories from the Bible, and, especially, the latest gossip—the more sensational, the better. (Murders, fires, and reports of "monstrous births" were perennial favorites.) These ballad sheets were an industry in their own right; as one observer noted, "Scarce a cat can look out of a gutter but out starts a half-penny chronicler."

Among these publications were a few dozen works written by women, including, from 1589, the first full-length defense of women's rights, penned by a woman in England (or at least the first such work to have survived). Written in response to a misogynist tract by a man named Thomas Orwin, the pamphlet's author, who calls herself Jane Anger, chides men for their illogic and argues for female sexual autonomy. A little over a decade later, a woman named Aemilia Lanyer,* who supported herself through her poetry, denounced "evil disposed men, who forgetting they were born of women, nourished of women, and that if it were not by the means of women, they would be quite extinguished out

---

* It is unfortunate, though perhaps not surprising, that Lanyer is better known not for her own writing but for (allegedly) being the "dark lady" of Shakespeare's sonnets. (There is no hard evidence for this theory, and the idea is not widely accepted.)

of the world, and a final end of them all, do like Vipers deface the wombs wherein they were bred, only to give way and utterance to their want of discretion and goodness." It is from this period that we also find the first original (as opposed to translated) play written by a woman in English, *The Tragedy of Mariam, the Fair Queen of Jewry*, by Elizabeth Cary, Viscountess Falkland. It was published in 1613—by coincidence, the approximate year of Shakespeare's retirement.

Shakespeare must have flipped through far more books than he actually purchased; still, the works that provided the backbone of his plays—Holinshed's *Chronicles*, Plutarch's *Lives*—he surely bought; perhaps Ovid, Virgil, and Horace, too. His grammar-school Latin was more than adequate for enjoying his favorite classical writers. His French and Italian, picked up through friends and from casual reading, were likely passable, though he probably preferred an English translation, when one was available (as it often was). He likely also borrowed books from friends. "From what we know of Shakespeare's insatiable appetite for books," writes James Shapiro, "no patron's collection . . . could have accommodated his curiosity and range. London's bookshops were by necessity Shakespeare's working libraries. . . . It's hard to imagine anyone in London more alert to the latest literary trends."

If we were lucky enough to cross paths with Shakespeare, would we recognize him? Today we are so accustomed to seeing pictures of the playwright that we tend to assume that we have a reliable notion of his appearance, but, as mentioned in the introduction, only two images have a reasonable claim to authenticity. The first is Martin Droeshout's famous engraving in the First Folio (figure 0.2); the second is the funeral effigy in Holy Trinity Church in Stratford. Although both date from several years after Shakespeare's death, they were at least carried out under the guidance of those who knew him; as such—as bland as they are—they are our best guess at his appearance. (The funeral bust has been famously described as looking like a "self-satisfied pork butcher.") A runner-up is the "Chandos portrait" that hangs in the National Portrait Gallery. Dating from about 1610, it has the advantage of being painted while the playwright was still alive; unfortunately, we don't know for sure that it actually *is* Shakespeare. Its subject certainly resembles the man in the Droeshout engraving, and the painting is from the right time period; unfortunately, its provenance is a blank prior to 1747—more than 130 years after Shakespeare's death—when it came into the possession of the Chandos family. The middle-aged man in the portrait, with slightly unkempt facial hair and sporting an earring, has something of the bohemian look that, warranted or not, we seem to expect in an artistic genius.

We can increase our chances of bumping into the playwright if we know where to look. Where, exactly, did Shakespeare live? He seems to have moved several times during his London years; by one account, he lived for a time in Shoreditch, and later in Bishopsgate, where, according to tax records, he was residing by 1596.* Later, toward the end of the decade, he resided in Southwark—a logical move, as by this time it was the heart of London's theater scene. (Today's visitors are of course drawn to the reconstructed Globe Theatre; but one must also head for Southwark Cathedral, where one can see a memorial plaque to Shakespeare's younger brother Edmund, an actor, who was buried in the church in 1607. The dramatist John Fletcher, Shakespeare's collaborator for his final plays, rests there as well. Stained-glass windows, meanwhile, illustrate scenes from Shakespeare's plays.)

The playwright would still have been living north of the Thames when his work first appeared in print. Shakespeare's epic poems, *Venus and Adonis* and *The Rape of Lucrece*, date from 1593 and 1594 respectively. But we know he had already made a name for himself as an actor and a playwright in London by 1592, because of a reference to him in a pamphlet by a poet and playwright named Robert Greene. In a snarky commentary on London's theatrical scene, Greene attacks Shakespeare as an "upstart crow, beautified with our feathers" and goes on to parody a scene from *Henry VI, Part 3*. Greene refers to the play's creator as someone "in his owne conceit the onely Shake-scene in a countrie"—clearly a put-down (as well as a not very clever pun on Shakespeare's name). Thanks to Greene, we know not only that Shakespeare was working in London by this time (and had at least one set of history plays under his belt), but that he was successful enough to make his colleagues jealous. (We know from independent sources that *Henry VI* was performed early in 1592—possibly with Shakespeare himself among the cast.)

Shakespeare was prolific, and he had to be: The demand for new plays was high, and there was a good living to be made for a playwright

---

* For whatever reason, Shakespeare chose not to pay his taxes on that occasion. As Charles Nicholl notes, "This is not remarkable—the system was chaotic, and evasion was common—but it is piquant to find that the first actual documentation of Shakespeare in London is as a tax-dodger" (Nicholl, p. 41).

capable of filling that need. It's been estimated that one-third of London's adult population saw a play at least once a month, with single performances drawing an audience of up to three thousand. As Shapiro notes, Shakespeare and his fellow dramatists were writing for "the most experienced playgoers in history." The playwright's colleagues were men of remarkable talent in their own right: This was the London of Thomas Kyd and Christopher Marlowe, both of whom were renowned for their poignant tragedies. Marlowe's life was cut short in the spring of 1593 when he was fatally stabbed during a bar-room brawl. We don't know if Shakespeare mourned for his colleague, but there is no doubt that Marlowe's death left a vacuum in London's theatrical world, and that Shakespeare helped to fill that void. He wrote, on average, two new plays every year, and kept up that pace until nearly the end of his career. Shakespeare was also a shrewd businessman. Together with his fellow actors, he became a shareholder in the construction of a grand new open-air theater, the Globe, to be built in Southwark, on the south bank of the Thames; Shakespeare technically owned one-eighth of the new building. Theaters, deemed morally dangerous, were forbidden within the city proper; but anyone who could afford a ferry ride, or who didn't mind a brisk walk across London Bridge, could pay one penny to see a performance of *Julius Caesar* or *Hamlet*.

Shakespeare's writing career can be divided very roughly into two halves, centered on the year 1600. In the seven or eight years up to this point, he produced a mixture of historical dramas and comedies, including several that are still among his best-loved works, including *A Midsummer Night's Dream*, *Romeo and Juliet*, and *The Merchant of Venice*. From early in this period we find the bloody revenge tale of *Titus Andronicus*; by its end we have one of Shakespeare's most refined comedies, *As You Like It*, and his first great tragedy, *Julius Caesar*. It was at about this time that the published versions of Shakespeare's plays began to routinely bear his name; previously, a playwright's name was hardly worth mentioning, but by this stage Shakespeare was famous enough that his publishers knew it would boost sales.* Indeed, Shakespeare was now earning a decent living from the theater, perhaps making two hundred pounds a year—

---

* Shakespeare had become something of a celebrity by this time. A remarkable booklet by a man named Francis Meres, dated 1598, lists a number of his plays (and is thus invaluable for deducing their chronology), and compares the playwright to the greatest of the ancients: "As Plautus and Seneca are accounted the best for comedy and tragedy among the Latins; so Shakespeare among the English is the most excellent in both kinds for the stage . . ." (quoted in Chute, p. 179).

"at least ten times what a well-paid schoolmaster could hope for," as Samuel Schoenbaum points out. He was, as Jonathan Bate and Dora Thornton note, "the first Englishman in history to make a serious living by his pen."

After 1600, Shakespeare produced a string of powerful tragedies, beginning with *Hamlet* and continuing with *Othello*, *King Lear*, *Macbeth*, and *Antony and Cleopatra*. And yet it is not as though he had forgotten how to be funny; *Twelfth Night* dates from the beginning of this period, and there is plenty of humor both in the tragedies and in the so-called romances, which include *The Winter's Tale* and *The Tempest*. All this time, however, Shakespeare was investing in property back in his native Stratford, and, sometime around 1613, he returned to the city of his birth. He died there on his fifty-third birthday, and, as with so many other details of Shakespeare's life, the cause is unknown; syphilis, typhoid, and influenza have all been suggested.

WOULD THE REAL WILLIAM SHAKESPEARE PLEASE STEP FORWARD?

This condensed biography gives us the bare bones of Shakespeare's life and career. We may feel we have almost gotten to know the man—but certainly not nearly as well as we would like to. How well *can* we know William Shakespeare? The documentary evidence is scant enough to count on the fingers of two hands—the parish records of the baptism, the marriage, and the birth of his children; a smattering of legal documents and investment records; and the famous will in which he leaves his "second best bed" to his wife. There are a mere fourteen words known to be written in his own hand (comprising six signatures and the additional words "by me" on the famous will).* It's not much to go on—and it puts biographers in a bit of a pickle, as they strive to reconstruct Shakespeare's thoughts and actions.† (And, inevitably, it encourages the anti-Stratfordians.) Of course, most books about Shakespeare—your local university library will likely have thousands—aren't biographies at all; instead, they examine Shakespeare's writings, a far more fruitful subject. The vast majority of Shakespeare scholarship focuses on what he wrote, not who he was.

---

* Although still the subject of some controversy, a manuscript for the play *Sir Thomas More*, now in the collection of the British Library, may contain a longer sample of Shakespeare's handwriting. The play was a collaborative effort, and a three-page section is widely believed to be in the playwright's hand.

† And yet it can be done: The best of the no-nonsense biographies is *Shakespeare: A Compact Documentary Life* by Samuel Schoenbaum (1987).

Still, the urge to "meet" Shakespeare is, for many of us, inescapable—just as we might fantasize about getting to know Mozart or Einstein. And so the biographical gaps gnaw away at us. Consider, for example, the famous "lost years"—the period following the birth of Shakespeare's twins, in Stratford, in 1585, and the first known reference to the playwright as an active Londoner (from Greene's pamphlet of 1592). Where was he during those seven years, and what was he up to? One of his earliest biographers asserted that he was "a schoolmaster in the country"; others imagined him working as a law clerk or serving his country as a soldier. (As Jonathan Bate notes, schoolmasters seem to like the schoolmaster theory, while lawyers tend to favor the law-clerk theory.) The various suppositions go in and out of fashion, and none are backed by any hard evidence. More recently, as mentioned, a BBC television documentary argued (as others had previously imagined) that he traveled to Italy during this time. Again, no actual evidence.

Even for the years in which we *do* know Shakespeare's whereabouts, there is much more we would like to know. Take his marriage to Anne, for example. Were they happy? Did he love her, or was he simply doing what needed to be done—"a bow and arrow wedding," as one of my tour guides put it—when he took her to the altar, seven months after his eighteenth birthday? The short answer is that we don't know. Here's one biographer's attempt at a longer answer:

> In common with most women of her class, [Anne] did not read or write, and may well have been quite willing to play the role of stay-at-home housewife and mother, while her husband acted and wrote in London. She does not seem to have had any great hold over his affections—Shakespeare was not a dissipated man, but nor was he a model of virtue. The most we can say is that he made "an honest woman" of her.

How much does this paragraph tell us? Very little, in fact. Aside from the brief reference to Anne's lack of schooling, it adds almost nothing to the smattering of data that we began with. Add a "may well have" and a "seem," and we have the skeleton of a character sketch of a young man married to an older woman who bore his children. This isn't meant to be a criticism; when facts are in short supply, we rely on educated guesswork. There is no other way. Indeed, pick up any Shakespeare biography, even one of the very good ones, and you will find a story laced with maybes: "he would have"; "he would likely have"; "one can imagine"; and so on. "It is possible" that Shakespeare's mother took her young son to Wilmcote to avoid the plague (Donnelly and Woledge); "It is easy to imagine" Shakespeare as a young law clerk (Greenblatt); "It is likely" that Shakespeare's old schoolmate, Richard Field, helped him find accommodations on his arrival in London (Day); Shakespeare "must have been" a familiar presence in the London bookshops (Shapiro); "we may plausibly imagine" Shakespeare haunting the bookstalls (Ackroyd). Frank Kermode, in his own very good Shakespeare biography, asks readers to indulge him as "these speculations grow more and more far-fetched as one 'might have' succeeds another, or a 'may well have' or a 'surely.'" (Kermode uses such constructions judiciously—and unavoidably: "We must assume" that Shakespeare attended the local grammar school; Shakespeare "could well have" attended the so-called mystery plays in Coventry. Indeed, the reader may have noticed them in the present work.)

NOBLE WEEDS AND OTHER GRAINS

Our picture of Shakespeare may seem more like a collection of fragments than a unified whole. There is so much more that we would like to know: We don't know if any of Shakespeare's children ever visited him in London, or if he ever gave them any fatherly advice; we don't know if he cried when told of the death of his son, Hamnet, in 1596. We don't know if he was a good archer, or if he gambled at dice or cards; we don't know who he went to the taverns with, or what they spoke about after a few pints of ale; we don't know if he was a regular at the Southwark whorehouses or if he remained faithful to his wife back in Stratford. And speaking of Shakespeare's sexuality, biographers (and ordinary readers) have often wondered if he was bisexual.* We yearn for any clue,

* Of Shakepseare's 154 sonnets, 126 appear to be addressed to a young man (referred to as a "fair youth"), including sonnet 18 ("Shall I compare thee to a summer's day? / Thou art more lovely and more temperate . . ."). The sonnets hardly constitute proof of homosexuality, as men routinely spoke of "loving" other men in Shakesepare's time, whether there was a sexual component or not. Even so, a number of candidates for the playwright's possible gay lover(s) have been put forward, including his early patrons, Henry Wriothesley, the Third Earl of Southampton, and William Herbert, the Third Earl of Pembroke.

no matter how minuscule. When one is offered, we bite: When archaeologists discovered traces of cannabis in the garden of one of Shakespeare's properties in Stratford, about a decade ago, it was big news. "Did cannabis fuel Bard's genius?" asked a headline on the BBC News website. The analysis was carried out on some two dozen clay pipes found on the site. The lead archaeologist also cited the reference to a "noted weed" in Sonnet 76. Could this be an allusion to the poet's predilection for marijuana? It's a stretch, to say the least. (As is so often the case, the theory was in the news for about a day, and then promptly forgotten.) But it does show how desperate we are for any glimpse, no matter how speculative, into the playwright's life.

More recently, in March 2013, a team of researchers uncovered documents that suggest that Shakespeare, during a time of famine, hoarded grain for profit and chased down those who could not (or would not) repay their debts. Again, it made headlines: "Bad Bard: a tax dodger and famine profiteer," trumpeted the *Sunday Times*; "Shakespeare the 'hard-headed businessman' uncovered," declared the *Independent* (although, as we've seen, we *already* knew he was a savvy businessman). As usual, the urge to reinterpret the plays in light of the new "evidence" was irresistible: The lead researcher, Jayne Archer of Aberystwyth University, suggested the newly revealed facts are reflected in *Coriolanus*, set in ancient Rome during a time of famine, and were perhaps inspired by the real-life uprising by peasants in the English Midlands in Shakespeare's time. What does it mean if Shakespeare was against grain hoarding in the play, but all for it in real life? Perhaps, Archer speculates, *Coriolanus* was the playwright's attempt "to expunge a guilty conscience." She also sees famine as central to the story of *King Lear*, in which the king's unfair distribution of resources to his daughters triggers a war.

We can't blame scholars for building theories around their data, scant as it may be. But that doesn't mean that "anything goes." And that applies whether you're building your theory from the ground up, from bits of "evidence" (like the grains of cannabis), or starting from the standard narrative and chipping away at it, as the anti-Stratfordians are wont to do. And of course, absence of evidence is not evidence of absence. Consider Shakespeare's lack of a personal library: As a skeptical *New York Times* article recently noted, no books are mentioned in the playwright's will, and there is no record of him paying tax on such a collection. What can we make of this? Very little, actually—and Bill Bryson has a humorous rebuttal to those who try. He reminds us that we know nothing of

Shakespeare's incidental possessions one way or another. The author of the *Times* article, Bryson says, "might just as well have suggested that Shakespeare never owned a pair of shoes or pants. For all the evidence tells us, he spent his life naked from the waist down, as well as bookless, but it is probable that what is lacking is the evidence, not the apparel or the books."*

We should also note that while Shakespeare is somewhat of an enigma, he is *no more of an enigma* than others from his social rank living in England at that time. Indeed, the problem with the surviving traces of Shakespeare's life, as Stephen Greenblatt has put it, "is not that they are few but that they are dull." Indeed, compared with Marlowe, a spy who faced accusations of brawling, sodomy, and atheism, Shakespeare seems to have been something of a couch potato. When I spoke with Greenblatt recently, he pointed out that Shakespeare is actually better documented than many of his contemporaries—at least, better than most of his artistic or literary contemporaries. "And certainly [he's] better known than contemporaries of his social class—unless they got in horrendous trouble with the police," Greenblatt said, referring to the body of evidence concerning Christopher Marlowe's activities. "Marlowe had a spy assigned to him, and the spy wrote reports. We could wish that Shakespeare had a spy assigned to him and left reports, but as far as we know that didn't happen."

"WORDS, WORDS, MERE WORDS"

There is another trap to avoid: We have Shakespeare's plays, poems, and sonnets—but we must resist the urge to read them autobiographically. Usually, that is easy enough: When Shakespeare writes about the assassination of Julius Caesar, we don't imagine either that he himself had plotted assassinations, or that he time-traveled to ancient Rome to witness the big moment. (He didn't have to: He could just read Plutarch's account, readily available in a recent English translation by Thomas North.) Few would suggest that Hamlet's indecision means that Shakespeare was indecisive, or that Iago's scheming implies that the play's author was a devious manipulator. But certain scenes—parts of *As You Like*

---

* At a recent lecture, I heard a well-known Shakespeare scholar respond to the "no books" issue as follows: Shakespeare was no dummy; knowing that the tax auditors were in the neighborhood, couldn't he have simply hidden the books? The audience seemed satisfied, but to my mind it's not much of a retort. (If I were doing battle with the anti-Stratfordians, I wouldn't make too much of the book-hiding theory.)

It and *The Winter's Tale*, for example—do conjure up something of life in a small English town, while the epilogue from *The Tempest* strikes many readers as offering at least a glimpse of the real Shakespeare. Another oft-mentioned example is the scene in *King John* in which Shakespeare writes poignantly of a mother's grief at the loss of her son, a play written very close to the time that Shakespeare lost his own son, Hamnet. The temptation to see the author through his works is even greater in the Sonnets, in which the words even play on the author's name (for example, Sonnet 135, which includes the line, "Whoever hath thy wish, thou hast my Will"). Are we really seeing the inner life of the author behind these lines? Caution would seem to be in order. As James Shapiro puts it, "Since I don't know when or where Shakespeare is speaking as himself, I steer clear of reading [the sonnets] as autobiographical. I'm not denying that there are elements of Shakespeare's personal experience woven into the fabric of these remarkable poems. But I am insisting that it is impossible to know how or when such personal elements appear. . . . It seems rather circular to me to construct the life out of the works and then read the works as autobiographical."

That is good advice—but even so, the Shakespeare enthusiast can be forgiven for *looking for* glimpses of the author behind the words. It is a natural urge, especially when those words are so powerful that they seem to speak directly to us, even four hundred years later. As Greenblatt told me:

> The longer you muse over and ponder [Shakespeare's works], the more you enjoy them, the more you feel that you're in contact with something important—you *do* wonder, 'Who was this person?' And that would be just as true if a message appeared on the beach in a bottle, and you opened it up. Even if you had no access whatsoever to whomever launched that, you would, if only idly, think, 'Who sent me this message?'—particularly if the message seemed actually addressed, in a strange way, to you. . . . This is not unique to Shakespeare. If Jane Austen reaches you, if Kafka reaches you—if you feel you're in contact with something powerful that is speaking to you deeply and personally—it is an absolutely natural human response to want to know who sent you this message and what to do about that response.

And so we yearn to know the "real Shakespeare"—while at the same time we learn to live with the biographical gaps. We have the most crucial elements of Shakespeare's life, and from those elements—after much scholarly research—a portrait, albeit an imperfect one, emerges.

There is a reason why I paused to look at the challenges of piecing together the details of Shakespeare's life. I want the reader to be conscious of just how tricky it is to separate the probable from the plausible in the world of Shakespeare; to judge how speculative is too speculative. Every discipline has its quacks: Biology has "intelligent design"; psychology has parapsychology and phrenology; medicine has homeopathy; geology has (or at least had) its flat-Earthers and (amazingly) hollow-Earthers. And though it doesn't have a catchy name, physics has its "Einstein was wrong"ers. Isaac Asimov, one of the great science communicators of the last century, spoke of his "built-in doubter," which he called on when confronted with controversial claims. The more radical the claim, the more skepticism was needed. Carl Sagan expressed a similar sentiment when he said, "Extraordinary claims require extraordinary evidence." A theory that claimed to overthrow four hundred years of established physics, for example, would demand the highest possible level of skepticism. One must doubt, but one must doubt intelligently.

Shakespeare studies has, of course, the anti-Stratfordians, "the Shakespeare wasn't Shakespeare" crowd. But a theory doesn't have to be off-the-deep-end crazy to warrant suspicion. Nor is the reverse the case: A claim need not have tangible, physical proof to be plausible or even probable—as with the supposition that Shakespeare, as a youngster, attended his local grammar school, or that he owned at least a few books, which the majority of scholars accept. I sometimes wish I had a "nonsense detector," analogous to Asimov's doubter, that could do the work for me: When someone mentions young William's attendance at the grammar school, it would emit a pleasing hum; perhaps a green light would go on. The discussion turns to Shakespeare's Catholic sympathies, or his sexual orientation, and a yellow warning light might come on. A visit to Italy? The yellow light begins to flash, and a buzzer comes on. Shakespeare was the Earl of Oxford? The light turns red, the buzzer gets louder, and smoke begins to stream out of the machine. . . . (Not as much fun as a time machine, to be sure, but still rather handy.)

We don't have such a device, so we must make do with our common sense, and the opinions of those scholars who have dedicated their careers to the study of Shakespeare and his work—exercising caution when the experts disagree with each other. In the chapters ahead, we'll face a task even more difficult than working out where Shakespeare was or who he was with at some particular time. We want to know *what he was thinking.* More specifically, we want to know what—if anything—he thought about the scientific discoveries of the day; the so-called "new philosophy." Along the way we will hear a variety of theories and opinions, some of them very plausible and some perhaps not-so-plausible. I hope by this point I have laid the groundwork to help the reader navigate through the coming arguments.

If this were a traditional Shakespeare biography, we would plod through obligatory chapters on Elizabethan drama, the layout of the Globe Theatre, Shakespeare's use of language, and so forth—but we must move on. As we've seen, the final decade of the sixteenth century was a remarkable time in English history. This was the age not only of Shakespeare, Marlowe, Jonson, and Kyd, but also the age of John Dee, Thomas Digges, and Thomas Harriot. Nobody could have known it at the time, but a new age—the Age of Science—was nascent. We are now ready to ask how much Shakespeare may have known about such developments. Did he ever meet any of the great scientific thinkers of the day? Did he hear about their work, or read about their ideas? And if he did, how did that knowledge shape his own work? We could begin with any one of the playwright's beloved works, but why not start with the most famous of all. Let us head for the battlements of Elsinore.

7. *"More things in heaven and earth . . ."*

THE SCIENCE OF HAMLET

More than four hundred years after its debut on the London stage, *Hamlet* remains Shakespeare's most famous work, his most frequently produced play, and, arguably, his greatest artistic achievement. (A significant number of critics, especially since the early decades of the twentieth century, have voted for *King Lear* over *Hamlet*—we'll look at that contest a bit more in Chapter 14—but for now, let's not quibble; for the sake of argument, let's say they're both works of the highest order of literary genius.) *Hamlet* is also Shakespeare's longest play—staged uncut, it would run for more than four hours—and one of the most problematic.* Incredibly, as the centuries pass, *Hamlet* seems more and more relevant: It is said to define what it means to be modern; to be self-aware; to be human. Prince Hamlet himself is Shakespeare's most complex character, and certainly the most thoroughly scrutinized figure in English literature. (He also loves to talk, speaking more than 1,500 lines, accounting for about 39 percent of the play.) Hamlet is the role most coveted by actors, in spite of (or perhaps because of) his obvious flaws. He is cowardly, narcissistic, indecisive, starkly misogynistic—the list goes on—and yet we can't seem to get enough of him. Perhaps, as William Hazlitt once put it, "It is *we* who are Hamlet." For a fictitious character, the prince and his inner turmoil seem all too real. His speeches, Hazlitt

* Cyrus Hoy, in the Norton Critical Edition of the play, begins his analysis: "Everything about *The Tragedy of Hamlet, Prince of Denmark* is problematic." Among the difficulties that Hoy notes are the uncertainty in dating the play, in choosing among its various texts (there are multiple editions from Shakespeare's time, and they differ significantly), in relating the play to its sources, and of course in comprehending the character of Prince Hamlet himself.

reminds us, are "but idle coinages of the poet's brain," and yet "they are as real as our own thoughts."

While *Hamlet* is rarely examined from the point of view of science, it is impossible not to think of the play, at least in part, as a reflection of its turbulent times—a period of remarkable intellectual upheaval. The time, many people surely felt, was indeed "out of joint." From the play's start to its finish, Prince Hamlet seems trapped between two worlds. In act 1 we find him "crawling between earth and heaven" (1.2.129); when his uncle asks him about his dark mood, he claims to have been, on the contrary, "too much in the sun" (1.2.67). Several acts later, we find him peering down at Ophelia's freshly dug grave while invoking the planets above; he notes that Laertes' grief "conjures the wand'ring stars" (5.1.249). And lest we imagine the stars are moving across the sky peaceably, the ghost has warned Hamlet that his tale will "make thy two eyes like stars start from their spheres" (1.5.17). The prince will soon be complaining that the world—"this goodly frame the earth"—is, for him, "a sterile promontory" (2.2.298–99).

"YOND SAME STAR THAT'S WESTWARD FROM THE POLE"

The action of the play is grounded in Denmark, but right from the opening scene we are asked to look upward. For the past two nights, the guards at Elsinore have been startled by the appearance of a ghost resembling the dead king (the recently deceased King Hamlet, father of the title character). Horatio, an old school friend of the prince, arrives on the scene, and Bernardo, one of the guards, explains the ghost's habits: He is prone to walking about the ramparts at night—not just at any time of night, but at one hour past midnight, when a particular star appears "westward from the pole":

> Last night of all,
> When yond same star that's westward from the pole,
> Had made its course to illume that part of heaven
> Where now it burns, Marcellus and myself,
> The bell then beating one—
>
> *Hamlet* (1.1.39–42)

Bernardo is struck silent at this point when (speak of the devil!) the ghost appears, as if on cue. Is Bernardo merely using this star as a way of marking the time? Perhaps—and, as we've seen, there are other occasions where Shakespeare's characters track the time by noting the positions of the stars (see page 16). But in *Hamlet*, the stars (and celestial happenings in general) seem to hold more gravitas for Shakespeare's characters than what we might associate with mere timekeeping. As Horatio explains, strange phenomena in the heavens are often accompanied by dire events on Earth. He refers to the murder of Julius Caesar, signaled by "stars with trains of fire"—a reference to meteors or a comet, perhaps—and "disasters in the sun"; the moon, meanwhile, was "sick almost to doomsday with eclipse" (1.1.120–23). And so the star "westward from the pole," we might surmise, holds more significance than some run-of-the-mill star that one might use to mark the hour.

But what star, exactly, are we talking about? Can we reconstruct the skies over Denmark at the time of *Hamlet* and find out?* Astronomer Donald Olson has attempted to do just that. Olson, who teaches at Southwest Texas State University, is sometimes described as a "forensic astronomer." He and his students analyze astronomical references in art and literature, in an effort to gain a deeper insight into the works in question. Over the years, he's tackled such diverse subjects as Julius Caesar's account of his invasion of Britain, Edvard Munch's painting *The Scream*, and the photographs of Ansel Adams. In the 1990s, he turned his attention to *Hamlet*, and, in particular, to the star seen from the ramparts of Elsinore.† He begins with the clues present in the text itself: We know the time of night (1 a.m.), as well as the star's location in the sky ("westward from the pole")—but to know which star that might be, we also need to know the time of year. Fortunately, there are further clues. Francisco complains that the night is "bitter cold" (1.1.8), and Hamlet, on the following night, agrees that "the air bites shrewdly, it is very cold" (1.4.1). Olson argues, quite reasonably, that this is suggestive of late fall or winter. Another reference makes it clear that we are not currently in "that season . . . Wherein our Saviour's birth is celebrated" (1.1.163–64), implying that the scene is not unfolding during Advent; thus most of December is ruled out. But we also know that two months have passed since old Hamlet's death—a murder, as it turns out—which happened

* Denmark and southern England differ in latitude by about four degrees—which means the height of a celestial object above the horizon would differ, slightly, between the two locations at any one time (which Shakespeare may or may not have been conscious of). This does not affect the positions of celestial objects relative to one another, or their distance to the celestial pole.

† Donald Olson's article, "The Stars of Hamlet," was published in the November 1998 issue of *Sky & Telescope*.

while he was taking a nap outdoors, in his garden. Putting these clues together, Olson concludes that King Hamlet died in September, and that the ghost's appearance on the battlements takes place in November. (So far, so good; a number of other scholars have also put forward November as the most probable time of the play's initial action.)

Now that we have the time of year as well as the time of night, what star might we find "westward from the pole"? Olson and his students used astronomical software* to try to answer that question, but as it turns out, there is no obvious candidate for such a star—at least, not at first glance. Olson considers, and then rejects, the various stars that make up Ursa Major and Ursa Minor (the Great Bear and the Little Bear); they aren't in the right part of the sky at that time of year; neither are the bright stars Vega or Deneb, which lie at the right declination (that is, the right distance from the pole star) but which again would not lie to the west of the pole in late fall. About all that's left, Olson argues, are the stars of Cassiopeia, which does happen to lie fairly close to the pole (in a direction roughly opposite to that of the "Big Dipper" asterism of Ursa Major). Unfortunately, Cassiopeia contains no particularly bright stars—none of the first magnitude†—and the second-magnitude stars that make up the familiar "W" (or "M") are all of about equal brightness. If one were keeping time by noting the position of Cassiopeia, one might just as well refer to the constellation as a whole than to single out any one of its virtually identical stars.

But as Olson points out—and as we saw in Chapter 3—there *was* a bright star in Cassiopeia, back in the days of Shakespeare's youth. It was, of course, Tycho's star, the supernova of 1572, which lit up the skies over Europe that fall, remaining visible for more than a year. It would have been in just the right part of the sky to lie "westward from the pole" at about 1 a.m. on a crisp November night, as seen from England or Denmark (or anywhere else of that approximate latitude). As we've seen,

---

* Although sky-simulation software ("planetarium software") is useful for its precision (and is necessary if one is interested in the positions of the planets as well as the stars), the basic task—determining which stars in which constellations are visible in which part of the sky, for an observer in the mid-northern latitudes, at a particular time of night at a particular time of year—can be tackled with little more than a cheap "planisphere," the rotating star map that you can buy in any planetarium gift shop for about ten dollars.

† In astronomy, "magnitude" is a measure of a star's brightness. The lower the number, the brighter the star. The very brightest stars actually have negative magnitudes: Sirius shines at magnitude −1.5, with Arcturus and Vega very close to zero. Capella, Aldebaran, and Antares are all about magnitude 1.0 ("first magnitude"), while the stars of the Big Dipper (part of Ursa Major) shine at about 2.0 ("second magnitude").

beginning with the Prologue, Shakespeare was just eight years old when Tycho's star appeared. The Prologue is fiction, of course, but it seems reasonable to expect that young William would have remembered such a sight from his childhood. We can't know how vivid the memory would have been more than twenty-five years later, when he sat down to write *Hamlet*, but I would suggest, as Olson does, that one's first sighting of a bright new star—a star that's *not supposed to be there*, which stays in the sky for months, and which people keep on talking about for *years*—is not something one would soon forget. Moreover, when Shakespeare was a young man, there would have been a reminder: As Olson points out, the historian Raphael Holinshed discusses the star at some length in his *Chronicles*—one of Shakespeare's key sources for his history plays. (The *Chronicles* also gave Shakespeare the plot of *Macbeth*, and bits of *King Lear* and *Cymbeline*.) Published in 1577 and reprinted in 1587, the *Chronicles* refer to a new star "in the constellation of Cassiopeia . . . [appearing] bigger than Jupiter, and not much lesse than Venus when she seemeth greatest." The star was "so strange, as from the beginning of the world never was the like." I tend to agree with Olson, that Shakespeare's "boyhood memory of the new star could have been reinforced [by Holinshed] at the time he was writing *Hamlet*."

Of course, Olson isn't the first to ponder the nature of the star "westward from the pole," but for some reason its identity has left Shakespeare scholars somewhat baffled. Many editions of *Hamlet* helpfully point out that "pole" means "pole star" (so far, so good) but leave it at that. Many also note that it is perfectly reasonable for the guards doing the night shift at Elsinore to mark the time by following the stars (even when clocks that strike the hour are present, as we are told they are in Shakespeare's play).* In the Penguin edition (1980, reprinted 1996), T. J. B. Spencer writes that the playwright "throughout the scene gives an impression of a clear, frosty, starlit sky." Fair enough. He adds, "Bernardo presumably points to the sky at one side of the stage, guiding the eyes of the audience away from where the ghost will enter," reminding us that the reference to the star may be motivated more by utilitarian stagecraft than by astronomical accuracy. Spencer then points to some of the other

* Shakespeare's plays are famously full of anachronisms. Striking clocks date from the mid-fourteenth century, making them rather out of place in *Julius Caesar*. But *Hamlet*, in spite of its medieval roots, seems to call for a Renaissance setting: The action can take place no earlier than the founding of the university at Wittenberg (established in 1502), so the presence of striking clocks at Elsinore is reasonable.

words that Bernardo uses to describe the star and its motion: "made its course"; "illume"; "burns." Taken together, Spencer says, this "seems to imply that the *star* is a planet" (italics in the original). Unfortunately, this cannot be: Planets are always located near the *ecliptic*—that is, near the imaginary line that runs through the constellations of the zodiac—and not near the pole.*

"THESE BLESSED CANDLES OF THE NIGHT"

It's remarkable how much confusion Shakespeare's astronomical references have wrought—with references to the pole star being (for some reason) among the most problematic. Bernardo's speech on the ramparts is one of at least three references to the celestial pole—the spot marked, roughly, by the "pole star" or "northern star." The most famous instance comes from the lips of Julius Caesar, who declares, ". . . I am constant as the northern star, / Of whose true-fixed and resting quality / There is no fellow in the firmament" (3.1.60–62). The star shines again in act 2 of *Othello*, where we find Montano, the Venetian governor of Cyprus, discussing the fate of a Turkish naval fleet caught in a violent storm at sea. Even watching from shore, they can discern the storm's fury. Montano asks his companions what the fate of the Turkish fleet will be. One of them (identified only as "Second Gentleman") replies,

> A segregation of the Turkish fleet:
> For do but stand upon the foaming shore,
> The chidden billow seems to pelt the clouds,
> The wind-shaked surge, with high and monstrous mane
> Seems to cast water on the burning Bear
> And quench the guards of th'ever-fixèd pole.
>
> (*Othello* 2.1.10–17)

In spite of the flowery language, the gist of the passage is clear enough: The storm is so bad it will surely be the demise of the Turkish ships. Owing to intense winds, the ocean spray has become so thick as to render the stars invisible—or, perhaps, the spray is rising so high that it seems to extinguish the stars. (Or, as George Costanza would have put

* The closest a planet can ever come to the pole is about sixty degrees. An observer in the northern hemisphere might say that a planet is "in the east" if it was rising or "in the west" if it was setting; if it was midway in its trajectory across the sky, one might say the planet was "in the south" (or "nearly overhead," if it was at the northern extremity of its orbit at that time).

it, "The sea was angry that day, my friends—like an old man trying to send back soup in a deli.") We might stumble on some of the particular words—like "chidden," for starters. (Footnotes to the rescue: Apparently it means "repelled by the shore.") Easier to gloss is the "burning Bear," presumably a reference to either Ursa Major or Ursa Minor—and since the next line mentions the "guards" of the pole, we can surmise that we're talking about the pole star and its neighbors: The guards are the two bright stars in the bowl of the Little Dipper, a part of Ursa Minor (the Little Bear), believed to guard or protect the north star.

The phrase "ever-fixèd pole" seems straightforward, but, as with so many passages in Shakespeare, it suffers from textual ambiguity. The quarto (1622) and folio (1623) editions of the play differ in a number of places, and this is one of them: It is *fired* in the former but *fixed* in the latter. Were the folio editors simply correcting a typo in the quarto? If so, the line expresses the most familiar property of the north star, which remains fixed in the sky while the other stars appear to circle around it. If we instead read it as "fired," then it presumably has something to do with "burning" in the previous line, and similar phrasings do crop up, occasionally, elsewhere in the canon. (The Arden edition goes with "ever-fired"; the Oxford, quoted above, goes with "ever-fixèd.") At least everyone seems to agree on the significance of these particular stars: The scene describes the fate of warships on the high seas, so there is little surprise that it contains a reference to stars that were of particular interest to navigators. In fact, their utility is twofold: The direction of the pole star indicates north, while the orientation of the "guards" relative to the pole can be used to determine the hour.

Still, it is possible to stumble. In the Oxford edition (2006), Michael Neill writes, "The Pole Star's usefulness to navigators seeking to take their bearings was that it was one of the so-called 'fixed stars.'" Unfortunately, this is a conflation of two senses in which a star might be "fixed": In the broad sense, all of the stars in the sky are "fixed stars," in contrast to the planets, known as "wandering stars"; that is, although they move across the sky, the stars (unlike the planets) maintain the same relative positions to one another. But the usefulness of the pole star comes from the fact that it doesn't join in with this collective motion: It remains in the same part of the sky, while all of the other stars appear to revolve around it. If modern editors seem strangely confused about this distinction, or about celestial mechanics in general, it may simply be because we spend less time looking up than we once did. And yet even in the nineteenth century editors were having trouble with Shakespeare's astronomical references. When Horace Howard Furness was compiling his massive "Varorium" edition of *Othello* in 1886, he waded through the various competing glosses from the preceding hundred years or so, assessing their merits. The scene that begins act 2 had been particularly troublesome, with critics recruiting a bewildering array of stars and constellations in order to make sense of it. Furness, cutting through the clutter, chose "fixed" over "fired," and concluded that the stars in question are indeed the three brightest stars of Ursa Minor—the pole star and the two stars which guard it. "Shakespeare," he concluded, "knew better than his commentators what he was talking about when he spoke of the guards of the pole." (Furness went on to cite a number of sixteenth-century astronomical manuals that describe how one can use the north stars and its companions for navigation.)

Let's return now to the opening scene of *Hamlet*, where once again the relationship between the north star and other celestial objects—in particular, the star "westward from the pole"—is paramount. The Arden editions are considered by many to be the gold standard for Shakespeare commentary, so it is illuminating to see how they treat the star mentioned by Bernardo on the battlements. In the second-to-last Arden edition, from 1982, editor Harold Jenkins notes that the reference need not be to a particular star—that much is certainly true—but then adds that "Shakespeare had presumably seen the brilliant star Capella, which would appear in the winter sky 'westward from the pole.' " Unfortunately, Jenkins is way off: At 1 a.m. in November, Capella, as seen from mid-northern latitudes, lies nearly overhead. One might describe it as being "above the pole," but certainly not "westward from the pole." (Having said that, it's still a better guess than suggesting the object is a planet.) But the latest Arden edition—the hefty 2006 text edited by Ann Thompson and Neil Taylor—brings a fresh start to the guards' nocturnal observations, and provides a revised take on the star's identity: Capella is out and the supernova is in, with a reference to Donald Olson's article from *Sky & Telescope*.

I don't want to make too much of the Arden revision; it's just a few lines of commentary in a 613-page book—and, as Jenkins reminds us, Shakespeare may not even have had a particular star in mind. But still: Here is one of the few cases of an astronomer telling the community of

Shakespeare scholars, politely but firmly, "Hey, you missed a spot." And at least a few of those scholars have said, "You're right, we did."*

FROM HAMLET'S CASTLE TO TYCHO'S ISLAND

We have already examined (in Chapter 3) the impact of the new star of 1572, one of the key events in the demise of the Ptolemaic model of the cosmos. As we noted, Tycho Brahe made detailed observations of the star from his Danish island, while Thomas Digges and John Dee, observing from England, were similarly captivated by its appearance. But there is more to connect *Hamlet* and Tycho than just the supernova (that is, assuming the star in question *is* the supernova). To begin with, there is the play's setting. It's no surprise that Shakespeare chose to locate his play in Denmark. One of his sources was a medieval story about a Scandinavian prince named "Amleth," dating from the twelfth century. A written account of this tale, by an author known as Saxo Grammaticus ("Saxo the Grammarian"), was first set to print in the early 1500s. Shakespeare may not have read Saxo's version, but he surely read a more recent French version by François de Belleforest, published in 1570. (By the 1580s, the story had been adapted for the stage, and was being performed in London by Shakespeare's own company. Possibly written by Thomas Kyd, the Ur-*Hamlet*, as scholars refer to it, has sadly been lost.) Some of *Hamlet*'s key elements, including the murder of the old king and the quest for revenge, go back to the story's medieval roots. (The ghost's origins seem to be more recent; he may have made his debut in the lost Ur-*Hamlet*.) But while both Saxo and Belleforest locate the story in Denmark, it was Shakespeare who specifically took the action to the royal court at Elsinore. Though we have no reason to imagine that Shakespeare ever visited Denmark—or, indeed, that he ever traveled beyond England—he would certainly have known of the castle at Elsinore, since some of his fellow actors had played there. (As noted, King James of Scotland—the future king of England—had visited there as well; apparently Tycho and his island were quite a draw.)

Elsinore—Helsingør in Danish—stands on the eastern shore of the

* And yet, even Thompson and Taylor have what I suspect is an astronomy-related stumble: When Laertes says that he has obtained a poison "So mortal that, but dip a knife in it, / Where it draws blood no cataplasm so rare, / Collected from all samples that have virtue / Under the moon, can save the thing from death / That is but scratched withal" (4.7.140–44), they suggest that "under the moon" simply means "anywhere on earth." I suspect it more likely means "sublunar," in the Aristotelian sense—that is, no *earthly* potion could serve as an antidote, though a *divine* remedy (from the heavenly realm above the moon) might do the trick.

Danish island of Zealand, overlooking the channel separating Denmark from present-day Sweden (though in Shakespeare's day, this was all part of the Kingdom of Denmark). Aside from hearing about the castle from some of his actor friends, might there have been any other reason for Shakespeare to locate the play at Elsinore? Here, Donald Olson draws our attention to another recently published book. Perhaps Shakespeare had been flipping through the pages of the *Atlas of the Principal Cities of the World*, a lavishly illustrated pictorial atlas printed in 1588. One of the book's engravings shows an oblique aerial view of the region surrounding Elsinore castle, including, not more than a few miles away, the little island of Hven—Tycho Brahe's island. The engraving even shows Tycho's observatory, Uraniborg ("Heavenly Castle"), labeled in Latin as Uraniburgum (figure 7.1).

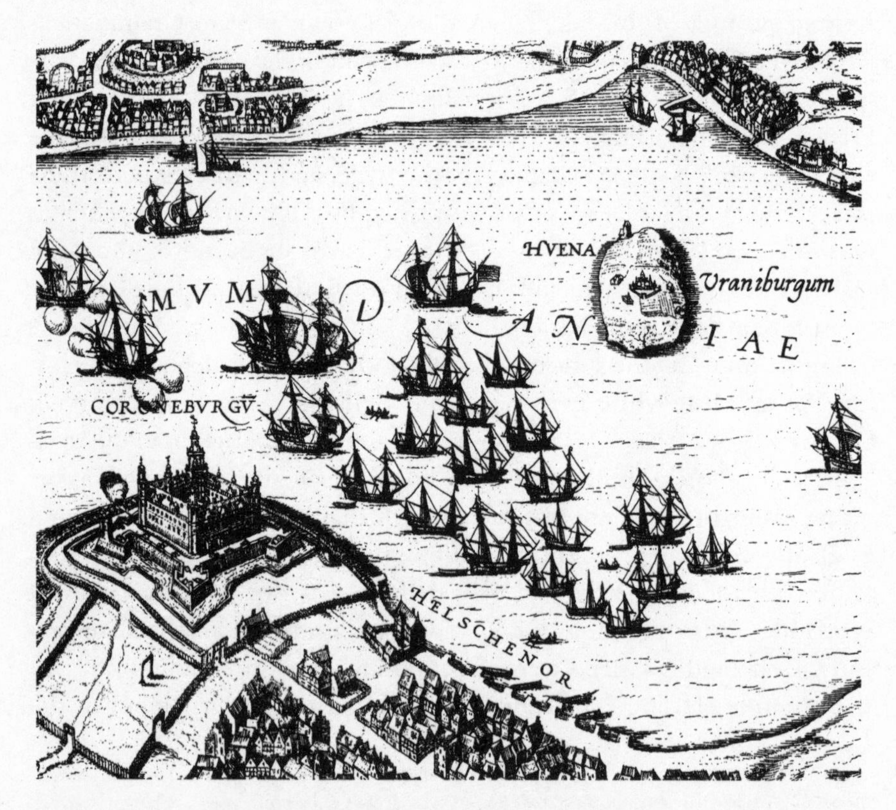

Fig. 7.1 The lavishly illustrated *Atlas of the Principal Cities of the World*, published in 1588, includes this depiction of the Sound of Denmark (separating present-day Denmark from Sweden). Tycho Brahe's observatory-castle, Uraniborg, on the island of Hven, is seen at the right; on the left is the castle of Elsinore (Helschenor), soon to be made famous by Shakespeare as the setting for *Hamlet*. Donald Olson

Fig. 7.2 In this engraving from 1590, Danish astronomer Tycho Brahe is surrounded by the crests of members of his extended family. Among the sixteen relatives, we find a "Rosenkrans" and a "Guildensteren." Copies of the engraving were sent to several English scholars, possibly including Thomas Digges, whose family had connections to Shakespeare. Donald Olson

---

At the risk of getting a little carried away, we can imagine one of Shakespeare's fellow actors getting the royal tour, so to speak, of the castle and its environs. ". . . And if you look out across the channel, you can just barely see the little island that the king gave to an eccentric astronomer. His name is Tycho. What he can see through all these clouds I don't know, but . . ."

We might also examine the names in *Hamlet*: While most of the key characters have fairly generic, more or less classical names ("Claudius," "Gertrude," "Ophelia"), Hamlet's old schoolmates, the courtiers Rosencrantz and Guildenstern, do in fact sound stereotypically Danish. What led

Shakespeare to choose these names? Again, we perhaps have a clue in the material that Shakespeare may have encountered in print—and once again it leads us back to the astronomer Tycho Brahe. In the 1590s, Tycho commissioned an engraving of himself—a portrait depicting the astronomer in a somewhat pompous-looking pose—surrounded by the crests of members of his extended family. When we look at the engraving closely, we find, among sixteen relatives, one named "Rosenkrans" and another named "Guildensteren" (figure 7.2).

There are, in fact, multiple versions of the engraving shown in the above figure. The one depicted here was made in 1590, but similar designs were printed several times in the 1590s and in 1601, and were included in various books, including the published collections of Tycho's astronomical letters, known as the *Epistolae*. Copies of those letters were circulated to men of learning across Europe, and the copy sent to an English scholar, Thomas Savile, has survived. Written in 1590, it includes Tycho's well-wishes for the two best-known English scientists of the day—John Dee and "the most noble and most learned mathematician Thomas Digges." We know that Digges kept up a correspondence with Tycho as well. As the historian of astronomy Owen Gingerich suggests, "it is entirely possible that Digges received a copy of the *Epistolae* directly from Tycho himself." The overall impression is that the "scientific community" of the time, to use an anachronistic term, was small and rather tight-knit. One imagines that everyone who was interested in the structure of the cosmos knew (or at least knew of) everyone else who was pondering similar questions. In his letter to Savile, Tycho adds, "I have included four copies of my portrait, recently engraved in copper at Amsterdam." He even suggests that one of the talented English poets (he does not mention any names) might like to compose a few lines in his praise. (If the request makes Tycho appear somewhat full of himself, it would seem to mesh with what we know of the astronomer's character from other sources—and perhaps also from the pose that he strikes in the engraving itself.)

Olson, incidentally, is not the first scholar to point out the seeming coincidence of names between Tycho's portrait and the courtiers in Shakespeare's play. It had been noted by Leslie Hotson in the 1930s, by A. J. Meadows in the 1960s, and probably by others. In the early 1980s, Owen Gingerich concluded that the coincidence with the names Rosencrantz and Guildenstern is "so striking that we may be sure that Tycho's portrait was one of the sources for *Hamlet*'s cast of characters." As

Olson puts it, "Shakespeare's imagination may well have associated the English astronomers, the new star, the Danish astronomer, and the Danish Hamlet."

WHAT'S IN A NAME?

How much should we make of these two names, found both in Shakespeare's play and among Tycho Brahe's relatives? The engraving, scholars have pointed out, isn't the only place to find these somewhat common Danish names. Apparently a Danish diplomatic mission to England in 1592 included two delegates bearing those same names, and it seems that the two men were inseparable. Moreover, they had been students at Wittenberg, just like Shakespeare's courtiers. (Although not the same men as the relatives found in the engraving, they, too, are believed to have been distant relations of Tycho.) So there is more than one way in which Shakespeare might have encountered the names Rosencrantz and Guildenstern. Still, a picture is worth a thousand words. What role might Tycho's engraving have played? In the Arden editions, both Jenkins (1982) and Thompson and Taylor (2006) mention the portrait as a possible source. Jenkins concludes that Shakespeare need not have seen the picture with his own eyes; simply hearing the names at some point was enough to give the play "an authentic touch of Denmark." He also cautions that both names "were common among the most influential Danish families," and that Frederick II, the Danish king who gave Tycho his island, had nine Rosencrantzes and three Guildensterns at his court. Even so, there are those who feel that Shakespeare's choice of these names is significant, and indicative of an important link between the playwright and the greatest of the pre-telescopic astronomers. Howard Marchitello notes "an admittedly striking series of coincidences that, from a certain perspective, can be said to connect Tycho's book to *Hamlet* (or *Hamlet*—or Hamlet—to Tycho's book)." A few scholars, including Scott Maisano, an associate professor of English at the University of Massachussetts–Boston, see the apparent *Hamlet*–Tycho link as an intriguing connection deserving of further study. "I would say that the Rosencrantz and Guildenstern connection to Tycho Brahe isn't just a coincidence," Maisano told me recently. "The Tycho Brahe connection is one of the most important ones, and probably one of the least explored ones, for Shakespeareans." Maisano suggests that the idea of a scientist on an island may have even more significance for *The Tempest* than for *Hamlet*, and sees the character of Prospero as more strongly linked to

Tycho Brahe than to John Dee, the scientist most often associated with Shakespeare's island magician.

It would appear, at a minimum, that Shakespeare had *some* awareness of Tycho Brahe's reputation. But the weight that we choose to lend to these connections depends on how much Shakespeare knew about the astronomical thinking of his day. It would be a stretch to imagine that Shakespeare had any direct contact with Tycho—but maybe he didn't need to. Perhaps what he knew of English astronomy, and English astronomers, was enough—and so we turn once again to Thomas Digges, the greatest of the English scientists of the Elizabethan age. As we saw in Chapter 3, Digges was the astronomer who first popularized the Copernican theory in England, and who went even further than Copernicus by daring to imagine an infinite cosmos. Because Digges died in 1595—within a few years of Shakespeare's arrival in London—it is unlikely (though not impossible) that the two men ever met. However, the Digges family had a number of connections to Shakespeare—connections that strengthened in the years following the scientist's death.*

We might begin in 1590, when an updated edition of Digges's book on military strategy, known as *Stratioticos,* was published in London. Overseeing the publication was Richard Field, an old childhood chum of Shakespeare's, from Stratford, who had settled in London a few years ahead of the playwright. A couple of years later, Field would publish Shakespeare's own *Venus and Adonis.* There is no reason to imagine that Shakespeare knew every book that his friend published—as voracious a reader as he surely was, he couldn't have plowed through *all* of them; and perhaps a book on military strategy wasn't all that enthralling. Or was it? You never know when you might have to write a good battle scene. Leslie Hotson found a number of similarities between Shakespeare's *Henry V* and Digges's *Stratioticos.* For Hotson, the character of Fluellen, the fiery Welsh captain, brings to mind Digges himself, especially in a number of lines in which Fluellen (and Digges) praises the military discipline of the Romans.

While we're talking about who printed what, we must briefly mention another publisher, William Jaggard. (We encountered his surname on the title page of the First Folio: Jaggard, together with bookseller Edward Blount, published the first collection of Shakespeare's plays in 1623; since

* A few scholars have noted these connections, but the most thorough account can be found in Leslie Hotson's book *I, William Shakespeare* (1937).

Jaggard died a month before the Folio's publication, however, it is his son Isaac's name that appears on the famous frontispiece.) We remember Jaggard for his connection to Shakespeare, but he also published many less-well-known works, among them a textbook on astronomy by Thomas Hill. The book went to press in 1599—around the time that Shakespeare was putting the finishing touches on *Hamlet*. By this time, Hill himself was dead, but a preface written by Jaggard pays tribute to the late author's skill. And though Hill rejected the Copernican theory, Jaggard's remarks are suggestive of the appetite for popular scientific works in Elizabethan England: With the nation having been blessed with four decades of peace, "Students have never had more liberty to look into learning of any profession"; as a result, "England may compare with any Nation for number of learned men, and for variety in professions."

Besides Richard Field and William Jaggard, who else was Shakespeare friendly with in the 1590s? Certainly his fellow actors, including Richard Burbage, to whom he seems to have been particularly close, as well as John Heminges and Henry Condell, the two actors who would later compile Shakespeare's works for publication in the First Folio.* (Shakespeare left a small amount of money to Burbage, Heminges, and Condell in his will.) There was Ben Jonson, of course—both a friend and a competitor. Adding to the list, perhaps just below Shakespeare's professional colleagues, we can put forward the name of Leonard Digges, son of Thomas Digges.† We know that Leonard, who became a poet, was a fan of Shakespeare: He contributed a dedicatory verse in honor of the playwright in the First Folio in 1623, seven years after Shakespeare's death. (A few more lines of praise were included with a collection of Shakespeare's poems published in 1640.) Shakespeare may well have known Leonard's older brother, Dudley, and their mother, Anne (Thomas Digges's widow). As Hotson's research showed, the Digges family lived in Cripplegate, an area noted for its weavers and brewers, located just north of the city's ancient walls. Digges, like Isaac Newton in the following century, was not just a scientist but also something of a politician, serving as a member of Parliament for roughly the last thirteen years of his

* Today, a memorial to the First Folio, capped with a bust of Shakespeare, stands as a tribute to the role that Condell and Heminges played in committing Shakespeare's plays to print (figure 7.3). The memorial can be found, appropriately, at the end of Love Lane, a couple of blocks from the site of the playwright's house.

† We mustn't get our Leonards mixed up: Thomas Digges's *father* was also named Leonard Digges; in Chapter 3 we heard about his purported development of an early telescope-like device.

life. He was also rich—probably one of the wealthiest men in the neighborhood. Cripplegate, incidentally, was also home to Shakespeare's close friend Heminges; so if Shakespeare was spending time with Heminges, and certainly if he was chumming with either of the Digges brothers, occasional visits to the stately Digges home seem plausible.

A brief geographical note: We know that Shakespeare, in his first dozen or so years in London, was a man on the move. As we've seen, he lived, at various times, in Shoreditch, Bishopsgate, and Southwark. None of these neighborhoods is particularly close to the Digges residence in Cripplegate. But, given that Cripplegate was just a stone's throw from the booksellers' stalls north of St. Paul's, and that Shakespeare's good friend Heminges had been living there for several years already, the playwright is unlikely to have been a stranger to the area. (Hotson says that Shakespeare was "a frequent and welcome guest" at Heminges' house, although I'm not sure how he deduced this.) However, Shakespeare's connection to the neighborhood grew even stronger a few years later, when he rented a room on Silver Street, in the heart of Cripplegate, just two blocks away from the Digges family.

LIFE ON SILVER STREET

Of all of Shakespeare's London residences, the one we can pin down with the greatest precision is the home on Silver Street in Cripplegate. There we find him, beginning about 1603, renting a room from a man named Christopher Mountjoy. Mountjoy was a Huguenot—a French Protestant—and like many Huguenots, he had fled France to escape religious persecution. Apparently he did quite well for himself in London, manufacturing ladies' ornamental headpieces and wigs from his workshop in his London home. He lived there together with his wife, his daughter, and a number of apprentices and servants.

Where, exactly, was the Mountjoy house? As is so often the case, legal documents provide the answer. It seems that Mountjoy's daughter, Mary, became involved with one of her father's apprentices. There was a courtship, then an engagement—and then something went terribly wrong. Mountjoy had been obligated to pay a dowry, but refused, and the matter ended up before the courts. (Maybe stinginess was something Mountjoy and Shakespeare bonded over? The playwright didn't like paying taxes; his landlord didn't like paying dowries.) The court case would be of minimal interest, except that Shakespeare, who apparently encouraged the pair in their romantic pursuits, was called on as a witness. At any rate, the court documents mention that the Mountjoys lived at the northeast corner of Silver Street and Muggle Street. Don't bother looking for either of these thoroughfares in your current *A-Z*; sadly, neither street exists today. And "Muggle" was actually just an alternative name for "Monkwell," a street that ran northward from Silver Street, near the ancient city walls. Charles Nicholl, who pieced together the clues in his delightful book *The Lodger* (2007), concludes that the Mountjoy house—likely a timber-framed structure—was just across the street from the church of St. Olave's, which stood on the south side of Silver Street. This is the church where Shakespeare would have worshipped, though as Nicholl reminds us, this says nothing about his religious views, as church attendance was mandatory, and those who failed to attend services could be fined. Unfortunately, the Great Fire of 1666 leveled much of the neighborhood, destroying the church as well as the Mountjoy home. Silver Street itself lingered on for another 274 years, until a German bombing raid during the Second World War reduced the entire neighborhood to rubble. Postwar redevelopment of the area yielded the sprawling Barbican complex, just to the north; the Museum of London, immediately to the west; and the major east-west thoroughfare known

Fig. 7.3 A memorial to the First Folio, topped with a bust of Shakespeare, stands just off Love Lane, not far from the Museum of London and the Barbican complex. Shakespeare once rented rooms in a house a few blocks away. Author photo

as London Wall (part of the A1211 highway). About all that's left from Shakespeare's time is St. Olave's churchyard, now a small park (where London Wall meets Noble Street—two streets that you *can* find in your *A-Z*). Because street level has risen over the years, Nicholl's best guess is that the Mountjoy house occupied what is now an underground car park beneath London Wall. It is hard not to think of the neighborhood's transformation as something of an indignity. "An underground car-park is unmistakably an underground car-park," Nicholl writes, "whether or not Shakespeare once lived on the site of it." Nonetheless, a visitor to the neighborhood today may want to pause on the pedestrian footbridge that straddles the busy road adjacent to the Museum of London, and gaze out at this depressingly ordinary corner of Europe's busiest city: As thoroughly transformed as it may be, this was, at least for a few years, the center of Shakespeare's world.

It was the Digges' neighborhood, too. As Leslie Hotson has noted, the Digges family home was on Philip Lane—two streets to the east of Monkwell Street. But they were more than just neighbors. As it turns out, the connections between Shakespeare and the Digges family, already strong when Shakespeare arrived at Silver Street, would only grow: A few years after Thomas Digges's death, his wife, Anne—now a wealthy and much-pursued widow in her mid-forties—married Thomas Russell, a landowner from the playwright's native Warwickshire. Shakespeare and Russell must have been close; after Shakespeare's death, Russell, together with Stratford attorney Francis Collins, would serve as the executors of Shakespeare's will. Meanwhile, the Digges's older son, Dudley (later Sir Dudley), would become a member of the Virginia Company, which established the colony at Jamestown in 1607. Although Dudley Digges never visited the New World himself, he would certainly have heard reports from those who did, and Shakespeare may have heard the tales from Dudley. By the same route, scholars have speculated, Shakespeare could have heard about the wreck of the *Sea Venture* off Bermuda in 1609, an event often seen as part of the inspiration for *The Tempest*. Much later, in 1655, Dudley's son, Edward Digges, would be appointed governor of the Virginia colony. But it is Leonard, not Dudley, whose life was transformed by his acquaintance with Shakespeare. As he would write in the First Folio, "This book, / when brass and marble fade, shall make thee look / fresh to all ages." The twenty-two-line verse concludes: "Be sure, our Shakespeare, thou canst never die, / But, crown'd with laurel, live eternally."

Let's recap the Shakespeare–Digges connection: Shakespeare was likely a regular visitor to Thomas Digges's neighborhood, and was friendly with at least one of his two sons. In the late 1590s, while searching for plausibly Danish-sounding names for two of his characters, he may have stumbled on the engraving of Tycho Brahe's relatives and has a flash of insight; the names are perfect, problem solved. Perhaps, as Leslie Hotson suggests, he saw the engraving on a visit to Thomas Digges's house. (As Hotson puts it, there is "little doubt that from 1590 Digges had a copy of his learned friend's portrait, bearing the names *Rosenkrans* and *Guildensteren*, at his house in Heminges' parish. Perhaps Shakespeare saw [the names] there.") The death of Thomas Digges in 1595 need not sink this theory: It is possible that Digges's sons, or his widow, kept the picture after his death, a treasured keepsake, perhaps, of Thomas Digges's far-reaching influence and interests. (We know the family continued to live in the same house, in Cripplegate, for several more years.) However the playwright may have chanced upon the image, Shakespeare scholars seem at least willing to entertain the notion that he saw the Tycho engraving with his own eyes. If Shakespeare could have seen Tycho's astronomical letters, might he have also seen Digges's own writings? In particular, might he have perused Digges's updated edition of his father's almanac—the one featuring the now-famous diagram of an infinite cosmos? If he did, perhaps it can illuminate yet another remarkable passage in *Hamlet*.

THE UNIVERSE IN A NUTSHELL

The scene in question comes in the play's second act, where we find the prince conversing with his old schoolmates, Rosencrantz and Guildenstern. Prince Hamlet, as usual, is feeling a bit melancholy. More specifically, he feels trapped in his own country, which he feels is like a prison:

GUILDENSTERN:

Prison, my lord?

HAMLET:

Denmark's a prison.

ROSENCRANTZ:

Then is the world one.

HAMLET:

A goodly one, in which there are many confines, wards, and dungeons, Denmark being one o'th' worst.

ROSENCRANTZ:

We think it not so, my lord.

HAMLET:

Why, then 'tis none to you; for there is nothing either good or bad but thinking makes it so. To me it is a prison.

ROSENCRANTZ:

Why, then your ambition makes it one: 'tis too narrow for your mind.

HAMLET:

O God, I could be bounded in a nutshell and count myself a king of infinite space—were it not that I have bad dreams.

(*Hamlet* 2.2.242–256)

We all have bad dreams on occasion, but where did Shakespeare come up with this striking phrase, "king of infinite space"? The scholarly editions are of little help; most of them simply let it pass without comment. Even the venerable Arden editions (Jenkins, 1982, and Thompson and Taylor, 2006) allow the phrase to slip by without a footnote—and the Ardens have footnotes up the wazoo. (Don't be alarmed if your own edition of *Hamlet* doesn't contain the "infinite space" remark—it occurs only in the folio text of 1623, and not in the various quarto editions, so it depends on which text your edition was based on.) Shakespeare does use the phrase "infinite" or "infinity" about forty times in the canon, but, with the exception of this scene from *Hamlet*, it is never invoked for the purpose of describing spatial extent.* Of the remaining cases, the most beguiling is the soothsayer's line from *Antony and Cleopatra*: "In nature's infinite book of secrecy / A little I can read" (1.2.10–11).

The phrase "infinite space" was certainly not in common use in Shakespeare's time—though, as we've seen, there were a handful of thinkers who were giving the matter serious consideration. Digges, of course; and before him, Nicolas of Cusa. And then there was Giordano Bruno, who lectured in England in the 1580s. Bruno tackled the subject in one of his dialogues, *On the Infinite Universe and Worlds* (1584), writing that "There is a single general space, a single vast immensity which we may freely call Void. . . . This space we declare to be infinite." Bruno was an ardent Copernican; so, too, was Digges. Might one of these thinkers have sparked a thought on the nature of the infinite as Shakespeare was working on *Hamlet*? Did he perhaps get the idea from looking at Thomas Digges's diagram of the cosmos (figure 3.3)—a diagram in which the stars, for the first time, are depicted as extending outward to infinity?

As we've seen, Digges's book, *A Prognostication Everlasting*, was published in 1576, and reprinted seven times before 1605, always with the fold-out diagram of the infinite cosmos. There were thousands of copies in circulation during Shakespeare's most productive years in London. Digges, as mentioned in Chapter 3, embraced Copernicanism wholeheartedly, and told his readers that he included excerpts of Copernicus's own treatise, translated into English, "so that Englishmen might not be deprived of so noble a theory."

It would be easier to tell what Shakespeare thought of this "noble theory" if his characters were a bit more plainspoken when discussing the cosmos. Consider Hamlet's peculiar love poem to Ophelia:

> Doubt thou the stars are fire,
> Doubt that the sun doth move,
> Doubt truth to be a liar,
> But never doubt that I love.
>
> (*Hamlet* 2.2.115–18)

---

* The critical resource for such matters is Open Source Shakespeare, a searchable online collection of the playwright's complete works: www.opensourceshakespeare.org.

The astronomy alluded to in the passage is that of Ptolemy, but is Hamlet endorsing it, or urging Ophelia to question it? The usual interpretation is that Hamlet is telling Ophelia that his love is more reliable and more dependable than whatever she had been taught about the workings of the cosmos. Roughly: "Question the unquestionable, but don't question my love." (The third line is the trickiest of the bunch: Here the meaning of the word "doubt" seems to change from "question" to "suspect.") The play was written at a time when a handful of philosophers were indeed questioning whether the Earth moved. Is there any trace of Copernicanism in Hamlet's poem? In *1599: A Year in the Life of William Shakespeare* (2005), James Shapiro writes, "The Ptolemaic science on which Hamlet's prognostications are grounded, as Shakespeare knew, was already discredited by the Copernican revolution. The stars aren't fire, the sun doesn't revolve around the earth. In such a universe, the truth may well turn out to be a liar." (Indeed, the sun doesn't revolve around the Earth, but I'm not sure about the other half of Shapiro's statement—after all, stars are certainly *like* fire.*) Harold Jenkins, in the 1982 Arden edition, seems to share Shapiro's view. The poem, taken at face value, refers "to the orthodox belief of the Ptolemaic astronomy that the sun moved around the earth," but Jenkins adds that there is also a hint of Shakespeare's possible awareness of Copernicanism: "Since each of the poem's first two lines assumes the certainty of what had now begun to be doubted, there is an irony of which Shakespeare (though not, I take it, Hamlet) must have been aware." Even T. J. B. Spencer, who suggested that the star "westward from the pole" was a planet, shares the suspicion that the shadow of Copernicus may loom over the play: Hamlet's poem "is a clever epitome of some of the poetical tendencies of the 1590s: cosmological imagery, the Copernican revolution, moral paradoxes, all illustrating amorous responses." Shakespeare's characters often speak in riddles, Hamlet above all. Perhaps the biggest riddle of the time was whether the universe was small, comfortable, and human-centered—or whether, as a handful of bold thinkers had suggested, it was enormous, with mankind a mere speck and our planet, on the cosmic scale, little more than a dot. No wonder Hamlet sees "this goodly frame the

* I suppose one can debate whether the process of "burning" hydrogen through a process of nuclear fusion should be classified as "fire." At any rate, Hamlet's poem is not Shakespeare's only mention of the subject: It comes up again in *Coriolanus*: A messenger delivers some news to Sicinius, who asks if he's certain, and the messenger replies, "As certain as I know the sun is fire" (5.4.46), and Macbeth pleads with the heavens, "Stars, hide your fires . . ." (1.4.50).

earth" as nothing more than a "congregation of vapours," a "sterile promontory" (2.2.298–303).

THE BRUNO CONNECTION

Digges had pondered the possibility of an infinite universe—and so had Giordano Bruno. The connections between Shakespeare and Bruno are perhaps more speculative that those that connect him to Digges, but they are worth exploring. The Shakespeare–Bruno link runs through a key middleman—a Londoner with Italian roots by the name of John Florio. We know that Florio became close friends with Bruno, having met his countryman while working as a tutor for the French ambassador. At a dinner party in Whitehall Palace, he is said to have told the assembled guests about Bruno's theory of multiple inhabited worlds. The Italian thinker, who was both a Copernican and an atomist, seems "to have left a mark on some of the most cultured people in England" living at that time, according to Hilary Gatti. The usual argument is that Shakespeare, eager to make the acquaintance of learned men from diverse backgrounds (and perhaps especially Catholics), would have befriended a number of Italians in London, with the eloquent (and bilingual) Florio at the top of the list. Shakespeare and Florio had at least one more connection: Florio had served as a tutor to one of Shakespeare's patrons, the Earl of Southampton. (Certainly Shakespeare knew of the Italian's work: He clearly read Florio's translation of Montaigne's *Essays*. More on that in Chapter 13.) It is through Florio and his circle of friends that Shakespeare supposedly became acquainted with Italian ways and customs of Italy—material that he would make much use of in his plays, thirteen of which were set partially or entirely in Italy. Perhaps he also acquired a basic comprehension of the Italian language, allowing him to gloss the Italian texts identified as source material for many of his dramas. (As we've seen, another theory—a more tenuous one—is that Shakespeare visited Italy, perhaps before settling in London, a claim for which there is no solid evidence.)*

The Shakespeare–Bruno connection has had its ups and downs: In the late nineteenth century, a number of Shakespeare scholars argued that the playwright was deeply influenced by Bruno's philosophy, acquired

---

* The extreme version of this line of speculation is the allegation that Florio *was* Shakespeare—that is, that Florio was the author of the plays attributed to Shakespeare. As with the other "alternative" authorship claims, few if any Shakespeare scholars take the idea seriously.

via Florio—a view that has since fallen out of favor. "The Bruno-Shakespeare discussion has become a historical curiosity," writes Gatti, "of which many Shakespearean scholars of today are no longer even aware." A degree of skepticism is in order. As Gatti notes, given the time frame of Bruno's visit to England, a meeting between Bruno and Shakespeare is "extremely improbable"—still, he may well have heard something of Bruno's ideas. (And, as mentioned in Chapter 5, we have a loose connection between Shakespeare and Bruno via the printers Richard Field and Thomas Vautrollier.) At the very least, Shakespeare would have read the favorable mention of Bruno in the preface to Florio's translation of Montaigne. Perhaps he read Bruno's only dramatic work, a comedy called *Candelaio*—echoes of which, Gatti argues, can be found in Shakespeare's *Love's Labour's Lost* as well as in Ben Jonson's *The Alchemist* and Christopher Marlowe's *Doctor Faustus*. Gatti concludes, "A convincing basis for knowledge of Bruno on the part of Shakespeare, probably mediated through John Florio, thus undoubtedly exists."

Looking for traces of Bruno's philosophy in *Hamlet*, one might begin with the prince's love poem, with its obsession over the issue of "doubt." Bruno writes passionately on the question of doubt in one of his final works, *De triplici minimo* (*The Triple Minimum*), published in Frankfurt in 1591:

> Whoever wishes to philosophise, doubting all things at first, must never assume a position in a debate before having listened to the opinions on all sides, and before carefully weighing the arguments for and against. He must judge and take up a position not on the basis of what he has heard said, according to the opinion of the majority, their age or merits, or their prestige. But he must form his own opinion according to how persuasive the doctrine is, how organically related and adherent to real things, and to how well it agrees with the dictates of reason.

Bruno, as we've seen, was as much a mystic as a scientist—but in this passage we see a glimpse, and perhaps more, of the modern scientific approach. It was also a way of thinking that was seen, at the time, as dangerous—which is part of the reason Bruno was eventually deemed "a particularly obstinate heretic . . . author of a number of enormously dangerous opinions." For Hilary Gatti, the parallel to *Hamlet* is profound, beginning with the doubts concerning the ghost's identity, and continuing with Prince Hamlet's quest to find the truth behind the story of his father's death while at the same time uncovering Polonious's scheming and the double-dealing of Rosencrantz and Guildenstern. "Thus Hamlet, like Bruno, finds comfort in a systematic exercise of skepticism," Gatti writes. "The truth must be pursued, and nothing lie hidden in the obscure shadows of deceit. . . . What lies at the center of Shakespeare's drama is not so much the murder of a king as the murder of truth itself."

Of course, Shakespeare's own characters frequently argue about what is true and what is illusory. In the opening scene, the guards put their trust in Horatio, the intellectual—the closest thing the play has to a scientist—to ascertain the ghost's true essence: "Thou art a scholar, speak to it, Horatio" (1.1.46). At first, he doubts that the apparition will even appear; later he says he wouldn't have believed it unless he had seen it with his own eyes. Hamlet is less amazed than his friend. In one of the play's most quoted lines, he says "There are more things in heaven and earth, Horatio, / Than are dreamt of in your philosophy" (1.5.174–75). If the play had been written in our own time, he probably would have said "science" instead of "philosophy" (recalling that the closest thing to "science" in Shakespeare's time was "natural philosophy"). (Another small complication is that it's "your" philosophy in the quarto editions, but "our" philosophy in the later folio: Thus the jab may not be directed at Horatio's scholarly learning in particular, but rather at all such learning.) Hamlet is looking for truth, even as he questions whether his friend's worldly knowledge—his science—is up to the challenge.

It is only natural to ask what *Hamlet*, which has so much to say about so many aspects of the human condition, might have to say about our place in the universe. As Donald Olson has shown—building on the work of Hotson, Meadows, and others before him—there is an argument to be made concerning Shakespeare's awareness of the work of Tycho Brahe and Thomas Digges. And as Hilary Gatti suggests, perhaps more tentatively, there may also be signs of Bruno's bold philosophy. Shakespeare covers a lot of ground in *Hamlet*, and quite possibly the physical nature of the universe was among his concerns. Next we will meet a scholar who takes the argument further. Is it possible that Shakespeare's most famous play is *all about* the structure of the cosmos?

8. *". . . a hawk from a handsaw"*

READING SHAKESPEARE, AND READING INTO SHAKESPEARE

It's the world's largest gathering of astronomers and astrophysicists: The American Astronomical Society meets twice a year, giving its members a chance to talk about their research and to announce the most newsworthy findings. I attended for the first time in January 1997, when, by luck, it was held in my home city of Toronto. As an aspiring science journalist, I was eager to take it all in: extrasolar planets, exploding stars, galaxies, black holes, the latest findings in astrophysics and cosmology—whatever was on offer. The AAS always picks out a handful of papers to publicize during the meeting, in the hope of garnering media attention, and one of these highlighted papers, titled "A New Reading of Shakespeare's *Hamlet*," caught my attention. What did Shakespeare have to do with astronomy?

According to the presenter, quite a lot. His name was Peter Usher, an astronomer from Penn State University, and his paper made some bold claims: "I argue that as early as 1601 Shakespeare anticipated the new universal order and humankind's position in it." The journalists at the press briefing listened attentively, if skeptically, as Professor Usher outlined his new interpretation of *Hamlet*; afterward, the professor answered a handful of questions. It is perhaps not surprising that the reporters from the British newspapers showed the most interest; after all, Shakespeare is "one of theirs." "Astronomer discovers cast of stars hidden in Hamlet" was the headline when the story ran in the next day's London *Times*.

* * *

Peter Usher became a Shakespeare enthusiast by accident. Born in South Africa, he taught astronomy for many years at Penn State, where he still holds the title of Professor Emeritus in astronomy and astrophysics. Often, while teaching introductory astronomy, he sought to engage his students by looking for connections across disciplines—for example, by connecting physics and astronomy with music or literature. Eventually he turned to Shakespeare, poring through the canon in search of astronomical references, and looking, in particular, for anything that might hint at the "new astronomy" of Copernicus.

At first, Usher came up empty-handed. It's not that there weren't any astronomical references in the canon; in fact they seemed fairly common in Shakespeare's work. As Usher has pointed out, happenings in the sky were simply a "bigger deal," so to speak, in Elizabethan England than they are now, partly because there was less light pollution, and partly because many of the trappings of our perennially distracted information-drenched culture hadn't yet been invented. But most of these astronomical references seemed to either reflect the medieval, Ptolemaic view of the cosmos, or to be phrased in such a way as to render them ambiguous. There didn't seem to be anything that pointed directly to the Copernican model of the heavens. This left Usher somewhat puzzled, given the profundity of the new discoveries unfolding at the time, and Shakespeare's obvious curiosity about the world. "It seemed to me that someone who lived through the beginning of the Scientific Revolution would have *something* a little more strongly to say about it, because this was a major upheaval in the worldview," he says. Or, as he puts it in the preface to his book *Shakespeare and the Dawn of Modern Science* (2010), "It is simply not credible that a poet of this stature could remain ignorant of the cultural impact that the New Astronomy was having during his lifetime—or that he would refrain from using the literary devices at his command to address the topic if he was not ignorant of its significance." He spent his spare time "hunting through the canon, to find whether Shakespeare did or didn't have any knowledge of heliocentrism." Once he began his search, there was no turning back.

Usher, now retired, lives in a leafy neighborhood a few miles east of downtown Pittsburgh. He is tall, slim, and sports wire-rim glasses; when he was a bit younger, he may have borne a passing resemblance to the actor Ed Harris. His knowledge of astronomy and its history served as a starting point for his literary quest. Soon he had read Leslie Hotson's account of the connections between Shakespeare and the Digges family, including the reference to Tycho Brahe's portrait with its "Rozencrans" and "Guildensteren" crests; and of course he considered the possible cosmological significance of Prince Hamlet's reference to "infinite space." Soon he was scrutinizing Shakespeare's most famous play scene by scene, line by line. Whatever Shakespeare might have known about the "new astronomy," he reasoned, *had to be in there somewhere*; it was, after all, his most ambitious play and certainly his longest and most complex work. The "aha moment" came as Usher pondered the name of Hamlet's villainous uncle, Claudius. Could the name be an allusion to Claudius Ptolemy, the Greek astronomer who had worked out the mechanics of the geocentric system? Soon, Usher was finding what seemed like other correspondences in the text. Gradually, he began to see the entire *dramatis personae* of Shakespeare's masterpiece in a new light. The play, he says, can be interpreted as an allegory about competing cosmological models.

THE PLAY WITHIN THE PLAY

In Usher's interpretation of *Hamlet*, nothing is quite what it seems—or, rather, nobody is quite who they seem to be. Nearly everyone in the play, he says, can be seen as "standing in" for an astronomer from Shakespeare's time, or from the annals of the history of astronomy—figures who, in one way or another, had a stake in the competing descriptions of the cosmos that were battling for acceptance in Renaissance Europe. According to Usher's interpretation, Prince Hamlet, the play's hero, represents the true picture of the universe—the heliocentric (sun-centered) model proposed by Copernicus and championed in England by the astronomer Thomas Digges. The correspondence applies also to the previous generation: The deceased king, Hamlet's father, is Leonard Digges, father of Thomas Digges—the man who, according to his son, may have invented a telescope-like device in the mid-sixteenth century, and whose work was continued by his son. ("Thomas and Hamlet are both compelled by the spirits of their deceased fathers to finish the job," Usher says.) The courtiers Rosencrantz and Guildenstern, meanwhile, represent Tycho Brahe, and serve as a surrogate for Tycho's "hybrid" model of the universe (in which the planets revolve around the sun, but the sun, in turn, revolves around the Earth). Laertes stands in for another English astronomer, Thomas Harriot. The lesser characters are vital, too. Bernardo, for example, is the medieval philosopher Bernardus Silvestris, an early proponent of a moving Earth. (Bernardus's major work, *Cosmo-*

graphica, "is an excellent fit to the *Hamlet* subtext," Usher writes.) In all, Usher finds such correspondences for twelve characters in the play (a list that includes all of the main characters except for the two women, Gertrude and Ophelia).

As Usher sees it, the battle between worldviews gets under way right from the play's famous opening scene, with the ghost of old King Hamlet (Leonard Digges) seeking revenge on his evil brother (Claudius Ptolemy). As the drama unfolds, some of Prince Hamlet's more enigmatic lines are seen in a new light. Consider his claim that he could be "bounded in a nutshell" while still counting himself "a king of infinite space." To Usher, this one line serves to highlight the essential difference between the old and the new models: A "shell" of fixed stars forms the universe's "bounds" not only in the outdated Ptolemaic model, but in those of Copernicus and Tycho as well; they were eliminated only when Digges put forward his vision of an unbounded cosmos—a world of "infinite space." In the end, although Hamlet dies, his ideas live on: When Rosencrantz and Guildenstern are killed, it represents the demise of the Tychonic system; and when Claudius is killed, it's the long-overdue downfall of Ptolemy and geocentrism. Finally we have the return of Fortinbras from Poland, and his salute to the English ambassadors—symbolizing the final triumph of Copernicus, the Polish astronomer.

Usher was also struck by the prominence given to the German city of Wittenberg, the university town where Hamlet and Horatio, as well as Rosencrantz and Guildenstern, are said to have studied, and to which Hamlet is apparently intent on returning. In Shakespeare's time, Wittenberg was renowned for its scholarship, and, as Harold Jenkins points out, it was already well known to the playwright's audience, having been mentioned in Marlowe's *Doctor Faustus*, which had premiered about eight years earlier. Wittenberg was also the seat of the Protestant Reformation: Luther studied there, and in 1517 he nailed his ninety-five theses to the door of the city's Castle Church. But as Usher points out, Wittenberg is also connected to Copernicanism: Rheticus, the only pupil of Copernicus, studied and later taught there, before supervising the publication of *De revolutionibus*. The city is mentioned four times in *Hamlet*, and was the home of the heliocentric theory's first ardent supporter; for Usher, it's not a coincidence. When Hamlet announces his intention to return to Wittenberg to resume his studies, Claudius declares that such a move is "most retrograde to our desire" (1.2.114). Usher sees this as both an allusion to the retrograde motion of the planets, which had originally motivated so much astronomical investigation in the first place, and to Claudius's (that is, Ptolemy's) opposition to the Copernican system, which was being taught at the famous German university. (Usher is not the first to make this connection; he notes that another astronomer, Celia Payne-Gaposchkin, suggested such a link in a textbook in the 1970s.)

THE VIEW FROM HVEN

If Shakespeare's decision to give Wittenberg such prominence in *Hamlet* is linked to Copernicus, then perhaps this connection can be used to make sense of some of the play's more puzzling lines, which, according to Usher, can now be read so as to support the cosmic allegory. Consider Hamlet's peculiar description of his apparent madness:

> I am but mad north-north-west. When the wind is southerly, I know a hawk from a handsaw.
>
> (2.2.374–75)

Needless to say, critics have puzzled over this line at great length.* The basic idea is plain enough: Hamlet is only mad at certain times—or rather, he is choosing to appear mad at certain times, while underneath he is quite sane. The proof of his sanity is that he can distinguish one object from another—although the two objects are not necessarily what they first appear to be. As the footnotes explain, one possibility (among many) is that "handsaw" may be a corruption of "hernshaw" (or "heronshaw"), referring to a kind of heron—in which case the passage refers to the ability to distinguish one type of bird from another. (However, this is still a bit odd: A prince, accustomed to hunting, would surely have little difficulty in telling a bird of prey from a wading bird, and wouldn't be inclined to brag about being able to make such a distinction.)

Usher sees the passage quite differently. His focus is not so much on the hawk and the handsaw, but on the two compass directions: For someone living on Tycho's island of Hven, the castle at Elsinore would lie to the north-northwest, while the German city of Wittenberg would lie to the south—and he feels confident that it's not just a geographical fluke. "We've got to think that there's something important going on here, because these are not mere coincidences," Usher says. "Shakespeare could have chosen other directions, but he happened to choose these

* It has been argued that the title of Alfred Hitchcock's 1959 film *North by Northwest* is an allusion to this passage.

particular two directions: one in which Claudius, namely Claudius Ptolemy, resides; and the other, where Rheticus is teaching Copernican astronomy. And he's contrasting the two." In other words, north-northwest leads to Elsinore, Ptolemaic astronomy, and madness; south leads to Wittenberg, Copernicanism, and sanity.

There is more to Usher's *Hamlet* theory, but that is enough to give the flavor of his argument. Perhaps not surprisingly, he encountered difficulty in getting his novel interpretation published. His first paper was rejected by "all the conventional Shakespearean outlets," he says. A turning point came in 1997, when a brief account of his theory ran in *Mercury*, the journal of the Astronomical Society of the Pacific, and, at about the same time, his paper on *Hamlet* was accepted for presentation at the AAS meeting in Toronto. More publications followed. Among the periodicals that have taken an interest in his work are the *Journal of the Royal Astronomical Society of Canada*, *The Elizabethan Review*, *The Shakespeare Newsletter*—along with the *Oxfordian*, the journal of the Shakespeare Oxford Society, a group that, according to their website, welcomes papers on "Shakespeare authorship issues" and is dedicated "to Researching and Honoring the True Bard."

While Usher didn't set about his research with the aim of wading into the so-called authorship question, he eventually found it impossible to separate the question of Shakespeare's astronomical knowledge from the question of his identity. Still, when he was writing *Shakespeare and the Dawn of Modern Science*, he managed to put off that problematic issue until the book's final chapter. Even so, one finds a hint of where things may be heading in the book's preface, where he refers to "the Stratford actor William Shakspere," who is "widely regarded as the poet Shakespeare. Some have wondered whether Shakspere wrote Shakespeare, and they have proposed alternate candidates for the authorship of the Canon—but the unique fact of this work is that the cosmic allegorical context provides an entirely new perspective on these propositions." Usher also had an earlier book, *Hamlet's Universe*, which he self-published in 2006, but *Shakespeare and the Dawn of Modern Science* takes the story much further, exploring astronomical references in *Love's Labour's Lost*, *The Merchant of Venice*, *The Winter's Tale*, *Cymbeline*, and of course *Hamlet*.

Usher is fully aware that his ideas are unorthodox, and that they may strike many Shakespeare scholars (and perhaps many ordinary readers) as far-fetched. And wading into the authorship debate certainly doesn't help. But acceptance, or lack thereof, doesn't seem to trouble him. "Change comes very, very slowly," he says. "But I'm not concerned about that. I'm a scholar, and I've written a scholarly book, and we'll see where it goes."

A few scholars are at least citing Usher's work; for example, Helge Kragh, in his book *Conceptions of Cosmos*, notes the Shakespeare–Digges connection, adding that "it has been argued that [Digges's] world picture enters allegorically in several of the Bard's plays," with an endnote crediting Usher.

That Shakespeare was influenced by Thomas Digges is one thing; that he was some kind of literary secret agent, slyly alluding to taboo subjects by means of allegory in his most famous plays, is another. The way Usher sees it, Shakespeare could discuss Copernicanism only allegorically, rather than tackling it head-on, because the heliocentric theory was supposedly a dangerous doctrine in Elizabethan England; that any mention of it could somehow jeopardize one's career or even one's life. ("The idea of having, for example, imperfections in the heavens was enough to have you decapitated, or at least disemboweled," he said during our interview. "You had to be careful what you were doing.") However, there seems to be scant evidence to support such a view. As Allan Chapman points out, while there were any number of ways to get into trouble for one's political beliefs in Elizabethan England, scientific views were considered harmless. "Nobody, as far as I am aware, got into any hot water whatsoever for their *scientific* beliefs," he writes. Still, we need to remember that science and philosophy were part of a single package at that time, and there were certainly "dangerous ideas." (Atomism, with its association with atheism, was one example—though I'm not aware of anyone in England facing persecution specifically for adhering to the atomic theory.) At the very least, there were perceived dangers—and a hint of these dangers (real or imagined) can be glimpsed in a letter that Harriot wrote to Kepler in 1608. Regarding his decision not to publish his astronomical findings, he says: "Things with us are in such a condition that I still cannot philosophize freely. We are still stuck in the mud. I hope Almighty God will soon put an end to it." How tangible was the danger? To be sure, anything that might be interpreted as an attack on the queen or her court might lead to a charge of treason, Chapman notes, but "as far as science was concerned, you could, if you did not mind being laughed at, believe that the moon was a lump of green cheese, and nobody would touch you for it." He concludes that "there was no persecution at all for scientific subjects."

Kragh and Chapman are historians of science—but what about the literary community? What do mainstream Shakespeare scholars think of Usher's work? In fact, this was hard to determine, for the simple reason that most of them aren't familiar with it. The vast majority of the scholars that I spoke with hadn't heard of Professor Usher and knew nothing of his investigation into Shakespeare and astronomy. On a few occasions I tried to summarize his work, or at least his *Hamlet* theory, though it is possible that my attempts to condense it into a few sentences failed to do it justice. However, there are a small handful of Shakespeare scholars who have taken the time to read Usher's work (or at least portions of it), and, while they may have doubts about his *Hamlet*-as-allegory thesis, they say that he's done something quite valuable.

Among those willing to lend qualified support to Usher's approach is Scott Maisano at the University of Massachusetts–Boston. "I'm really glad that he's done it," Maisano told me, referring to Usher's work on *Hamlet*. "I think the new science makes a difference in a play like *Hamlet*, and makes a difference in perhaps the ways that Peter Usher is suggesting." The problem, he says, is that Usher's proposal "feels a little too allegorical." (One Harvard professor was more blunt, stating flatly that "Shakespeare *doesn't do* allegory.") It's not that the *Hamlet* story *can't* contain allusions to the decline and fall of the medieval worldview—but for Maisano, the idea that this is the play's primary function is too much of a straitjacket for such a broad and complex work. "It's hard for me to reduce *Hamlet* to being just about, or primarily about, the Copernican or Ptolemaic models of the universe," he says. "There's so much going on in that play, so many ideas, so many angles from which one can approach that play." Maisano is highly suspect of the idea "that you can reduce it to some sort of allegory about the overthrow of the Ptolemaic system." And if it really is such an allegory, he says, it's a staggeringly elaborate one. "I would resist the idea that *Hamlet*, or any other Shakespeare play, is primarily an allegory encoding a set of ideas or beliefs."

Nonetheless, Maisano (who has read Usher's published articles but not his books) says that Usher has provided a valuable service by focusing attention on a neglected subject—the question of what Shakespeare might have known about the revolution that was unfolding in the sciences, and whether he took an interest in it. "You ask, 'Did he know and did he care?' I think he did. I think Usher's right. He did know, and he did care."

John Pitcher, a professor of English at the University of Oxford, has a similar assessment of Usher's contribution. "It's good work; it digs up stuff," he says. "But it's got the same kind of determination to prove something about Shakespeare that you find in people who want to prove that Shakespeare is Oxford."* Usher has "this same kind of over-resolve to make everything fit. And I don't think everything *does* fit, probably." Even so, Pitcher, like Maisano, believes that Usher is shining a spotlight on what had previously been a dark corner of Shakespearean studies. "I think the most important thing is to ask, 'Do these plays have running through them an exploration of different aspects of the physical world?' And I think there's no doubt about that at all."

THE VIEW FROM ITALY

Usher is not alone in his effort to link Shakespeare to the Copernican Revolution. An Italian scholar, Gilberto Sacerdoti, tackled the subject head-on in his 1990 book *Nuovo cielo, nuova terra: La rivelazione copernicana di "Antonio e Cleopatra" di Shakespeare* (*New Heaven, New Earth: The Copernican Revolution in Shakespeare's* Antony and Cleopatra). While Usher finds the crucial link between Shakespeare and astronomy in the figure of Leonard Digges, Sacerdoti finds it in the Italian philosopher and mystic Giordano Bruno; and while Usher focuses most intently on *Hamlet*, Sacerdoti homes in on *Antony and Cleopatra*. A great deal of his thesis hinges on a key line that occurs in the very first scene of the play. Antony, Rome's most powerful general, is already smitten with Cleopatra, the queen of Egypt; but just as things are heating up, he is called back to Rome. Torn between love and duty, he professes the depth of his feelings for Cleopatra. But she demands to know *how much* he loves her:

CLEOPATRA

If it be love indeed, tell me how much.

ANTONY

There's beggary in the love that can be reckoned.

---

* In Chapter 7 of *Shakespeare and the Dawn of Modern Science*, Usher argues that "Shakespeare" was in fact Leonard Digges (senior).

CLEOPATRA

I'll set a bourn how far to be beloved.

ANTONY

Then must thou needs find out new heaven, new earth.
(1.1.14–17)

Antony says his love is boundless; when Cleopatra insists on measuring it, he replies that one would run out of territory if one tried: You would need a new heaven and a new earth. The standard interpretation is that the line merely references a well-known passage in the Bible, occurring both in 2 Peter ("We, according to his promise, look for a newe heaven and a newe earth") and in Revelation ("I sawe a new heaven, and a new earth"), referring to the redemption of humankind at the Second Coming of Christ. (The phrasing is from the "Bishop's Bible" of 1572, the version most likely to have been known to Shakespeare.) But Sacerdoti sees something else. The play, he argues, reflects a vision of the universe "that is unmistakably the Brunian version of Copernicanism." Sacerdoti's work seems to have gone virtually unnoticed by English-speaking scholars.* I was able to find only a single review of the book, and, like the original work, it was in Italian. The reviewer, Henry Newbolt, had the same concern with Sacerdoti that Maisano and Pitcher have with Usher: He's just a bit too . . . enthusiastic. "If he keeps on like this," Newbolt writes, "Sacerdoti will end by attributing to Bruno the paternity of the Bard's works, if he does not end by identifying Shakespeare with Bruno himself (I hope not)."

SHAKESPEARE UNDER THE MICROSCOPE—AND THROUGH THE TELESCOPE

One thing that Usher and Sacerdoti have in common—and it is not unusual in the world of Shakespeare scholarship—is an eye for minutiae. If there is something within the plot or dialogue of one of the plays that might yield an astronomical allusion, Usher has probably found it. Sometimes these are references to particular celestial objects; sometimes they lead to specific astronomers or philosophers. In *Shakespeare and the Dawn of Modern Science*, Usher devotes nine pages to analyzing the most famous

* Stephen Greenblatt noticed it—and I'm grateful to him for bringing it to my attention.

stage direction in the canon—the "Exit, pursued by a bear" directive in the middle of *The Winter's Tale*. Usher eventually concludes that the bear represents Nicholai Reymers Baer, known as Ursus (Latin for "bear"), a sixteenth-century German astronomer who once visited Tycho Brahe at his island observatory. (Tycho himself, meanwhile, is represented by Antigonus in the play, Usher says.) Following a dispute, Tycho threw Ursus off his island and later tried to sue him for libel. Usher concludes that "the backwardness of Tycho's thinking and the Dane's haughty attitude to social inferiors would be sufficient reason to invent a vengeful bear."

Usher is also deeply concerned with passages that suggest not merely a knowledge of the night sky, but telescopic knowledge. Consider act 3, scene 4 of *Hamlet*: Prince Hamlet has confronted his mother, the queen, in her bedroom; in front of them are paintings of both King Hamlet and Claudius. He asks her to look at the images; to study the men's features. His father's image is clearly the more noble; he has, among other things, "an eye like Mars to threaten and command" (3.4.57). The standard interpretation is that the king's eye is suggestive of his power—that he could command an army like the god of war (Mars). Usher, however, sees it as a reference to the Great Red Spot on Jupiter. "The planet Mars . . . is not known for an 'eye,' only for its red color. What Shakespeare means is that Jupiter has an eye that is red; in short, it seems most reasonable that Shakespeare is describing Old Hamlet's face as like that of Jupiter which has an 'eye' that is red like Mars, in other words, a reference to the planet Jupiter's Great Red Spot"—discovered many decades ahead of schedule, so to speak. (The standard view is that the red spot was discovered in the 1660s by Giovanni Cassini; the oldest surviving drawings of it date only from the 1830s.)

Usher believes he's found references to the planet Saturn as well. In *The Merchant of Venice*, Graziano "associates Antonio's chronic melancholy with yellowness and that Bassanio is suitor number 7," he writes. "It happens that Saturn is yellowish and the seventh Ancient Planet. The recurrence of heptads raises the possibility that Saturn is at issue." Meanwhile, the ring that Portia gives to Bassanio, a crucial plot device in the play, could represent Saturn, the ringed planet. The same goes for Imogen's ring in *Cymbeline*—but with an important difference: When the ring is described as being cracked, in the play's fifth act, Usher sees it as a reference to the gap in Saturn's rings known as the "Cassini division." The standard view is that the gap was first observed by Cassini in 1675—

but Usher believes it was seen by Leonard Digges more than a century earlier.

Usher goes on to collect a vast array of astronomical references within the canon. *Hamlet*, not surprisingly, is the richest source of such information. Usher believes that within its pages, Shakespeare describes "properties of the Sun, Moon, planets, and stars that he could not have known without telescopic aid." (In an earlier paper, he writes, "In *Hamlet*, the Bard describes with relative clarity the phases of Venus, craters on the moon, sunspots, the stellar makeup of the Milky Way, the number of naked-eye stars, and the existence of stars lying beyond the pale of human vision. These data could only derive from telescopic observations.") The figure of Leonard Digges lurks in the background throughout; his early "perspective glass" would have been "a reasonable means of resolving such detail."

THE TUDOR TELESCOPE REVISITED

While Usher's *Hamlet*-as-allegory thesis can stand or fall on its own merits, his broader argument, focusing on Leonard Digges and Elizabethan telescopy, makes very specific claims about what historians call the "material culture" of early modern England. Usher asks us to consider the possibility that Digges had a fully functioning telescope some half a century before Galileo got his hands on such a device; and, for that matter, that Digges's telescope was significantly better than any of those employed by Galileo. But as we saw in Chapter 5, the case for the "Tudor telescope," though tantalizing, is hardly convincing. As noted, we have the case of Leonard Digges, who, according to his son, Thomas, used a "perspective glass" in the mid-1500s. And we have Thomas Harriot, who may have used a variety of devices, and, by around the time of Galileo's telescopic observations, was almost certainly using an instrument similar to Galileo's, with which he was able to view details of the lunar surface, sunspots, and the moons of Jupiter. But the existence of a telescope suited for astronomical observations at that date—around 1609–10—is not controversial. Peter Usher wants to push back the timeline of the telescope's invention and its use: In *Shakespeare and the Dawn of Modern Science*, and in a series of papers, he argues that Elizabethan telescopy was in an advanced state by the mid- to late 1500s. And he goes further, stating that Shakespeare himself had access to a telescope—or, more specifically, that whoever wrote the works of "Shakespeare" had such access. But, as we noted in Chapter 5, the evidence is thin. To sum up: In the case of Leonard Digges, all we have is Thomas's passing reference to a device his father supposedly used many years earlier, plus another secondhand account from Bourne; and Harriot—before his encounter with more sophisticated Dutch instruments in or around 1609—makes no mention of a device offering a magnified view of distant objects. If there really was a "Tudor telescope"—a device as good as instruments developed more than a century later—wouldn't more copies of the instrument have been made? Wouldn't its existence have been widely known and discussed? To be sure, Leonard Digges probably experimented with an array of optical devices, and, with difficulty, may have achieved reasonably good views of distant terrestrial objects. And, on occasion, he perhaps even glimpsed something in the night sky that no one else had seen up to that point. But observing the planets in fine detail is another matter. On the weight of the evidence, it doesn't seem likely. As David Levy puts it, the existence of an Elizabethan telescope capable of revealing Jupiter's Great Red Spot "strains credulity."

Of course, one could argue that the invention of the telescope was a state secret; that such a device had been developed, but word of its existence was kept quiet in the interest of national security. This is the line of reasoning Usher suggests: "Spyglasses had obvious military uses, and the lack of printed detail on optical devices no doubt stemmed from a need for a nominally Protestant England to guard its scientific and technical expertise against the Catholic theocracies that menaced her." As for the sights revealed by the telescope—these were perhaps not dangerous in their own right, but "would affect national security indirectly as discoveries that flew in the face of scholastic certainty could only inflame anti-Elizabethan sentiment at home and abroad and strengthen the resolve of theocracies to stamp out heresy on the island nation." However, historians are skeptical. Much like the claim that Copernicanism was a dangerous topic, the idea of a secret Elizabethan telescope doesn't quite hold up to scrutiny. Allan Chapman believes there is "a formidable argument against such a 'conspiracy' theory. . . . Frankly, I find the idea as implausible as Sir Winston Churchill suppressing radar in World War II, and then expecting no-one to talk or write about it thereafter!" Again, it is possible that Leonard Digges used a primitive telescope-like device; and John Dee and Thomas Harriot may have used one as well—but it was only with the improved Dutch inventions, circa 1609, that a new window on the heavens was truly opened. Only then did astronomers have, as Chapman puts it, "relatively good, clear images which conveyed a wholly new level of meaning in the universe."

The minutiae that catch a scholar's eye need not stem from Shakespeare's words: His numbers will do just as well. For Usher, every number in the canon is a potential clue. In *The Merchant of Venice*, he notes that Bassanio is Portia's seventh suitor, suggestive of Saturn, the seventh of the known planets. When the number ten thousand occurs in *Hamlet*, and again in *Cymbeline*, Usher says that this is approximately the number of stars visible to the unaided eye.* A scene of particular numerical interest comes in the middle of *The Winter's Tale*, in which the elderly Shepherd laments the foolishness of young men. He wishes that "there were no age between ten and three-and-twenty" (3.3.58–59). Men of that age, he says, do nothing but fight, steal, impregnate women, and disrespect their elders (Shakespeare says it with more poetic language, of course). They also lack common sense: "Would any but these boiled brains of nineteen and two-and-twenty hunt in this weather?" (3.3.62–64). Usher gets a lot of mileage out of these two pairs of numbers (10/23 and 19/22). With some deft numerical juggling, he pieces together a chronology within the play that parallels many of the key dates in the life and times of both Tycho and Ursus, including the years of their deaths. (The sixteen-year leap announced by the Chorus at the start of act 4 provides further temporal data for additional number crunching.) Another odd number is Hamlet's age: Though he's called "young Hamlet," and wants to return to university, he is apparently thirty.† Usher points out that this is the same age that Thomas Digges was when he published the supplement to his father's almanac—the one with the diagram of the infinite cosmos.

---

* The figure is likely too high. The standard number usually given is six thousand: This is the *total* number of stars that one could see, in theory, in the entire sky, from a dark-sky site, without optical aid. In practice, only a fraction of that total would be visible at any one time. Half would of course be below the horizon, and atmospheric dimming would obscure another thousand or so. Astronomers usually say that no more than about two thousand stars are visible to the unaided eye at any one time. Perhaps a person with exceptional vision, under exceptional viewing conditions, could see a total of ten thousand stars—but it seems like a stretch.

† The most direct evidence for Hamlet's age is found in act 5, scene 1, in which the gravedigger says he's had his job since King Hamlet defeated Old Fortinbras—which, he reminds us, also happens to be the day on which Prince Hamlet was born. He later adds: "I have been sexton here, man and boy, thirty years" (5.1.156–7). This figure also meshes with what we know about Yorick, the jester: Hamlet played with him as a child, but the jester has now been dead for twenty-three years. Nonetheless, there has been much hand-wringing over Hamlet's age; see, for example, Jenkins's lengthy discussion of the subject in the Arden edition.

WHAT'S IN A NUMBER?

What is intriguing about these arguments is that, if one removes the speculation about early Elizabethan telescopy, they begin to resemble the kind of "close reading" that has always formed a significant part of Shakespeare criticism; indeed, they echo many of the arguments that Shakespeare scholars have been putting forward for decades (if not centuries). Consider Usher's attention to numbers. At first, his number crunching might seem somewhat obsessive, like that of a Kabbalist—but there's always been a corner of Shakespeare scholarship in which obsession with numbers is de rigueur. A case in point is the work of Thomas McAlindon, a respected Shakespeare scholar who taught at the University of Hull, in England. In *Shakespeare's Tragic Cosmos* (1991), for example, McAlindon explored the "quadruple groupings" in *Julius Caesar*: two marriages; four plebeians responding to the speeches of Brutus and Antony; four plebeians attacking Cinna the poet ("no doubt the same ones"). The number two holds even more significance for McAlindon. The "dyad" crops up frequently in the canon—for example, in *A Midsummer Night's Dream*, where "unified duality is presented as incipient doubleness and confusion."

McAlindon finds numbers to be even more crucial to interpreting *Macbeth*, a play in which "number symbolism co-operates with nature symbolism in the process of signalling key ideas relating to the tragic theme of disunity and chaos. . . . Threes and twos, trebling and doubling, are closely linked throughout the play." This pattern of numbers "focuses sharply on the idea that 'doubleness' is the root cause of tragic chance and confusion, so that the witches' refrain, 'Double, double, toil and trouble; / Fire burn, and cauldron bubble' . . . might be taken as the play's epigraph." It is the three witches who naturally catch his attention first. (If three is a mystical number to Christians—think of the Holy Trinity—then why would Christians associate the number with witchcraft? McAlindon has the answer: "The explanation, of course, lies in the fact that witchcraft, like devilry, is a rival system which parodies what it seeks to overthrow.") The porter admits three imaginary sinners into hell; Macbeth hires three murderers. Banquo hallucinates a series of nine kings, but of course nine is "the witches' favourite multiple of three." (Wait a minute, is it really *nine* kings? The text says "a show of eight kings" with "Banquo following" [4.1.3]—eight plus one; there's your nine.) Duncan is murdered at 3 a.m. (there's a tradition that roosters crow three times—at midnight, at 3 a.m., and an hour before dawn); and the porter admits to "carousing till the second cock" (2.3.23–24).* And, of course, three "tomorrow"s in Macbeth's famous soliloquy.

That's a lot of threes (and I haven't listed them all). Were these triads as important to Shakespeare as they are to McAlindon, or is he reading into the plays a subtext that the author never intended? McAlindon insists that these number patterns "tell us something important about Shakespeare's beliefs and motivation." His findings show that *Macbeth*, for example, is "a far more intricate and artful play than has customarily been thought" and provide us "with firm clues as to its meanings. Its special relevance in this context lies, of course, in the fact that number symbolism is part of the language of cosmology. . . ."

Another scholar struck by Shakespeare's use of numbers is Shankar Raman at the Massachusetts Institute of Technology. In a paper titled "Specifying Unknown Things: The Algebra of *The Merchant of Venice*," he examines Bassanio's and Portia's argument over the status and value of unknown things—a quarrel that "parallels a shift in the history of algebra." He says that "the language of proportionality and algebraic equations permeates in particular Bassanio and Portia's responses to the 'hazard' of choosing the right casket." The play expresses "a fundamental connection between law and mathematics." In another paper, "Death by Numbers: Counting and Accounting in *The Winter's Tale*," Raman focuses on a "deep and abiding . . . connection between the language of Renaissance arithmetic and the (at first glance) unmathematical world of Shakespearean romance." The most important numbers in the play, he says, are zero and one, with much of the drama rooted in an "overarching tension" between the two: "The difference between the zero and the one in *The Winter's Tale* bespeaks a tension between the two different numbering systems, Arabic and Roman, to which the early modern era is heir," and offers the playwright a chance to create "dense ruminations on the finititude of existence."†

---

* However, timekeeping-by-rooster has its perils. It's true that there's a similar reference in *Romeo and Juliet*, in which Capulet states that ". . . the second cock has crow'd! . . .'tis three o'clock" (4.4.3–4)—but editors are hesitant to apply the same logic to *Macbeth*. A. R. Braunmuller (in the New Cambridge edition) and Nicholas Brooke (in the Oxford edition) shy away from equating "second cock" with "three o'clock," with Brooke interpreting the porter's phrase loosely as "early dawn" (Brooke, p. 131).

† At the far end of the Shakespeare-and-numbers plausibility spectrum is the theory that Shakespeare was responsible for some of the translation work in preparing the King James version of the Bible, printed in 1611—work that had to be done in secret, of course, since it would have been inappropriate for a lowly actor to contribute to such a lofty work. The "clues" can be found in Psalm 46, where the forty-sixth word from the beginning is "shake" (verse 3) and the forty-sixth

* * *

These highly analytic studies of Shakespeare's use of numbers are certainly intriguing, and no doubt Shakespeare knew what he was doing in his use of numbers *most of the time*. But occasionally he seems to have been downright sloppy. In *Hamlet*, the prince ponders the imminent battle between Norwegian and Polish forces, and the "two thousand souls" (4.4.25) that will likely perish; a few dozen lines later it is "twenty thousand men" who will die (4.4.60).* In *Henry V* (1.2), the archbishop of Canterbury rattles off a long list of names and dates in order to justify the English king's claim to France, and Shakespeare—repeating an error in Holinshed's arithmetic—subtracts 426 from 805 and gets 421 (rather than 379). In *Julius Caesar*, Octavius laments Caesar's "three and thirty wounds" (5.1.52), although Plutarch clearly has the number at twenty-three, not thirty-three. In *The Winter's Tale*, both Leontes and the Chorus ("Time") give the duration of the gap between the third and fourth acts as sixteen years, but Camillo gives it as fifteen (4.2.4). These few examples may not prove that, as Harold Jenkins has put it, Shakespeare "was often lax with numbers"; still, we should perhaps be cautious about attaching a deep significance to every number in the canon.

As we've seen, Usher is not the first to focus on Shakespeare's use of numbers; nor is he the first to look for hidden treasure in the curious words and phrases that one finds throughout the canon. Consider, for example, that peculiar line from *Hamlet* that we looked at earlier, about knowing a hawk from a handsaw. Recall that Usher's explanation involved geography, and in particular the relationship between Elsinore, Tycho Brahe's island of Hven, and the German university city of Wittenberg. It may have sounded like a stretch—but consider this explanation, offered by a nineteenth-century critic. After a reminder that "handsaw" may refer to a kind of bird, we are told that

> The meaning generally given to this passage is, that birds generally fly with the wind, and, when the wind is northerly, the sun dazzles the hunter's eye, and he is scarcely able to

---

word from the end is "spear" (verse 9). Oh, and Shakespeare was forty-six at the time (turning forty-seven in the year it was published).

* Another example from *Hamlet*: When the Player King speaks of "thirty dozen moons" (3.2.150) having passed since he married the queen, it seems like a poetic way of saying "thirty years"—except that the first quarto edition of the play gives the figure as forty years, not thirty—but then, the first quarto is not considered to be a very authoritative text.

> distinguish one bird from another. If the wind is southerly, the bird flies in that direction, and his back is to the sun, and he can easily know a hawk from a handsaw. When the wind is north-north-west, which occurs about ten o'clock in the morning, the hunter's eye, the bird, and the sun, would be in a direct line, and with the sun thus in his eye he would not at all be able to distinguish a *hawk* from a *handsaw*.

Who could have missed *that*? Still, if "handsaw" means "heron" in *Hamlet*, one might wonder why it appears to straightforwardly mean "handsaw" in *Henry IV, Part 1*. The relevant scene comes in act 2, where Falstaff claims to have been attacked by a horde of bandits. (The number of attackers grows with each telling of the tale, but in reality it was just two men—Prince Hal and his accomplice, Poins, in disguise.) After intentionally damaging his own sword to make it appear as though he had used it to fight off his assailants, Falstaff laments that his weapon is "hacked like a handsaw" (2.4.161). Maybe, just maybe, Hamlet's handsaw was a handsaw all along. As Freud is supposed to have said, "Sometimes a cigar is just a cigar."

WHEN SHALL WE 1,250 MEET AGAIN?

The Shakespeare Association of America is to Shakespeare studies what the American Astronomical Society is to astronomy. The SAA* is the world's largest professional association for Shakespeare scholarship and allied studies, boasting some 1,250 members from 36 countries. I had the privilege of sitting in on a number of sessions at their annual conference in 2012, when it was held in Boston, and again in 2013, when it was held in Toronto. The breadth of topics tackled at a typical SAA conference is truly staggering, encompassing not only Shakespeare's writings but also those of Marlowe, Jonson, and any other writers active in early modern England, as well as analyses of the rapidly changing material, social, and intellectual environment in which these writers lived and worked.

Scholars come to present their papers, listen to other scholars' papers, and to discuss research areas of mutual interest. That much it probably has in common with any other arts or humanities conference. But one difference stood out: In each seminar room, some fifteen or twenty

---

* Perhaps it is appropriate that the two conferences mentioned in this chapter, AAS and SAA, have opposite acronyms. The two conferences could hardly be more different.

chairs are placed around a central table, as in a corporate boardroom; people actually presenting a paper in the session are invited to sit in this "inner circle." Surrounding them one finds another thirty or forty chairs lined up against the outer perimeter of the room, facing inward. This is where the "auditors" sit—anyone who's not presenting a paper. (These are often graduate students, but they're just as likely to be tenured professors who are attending the session to listen rather than to present.) This arrangement, which one senses hasn't changed in many, many years, is clearly seen as normal by the attendees—but it seems to create an unnecessarily harsh divide between the presenters and the auditors. (The view from the auditors' circle is a bit odd, as one peers at the faces of half of the presenters, and the backs of the heads of the other half. Even when you can see a speaker's face, it is always partially obscured by the back of someone else's head.)

Topics that came up for discussion in 2012 and 2013 ranged from the predictable to the esoteric. A seminar titled "Reading Shakespeare and the Bible" is hardly surprising; in the twenty-first century, neither is "Early Modern Queer Colonial Encounters" or "iShakespeare: New Media in Research and Pedagogy." For myself, on the lookout for science-related topics, a panel on " 'The Famous Ape': Shakespeare and Primatology" was a highlight of the 2012 conference; so was a seminar on "Matter, Perception, and Cognition in the Renaissance." (I tried not to worry too much when a seminar leader referred to philosopher and cognitive scientist Daniel Dennett as "David Dennett"; I suppose anyone can make a one-time slip.) Sometimes what seem like minor things become major discussion topics: The question of how many people have to be able to see a dagger before it should be considered "real"; how sleep deprivation affects Macbeth's cognitive abilities; whether the names of Shakespeare's characters have scatalogical significance, reflecting (as one scholar suggested) "aspects of anality and flatulence." At the 2013 conference, I was intrigued by a listing for a two-hour session on *The Tempest*, one of my favorite plays. As the seminar unfolded, the presenters pondered questions that likely escaped casual readers of the play. For example: What language, exactly, does Prospero teach to Caliban? Should we think of Caliban and Miranda as grammar-school students, with Miranda the more advanced pupil? And is Caliban *still* a student of Prospero's, or was he expelled for attempting to rape Miranda? And why is Alonso so certain that he will never see his daughter again, just because she's moved from Milan to Tunis—when there was already a centuries-old trade route established between the Italian city and the North African port? Later, someone made a case for interpreting Caliban's log gathering as a symbol for deforestation and environmental destruction. An older gentleman along the outer perimeter of auditors chimed in: "Or at least for log gathering," he said dryly.

For anyone not up to speed on twenty-first-century Shakespeare scholarship, some of the topics covered may seem to come from somewhere just beyond the left-field wall. Some sample titles from the last two conferences:

1. "Diagnosing Hamlet: The mad prince and the autism spectrum."
2. "Dis-Eating *Macbeth*: Macbeth's indigestion and the matter of milk."
3. "The Ecology of *The Tempest*: Was Prospero's island carbon-neutral?"
4. "'Exhalations Whizzing': Meteorology, melancholy, and moral action in *Julius Casear.*"
5. "The Georgic Contract: Agrarian bioregionalism and eco-cosmopolitanism in *Henry IV, Part 2*."
6. "Head in the Clouds: Historicism, *Hamlet*, and neurophenomenology."
7. "Shakespeare's Quantum Physics: *The Merry Wives of Windsor* as a feminist 'parallel universe' of *Henry IV, Part 2*."
8. "The Unbearable Lightness of Being Ariel: Is Prospero's little helper a hologram?"

Okay, I confess: Three of these eight titles are made up. But can you tell the fake titles from the real ones? It's not easy, is it? (The answers can be found at the end of the chapter.)

I mentioned that Usher devotes nine pages to the famous "Exit, pursued by a bear" stage direction in *The Winter's Tale*. Perhaps his interpretation wouldn't have been too wildly out of place at the 2013 conference, where a paper examined "the bear's disruptive (and even desirable) queer effects that destabilize binary systems of difference. Shakespeare's onstage bear—whether live in the flesh or represented by a bearskin (made of bear-that-once-was)—materializes 'transspecies' connections that refuse the (false) separation between human/animal and nature/culture, preferring inter-actions and inter-dependence to autonomy and antagonism."

When scholars like Peter Usher pick over a Shakespeare play line by line, looking for nuggets of hidden meaning that have slipped by unnoticed over the years, they are hardly alone. They are merely the latest in a long line of scholars who have reached as far as plausibility may allow—or perhaps slightly farther—in discerning a hidden meaning in Shakespeare's centuries-old words.* They have mastered the peculiar blend of art and science that allows one to bend Shakespeare's words perilously close to the breaking point in search of a fresh insight, a novel interpretation. If nitpicking each line of a Shakespearean passage were a crime, the last SAA conference would have to have been held in the Don Jail rather than the Royal York hotel.

Whatever one may think of Usher's more extravagant claims, his work touches on a number of areas where the Shakespeare community is—gradually, perhaps—beginning to see things his way: More and more scholars are acknowledging that the connections between Shakespeare and Thomas Digges have been neglected, and that there is *some sort of link* between the characters and setting of *Hamlet* and the Danish astronomer Tycho Brahe. If his book manages to push a handful of researchers to more closely examine those issues, he will likely be satisfied. If they happen to embrace his other claims, I'm sure he'd be delighted—but he does not seem to be betting on it, or to be in any rush. "I think that good wine sells itself," he says. "I'm retired, and I'm not in a hurry. After all, the canon has been around for four hundred years; another couple of centuries is not going to make a difference."

As it turns out, from among Usher's myriad claims about Shakespeare and science one particular idea is gaining traction. It involves astronomy and one of the late plays, and it resonates particularly well with both Scott Maisano and John Pitcher. In fact, it seems that the three men all

* Note that even the *surface* meanings remain the subject of debate. In the new Arden *Hamlet*, for example, Thompson and Taylor point out that even the most famous speech in English literature is ambiguous: Hamlet says, "To be, or not to be—that is the question"—but what *is* the question? "Perhaps surprisingly after so much debate, editors and critics still disagree as to whether the question for Hamlet is (a) whether life in general is worth living, (b) whether he should take his own life, (c) whether he should act against the King." They add in their introduction that "it is still possible to disagree about almost every aspect of this play . . ." (Thompson and Taylor, p. 284; p. 137) In fact, one need not have an entire sentence, or even a whole word, to have a controversy: Certain letters, and even punctuation marks, have triggered much debate.

came upon the same idea, at about the same time, by chance. The play in question is *Cymbeline*, dating from the final few years of Shakespeare's career. It also involves the development of modern astronomy, and the invention of the telescope in particular. That's a subject that we have touched on only briefly, in connection with Usher's provocative claims regarding Leonard Digges, who purportedly used such a device in the mid-1500s. Now we must examine the work of the Italian scientist who quite definitely *did* aim such an instrument skyward, more than half a century later. And so we are ready to meet that *other* great mind who came into the world in 1564.

Answers to the quiz on page 189: Titles 1, 3, and 8 are fake. Paper number 7, which imagines *Henry IV, Part 2* and *The Merry Wives of Windsor* as taking place in parallel universes, is quite real. As noted in the paper's abstract, it invokes not only quantum theory but also string theory: "I posit that the parallel universe is one that contemporary quantum physics has demonstrated as a logical product of string theory." The essay notes, with explanations from quantum theory, how the transportation from one universe to another occurs, and it argues that Shakespeare's purpose in the creation of *Merry Wives* was to demonstrate "that female-determined justice against male abuses could indeed ultimately or even simultaneously transpire." (http://www.shakespeareassociation.org/abstracts/41.pdf) The author issued a caveat during the seminar, admitting that Shakespeare may not have been "consciously thinking of string theory" when he wrote the plays.

9. *"Does the world go round?"*

SHAKESPEARE AND GALILEO

While Shakespeare's birthplace is a major tourist attraction, the house where Galileo entered the world—a four-story, pinkish-brown town house in the northern Italian city of Pisa—is to this day a private residence, marked only by a small plaque and an Italian flag. It stands on a quiet street in a neighborhood known as the San Francesco Quarter. In Galileo's time, the area was home to artists, craftsmen, and shopkeepers. His father, Vincenzo Galilei, had settled there, with his wife, Giulia, just a year before the birth of their first son.

Vincenzo was a skilled musician, teacher, and music theorist. In spite of his talents, money was tight, and he traded in wool to make ends meet. Giulia was an educated woman who could claim a cardinal among her relatives. Galileo was the first of their seven children; as was the custom in Tuscany at that time, he was given a Christian name that reflected the family name—hence the echo-like "Galileo Galilei," the father of modern science, known to history simply as Galileo.

Vincenzo had hoped his son would become a doctor, and the youngster was duly enrolled in the university at Pisa to study medicine. Instead, he developed an interest in mathematics. He left the university without a degree, although he would later return; it was at Pisa that he landed his first teaching job. To call Galileo a misfit might be too harsh, but from an early age he was known for his argumentative nature. We know that he irked some of the more senior faculty members by refusing to wear the school's official robes, which he considered pretentious (and for which the university docked his pay).

It was also in Pisa that Galileo first became fascinated by motion, and began to investigate the way that objects move in response to a steady force, like the force of gravity (although no one called it gravity at that time). One example of such movement is the swing of a pendulum. Galileo's first thoughts on the matter are said to have been triggered by the sight of the massive chandelier in Pisa's cathedral, gently swaying in the breeze; eventually, he worked out the mathematical formula for the duration of a pendulum's swing. The movement of falling or rolling bodies intrigued him as well. One way to study such motion was to roll different kinds of objects down an inclined plane, carefully measuring how far the objects moved in a given interval of time. He also wondered about the special case of objects falling straight down. Suppose you had a ball of iron, like a cannonball, and another ball of the same size and shape, but made of wood. You might guess that the cannonball would fall faster, just because it's heavier. That's what Aristotle thought. He said that the heavier body would fall faster than the lighter one, with a speed proportional to its weight. That certainly sounded plausible—but Galileo had his doubts. He would later give a detailed argument in what was to be his final book, the *Discourses and Mathematical Demonstrations Relating to Two New Sciences* (1638). Written in the form of a dialogue, Galileo's line of reasoning is voiced by a character named Salviati:

> Aristotle declares that bodies of different weights, in the same medium, travel . . . with speeds which are proportional to their weights [and thus] a stone of twenty pounds moves ten times as rapidly as one of two [pounds]; but I claim that this is false and that, if they fall from a height of fifty or a hundred cubits, they will reach the earth at the same moment. . . . Aristotle says that "an iron ball of one hundred pounds falling from a height of one hundred cubits reaches the ground before a one-pound ball has fallen a single cubit." I say that they arrive at the same time.

According to his first biographer, Vincenzo Viviani, Galileo tested his hypothesis by dropping objects of varying weights from the top of the cathedral's famous bell tower, the Leaning Tower of Pisa. Historians, however, suspect that the story may be more legend than fact. (Viviani's account was driven largely by hero worship, and much is exaggerated—a common practice in biographies of the time.) Galileo certainly *could* have performed such an experiment; but he likely had already deduced the answer from his studies of motion on an inclined plane, in which the same principles are at work (in either case, the distance covered increases with the square of the elapsed time). Galileo also studied projectile motion, showing that a cannonball must follow a parabolic path. All of these findings contradicted Aristotelian physics, whose failings were becoming increasingly evident. Rather than relying on ancient wisdom, Galileo favored an experimental approach. At the same time, he was discovering the power of mathematics in describing the natural world. As he would put it in a short book called *The Assayer* (in Italian, *Il Saggiatore*), in 1623:

> Philosophy is written in this grand book, the universe, which stands continually open to our gaze. But the book cannot be understood unless one first learns to comprehend the language and read the letters in which it is composed. It is written in the language of mathematics, and its characters are triangles, circles, and other geometric figures without which it is humanly impossible to understand a single word of it; without these, one wanders about in a dark labyrinth.

Nature displayed an underlying order, and through careful observation and mathematical representations—what we would today call mathematical modeling—that order can be understood and investigated, and predictions can be made. This combination of experiment and mathematical analysis, taken together, would serve as the backbone of science in the centuries ahead.

Galileo moved to the northeastern city of Padua in 1592, and would teach at the university there for the next eighteen years. He was involved with, but did not marry, a woman named Marina Gamba, who bore him two daughters and a son.* He would later look back on that time as the happiest and most productive of his life. Even so, he struggled financially, especially following the death of his father. Happily, his teaching duties left ample time for tinkering. He was desperate to invent something he could patent—a machine or instrument whose utility would guarantee him some measure of financial security, perhaps by winning the pa-

* Another coincidence between Galileo's life and that of Shakespeare, who also had two daughters and one son.

tronage of a prince or duke. ". . . I have many diverse inventions," he wrote, "only one of which could be enough to take care of me for the rest of my life . . . if I can only find a grand Prince who would like it. . . . Then he could do with this invention and with its inventor whatever he likes. I would hope that he would accept not only the stone but the quarry."

Galileo invented a primitive thermometer and, more lucratively, a military compass designed to aid artillery officers in battle. (He made money not only by having his assistants mass-produce the instrument in his workshop, but by offering tutorials in its use.) By his early thirties, Galileo had taken an interest in astronomy. When a new star appeared in the sky in 1604—the supernova now known as "Kepler's star"—he delivered a series of lectures on the remarkable object, speculating on its significance; the university's auditorium was filled to capacity for each of his talks. He also corresponded with Kepler, after receiving a copy of the German scientist's book *Mysterium Cosmographicum* (1596). The book was a blend of science, mysticism, and numerology—but it also praised the Copernican model, which Galileo approved of. "It is really pitiful that so few seek the truth," Galileo wrote to Kepler. He noted that he himself had been "an adherent of the Copernican system for many years. It explains to me the causes of many appearances of nature which are quite unintelligible within the commonly held hypothesis." It was the first sign that Galileo, too, saw the Ptolemaic model as outdated.

"A CERTAIN FLEMING HAD CONSTRUCTED A SPYGLASS"

Up to this point, Galileo—like all skywatchers since the dawn of history—had only the unaided eye with which to observe the heavens.* That would soon change. It was in Padua that he first learned of a curious optical device from Holland—an instrument that was said to make distant objects appear nearby:

> About ten months ago a report reached my ears that a certain Fleming had constructed a spyglass by means of which visible objects, though very distant from the eye of the observer, were distinctly seen as if nearby. Of this truly remarkable effect several experiences were related, to which some persons gave

* Or perhaps all except for Thomas Harriot. As we saw in Chapter 5, Harriot appears to have aimed a telescope at the night sky a few months ahead of Galileo.

> credence while others denied them. A few days later the report was confirmed to me in a letter from a noble Frenchman at Paris, which caused me to apply myself wholeheartedly to inquire into the means by which I might arrive at the invention of a similar instrument.

As his own testimony shows, Galileo did not invent the telescope. In fact, we can't be sure who did, although credit usually goes to a Dutch spectacle maker named Hans Lipperhey (sometimes spelled Lippershey), who applied for a patent for a telescope-like device in October, 1608. It was an instrument, he claimed, "by means of which all things at a very great distance can be seen as if they were nearby." It apparently magnified distant objects threefold. That may not sound like much, but even so, the military applications must have been obvious. Still, his application was turned down on the grounds that the design was already well known, and indeed two other Dutchmen are thought to have independently come up with a similar device at about the same time.

Before long, Galileo had improved on the original Dutch invention. Soon he had a telescope—or as he called it, a "perspicillum"—that could magnify twenty or even thirty times, compared with what one would see with the unaided eye.* Galileo had managed, as Owen Gingerich has put it, "to turn a popular carnival toy into a scientific instrument." Galileo immediately recognized the potential of this new instrument. But his first thoughts had nothing to do with astronomy; instead, he saw the telescope's value as a military tool. He arranged a meeting with senior Venetian statesmen, leading them to the top of the bell tower at the Piazza San Marco. Galileo urged them to aim the telescope at ships in the harbor. The device worked so well that they could identify ships a full two hours before they arrived in port. The officials were suitably impressed, and offered to double Galileo's salary at the university. In the end, he used their offer as a bargaining chip to land an even better job in his native province of Tuscany. Galileo would soon head for Florence with a lofty new title: In July 1610, he was appointed Chief Mathematician and Philosopher to the Grand Duke of Tuscany. But he was still in Padua when he aimed his telescope skyward, and began to scrutinize the night sky with the new device. What he saw would change the world forever.

---

* Stillman Drake notes that the word "telescope" was coined in 1611.

* * *

By modern standards, Galileo's telescope was a joke: Today you can go to any department store and get a beginner's telescope that has vastly superior optics to anything Galileo would have been able to construct. Modest as it was, however, Galileo accomplished a great deal with it. Beginning in the fall of 1609, Galileo aimed his telescope at the night sky—and was amazed at what it revealed. He found the moon to be covered with mountains and craters, and calculated the size of these features by carefully observing their shadows.

He showed that the stars visible to the unaided eye are outnumbered perhaps ten to one by dimmer stars, too faint to show themselves without a telescope. He went on to deduce that the Milky Way itself must be made up of countless faint stars, too dim to see with the unaided eye. But the biggest surprise came when he aimed his telescope at the planet Jupiter:

> On the seventh day of January in this present year 1610, at the first hour of night, when I was viewing the heavenly bodies with a telescope, Jupiter presented itself to me; and because I had prepared a very excellent instrument for myself, I perceived (as I had not before, on account of the weakness of my previous instrument) that beside the planet there were three starlets—small indeed, but very bright. Though I believed them to be among the host of fixed stars, they aroused my curiosity somewhat by appearing to lie in an exact straight line parallel to the ecliptic, and by their being more splendid than others of their size. There were two stars on the eastern side and one to the west.

The following night he found "a very different arrangement": The three little stars were now all to the west of Jupiter, and closer together than they had been the previous night. A couple of days later, "there were but two of them, both easterly, the third (as I supposed) being hidden behind Jupiter."

Galileo was mesmerized by this dynamic display, this string of little stars that seemed to follow Jupiter across the sky from night to night. Eventually he discovered that there were not three but four of these peculiar stars. He observed every night in which the skies were clear, and gradually came to understand what he was seeing. "There was no way in

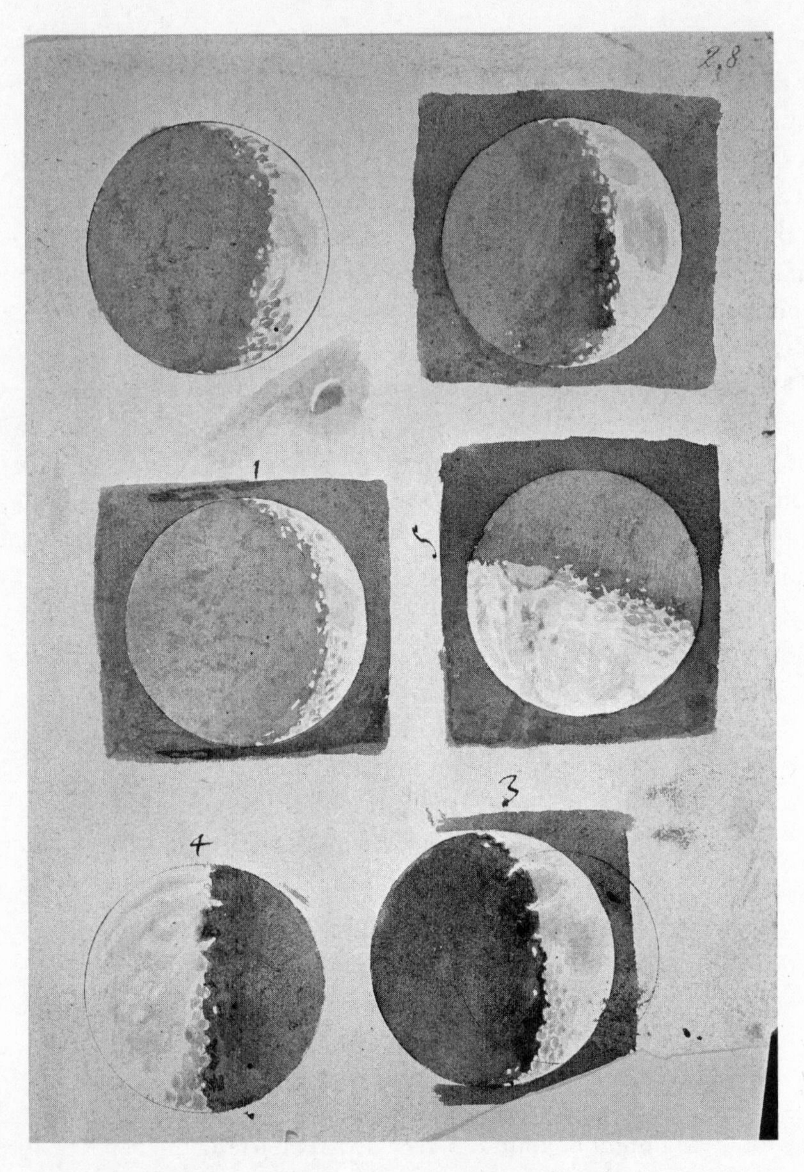

Fig. 9.1 Aided by the newly invented telescope, Galileo sketched the moon (these images date from November and December of 1609), showing it to be covered with mountains and craters. Engravings based on these drawings would help make his book *Siderius Nuncius* (*The Starry Messenger*) a bestseller. Scala/Art Resource, NY

---

which such alterations could be attributed to Jupiter's motion," he would later write, "yet being certain that these were still the same stars I had observed, my perplexity was now transformed into amazement." Soon he had "decided beyond all question" that there were four stars

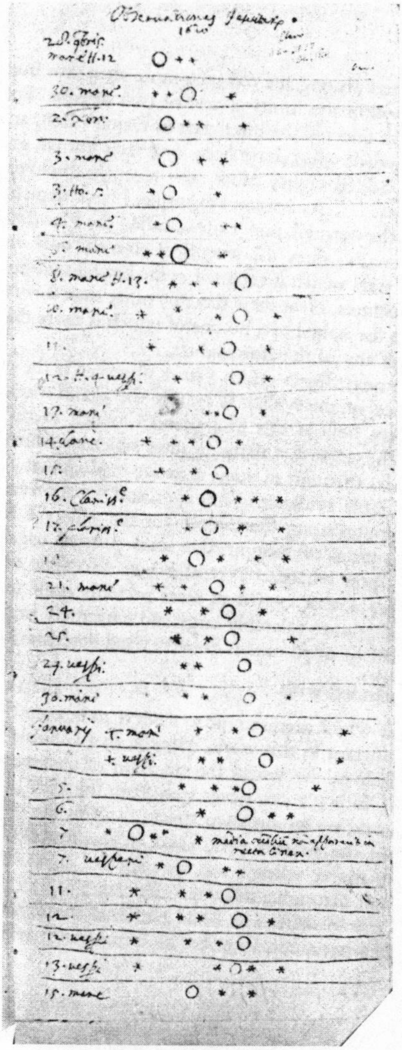

Fig. 9.2 In January of 1610, Galileo aimed his telescope at Jupiter—and found it accompanied by four previously unknown "stars." It soon became clear that these were moons revolving around Jupiter, just as our own moon revolves around the Earth. In this page from his notebook, Galileo tracks their positions over several weeks. The Granger Collection, New York

"wandering about Jupiter as do Venus and Mercury about the sun, and this became plainer than daylight from observations on similar occasions which followed." There was no doubt that "four wanderers complete their revolution about Jupiter." He named them the "Medicean stars" in honor of his new patron, Grand Duke Cosimo II de' Medici, but it didn't stick; today we call them the Galilean moons after their discoverer.

Today, anyone who's been to a public observing night at their local planetarium or observatory knows what Jupiter and its four bright moons look like (in fact, they can be seen with a good pair of binoculars, if the conditions are favorable and one has steady hands). But until the winter of 1610, no one knew these moons existed. No one had imagined that such objects *could* exist. The Earth, being at the center of the universe, was presumed also to be the center of rotation. Bodies can revolve around the Earth, and only the Earth. But Galileo now had incontrovertible evidence that there was at least one other body that could also serve as such a center: The planet Jupiter was now seen to have not just one moon, like the Earth, but four moons, circling around it; no doubt they had been doing so since the beginning of time.

Galileo knew that before long, someone else would duplicate his observations—so he rushed to get his ideas into print. The result was a slim book called *Siderius Nuncius* (*The Starry Messenger*), published in Venice on March 13, 1610. Although written in Latin, this twenty-four-page pamphlet conveyed a remarkably straightforward message: Here was a blow-by-blow account of the wonders of the night sky revealed by Galileo's telescope, sights "never seen from the creation of the world up to our own time." Finally there was observational evidence that would lay to rest "all the disputes that have vexed philosophers through so many ages." The book was an instant best seller. The initial print run of five hundred copies sold out at once, with additional orders pouring in from across Europe. Within five years, Galileo's discoveries were being discussed as far away as Peking, where a Jesuit missionary had published a summary of Galileo's findings in Chinese.

THROUGH THE EYEPIECE

Galileo continued his examination of the night sky, soon discovering that Venus goes through phases, just like our own moon. Something was odd about Saturn, too, though it was hard for Galileo to say exactly what. He could tell the planet was elongated in a peculiar way, but his telescope lacked the resolution needed to show its rings. He also found that the sun, like the moon, was imperfect, its surface peppered with dark spots (a discovery that, by now, others had also made).*

Everything Galileo saw through the telescope seemed to accord with the Copernican model of the universe. As we've seen, Galileo was leaning toward Copernicanism even before he began his telescopic observations; what he saw through the eyepiece, however, seemed to clinch it. Consider the planet Venus: If the Ptolemaic model were correct, with Venus circling the Earth in a "lower" shell, or sphere, than the sun, then it would always display a crescent phase.† The fact that Venus displayed

---

* Christoph Scheiner, a Jesuit priest and astronomer working in Bavaria, observed sunspots at about the same time as Galileo, in the fall of 1611, and the two fell into a bitter dispute over who was first—neither of them apparently being aware that Johannes Fabricius, an astronomer at Wittenberg, had published a short treatise on sunspots the previous summer. They had also been observed in England by Thomas Harriot, as we saw in Chapter 5.

† The argument is a bit subtle because, in the Ptolemaic system, Venus was believed to move in an epicycle that was "synched," so to speak, with the sun's motion around the Earth; that is, the center of Venus's epicycle was thought to lie at all times on a straight line connecting the Earth and the sun. (This was necessary to account for the fact that Venus always appears in close proximity to the sun.) With this constraint in place, Venus could only display crescent phases. In contrast, in the Copernican (or the Tychonic) model, one would see a full set of phases, whether Venus moved in its own epicycle or not.

a full set of phases, from "new" to "full," only made sense if Venus was actually revolving around the sun. (By extension, the same argument must apply to Mercury.) Upon making the discovery, he sent a coded message to Kepler: Unscrambled, it read "The mother of love emulates the figures of Cynthia"—that is, Venus emulates the moon. A short time later he wrote to Christopher Clavius, a Jesuit astronomer working in Rome, that the sun is "without any doubt the center of the great revolutions of all the planets." In 1613 he was ready to go public with the discovery. He wrote, "With absolute necessity we shall conclude, in agreement with the theories of the Pythagoreans and of Copernicus, that Venus revolves about the sun just as do all the other planets."

And then there was Jupiter, behaving like a miniature solar system in its own right. If Jupiter had moons of its own, how could anyone say that the Earth was the center of the universe? One of the old arguments put forward against the Copernican theory had involved the moon: How could the Earth revolve around the sun—presumably at enormous speed—without our planet "losing" its moon in the process? And yet, whether one believed Ptolemy or Copernicus, Jupiter certainly moved around *something*—and it did so accompanied by *four* moons! This was "a fine and elegant argument" in support of the Copernican model, Galileo wrote, adding that

> now we have not just one planet rotating about around another while both run through a great orbit about the sun; our eyes show us four stars which wander around Jupiter as does the moon around the earth, while all together trace out a grand revolution about the sun in the space of twelve years.

The moons of Jupiter, along with the phases of Venus, seemed to directly contradict Ptolemy's cosmology, while the imperfections seen on both the sun and moon were an affront to the Aristotelian paradigm on which the Ptolemaic model rested. The ancient Greek model of the cosmos was under attack, first from Copernicus's chisel, and now from Galileo's sledgehammer. After fifteen centuries, a worldview that had seemed unassailable was giving way.

"THE STRANGEST PIECE OF NEWS"

Until Galileo's observations, the appeal of the Copernican model had largely rested on its utility for computing planetary positions. Now there was concrete observational evidence in its favor. The Copernican system, by now almost seventy years old, was clearly more than a mathematical convenience: It could now be regarded as providing a physical description of the cosmos. Galileo's observations, as Owen Gingerich writes, "made it intellectually respectable to believe in heliocentrism as a physical reality."

The news spread quickly. In the French court, the queen, Marie de Médicis, cousin to Grand Duke Cosimo, ordered a telescope to be set up at her window; before it was in place, she became so excited that she fell to her knees in delirious anticipation. Meanwhile her husband, King Henri, suggested that Galileo might name his next discovery in honor of himself. Word of the Italian's discoveries reached England just as quickly. As it happened, an English diplomat, Sir Henry Wotton, was visiting Venice when *The Starry Messenger* was published. Before the day was over he had made sure that a copy of the book was on its way to his king, James I. He included a cover letter informing the king that Galileo's discoveries had "overthrown all former astronomy." The letter reads, in part:

> . . . I send herewith unto his majesty the strangest piece of news (as I may justly call it) that he hath ever yet received from any part of the world; which is the annexed book (come abroad this very day) of the Mathematical Professor at Padua, who by help of an optical instrument (which both enlargeth and approximateth the object) invented first in Flanders, and bettered by himself, hath discovered four new planets rolling about the sphere of Jupiter, besides many other unknown fixed stars.

As mentioned in Chapter 5, the king's eldest son, Prince Henry, also requested Galileo's book; and we've noted that Sir William Boswell, at Cambridge, corresponded with Galileo and helped to spread word of the telescopic discoveries. Another copy of *The Starry Messenger* reached Sir William Lower, who wrote, "Me thinks my diligent Galileo hath done more in his three-fold discoveries than Magellan in opening the straits to the South Sea, or the Dutchmen that were eaten by bears on Novaya Zemlya. I am sure with more ease and safety to himself, and more pleasure to me." Within a year of the publication of *The Starry Messenger*, students at Oxford were asked to debate the question "*An luna sit habitabilis?*" ("Is the moon habitable?"). As Mordechai Feingold writes, "Thus began the Galilean craze. By the end of the decade, almost every prominent man of science in England . . . was conducting telescopic observations." From this point on, Galileo's book "became seminal to the on-going discussion of the heliocentric theory in England."

You didn't have to be a scientist to be caught up in "telescope fever." Thomas Hobbes, the political philosopher, sought out a copy of *The Starry Messenger* and was disappointed when all the bookshops he visited reported being sold out. (And he sensed that they would not be resold any time soon: "They that buy such bookes, are not such men as to part with them againe," he noted in a letter to a friend.) Barten Holyday, a clergyman and playwright, mentioned Galileo's observations in a play published in 1618. And a Scottish poet named Thomas Segett, roughly contemporary with Shakespeare, observed:

> Columbus gave man lands to conquer by bloodshed
> Galileo new worlds harmful to none. Which is better?

And what about Shakespeare himself? As we've seen, the usual view is that these discoveries, announced in the spring of 1610, come too late in Shakespeare's working life to have had much of an impact. But perhaps we shouldn't be so hasty. Shakespeare was not quite ready to retire in 1610; indeed, he would write at least two more plays on his own, plus a few more in collaboration with playwright John Fletcher. Of the works he penned by himself, the most famous is *The Tempest.* This was the last of the three "romances," and indeed the last of the plays he wrote without a collaborator. The first of the romances is *The Winter's Tale.* In between we find one of the most intriguing plays (and one of the most overlooked works) in the entire canon: *Cymbeline.*

"STAGEY TRASH OF THE LOWEST MELODRAMATIC ORDER"

There's no getting around it: *Cymbeline* is an odd play. Our only surviving text is the version included in the First Folio, where it comes at the very end, wedged in with the tragedies as though an afterthought. But *Cymbeline* is anything but a traditional tragedy. Rather than ending in a bloodbath, like *Hamlet* and *King Lear,* this play ends in peacemaking, both familial and political. Scholars today sometimes classify it as a "tragicomedy." Jonathan Bate, editor of the RSC Shakespeare edition, is only half joking when he describes it as "tragical-comical-historical-pastoral." The play contains "a highly self-conscious . . . array of favourite Shakespearian motifs: the cross-dressed heroine, the move from court to country, obsessive sexual jealousy, malicious Machiavellian plotting, the interrogation of Roman values." In other words, Shakespeare, nearing the end of his career, put every possible dramatic ingredient into the pot, and stirred. The plot is labyrinthine, even by Shakespearean standards: The action is driven by at least three separate but intertwined stories, while texture and mood seem to change with the wind. Samuel Johnson despised *Cymbeline*, saying that it contains "much incongruity" and that to list all its faults would be "to waste criticism upon unresisting imbecility. . . ." George Bernard Shaw, meanwhile, called it "stagey trash of the lowest melodramatic order . . . vulgar, foolish, obsessive, indecent, and exasperating beyond all tolerance."

More recently, however, Shakespeare scholars (and ordinary fans) seem to be warming to *Cymbeline*, unconventional as it may be. In the Cambridge edition, Martin Butler defends it as a work whose "narrative grips and compels, rising inexorably from a naïve tale of sundered lovers to a peripeteia of dazzling artfulness. . . . *Cymbeline* was produced by a dramatist working at the height of his powers."

If *Cymbeline* is a hodgepodge, it is perhaps because it draws on multiple sources. Shakespeare takes elements from Holinshed's *Chronicles* and Giovanni Boccaccio's *Decameron*, as well as an anonymous play called *The Rare Triumphs of Love and Fortune*, from 1589. King Cymbeline* rules over ancient Britain, but Roman legions, under Augustus Caesar, are targeting the island for invasion. This was also the time of the birth of Christ—an event perhaps of more significance to Shakespeare and his audience. It may be due to the unusual setting that the play is said to contain more anachronisms than any other in the canon.

I haven't bothered to offer plot synopses for the plays discussed so far, on the grounds that most readers are probably familiar with them; but, due to its relative obscurity, perhaps a one-paragraph outline of *Cymbeline* is in order. The central characters are King Cymbeline, his daughter,

---

* Shakespeare's King Cymbeline has virtually nothing in common with the real-life King Cunobeline (in Latin, Kynobellinus) who ruled over southeastern England in the first century A.D. We might also note that, while Shakespeare had been content to refer to his country as "England" in the history plays penned during Elizabeth's reign, in *Cymbeline* he says "Britain" (or "Britains") nearly fifty times. It was written during the reign of King James—the first king to rule over both England and Scotland—and its language may reflect the new king's ambitions for a unified "Great Britain" (See Bate and Thornton, p. 49).

Imogen,* and the man she has secretly married, a commoner named Posthumus Leonatus. Cymbeline disapproves of their union, and banishes Posthumus; meanwhile the queen, Imogen's stepmother, plots to have Imogen married to the dim-witted Cloten, her son by a previous marriage. Posthumus heads for Rome, where he and his new companions debate whose wife is the most chaste. An Italian nobleman named Jachimo makes a bet with Posthumus, wagering that when he travels to England, he will be able to seduce Imogen. He fails—but he manages to gain access to Imogen's bedchamber by hiding in a trunk. At night he emerges, observing every detail of Imogen's body and of the room, in order to convince Posthumus that he in fact slept with her. A second plot involves Cymbeline's long-lost sons, Guiderus and Arviragus, who are cared for by Belarius, a former lord, but now an outlaw, who believes they are his own offspring. While out hunting, the three encounter Cloten and kill him; then they meet Imogen (who is dressed as a boy, Fidele) when she becomes lost in a Welsh forest on her way to meet Posthumus in Rome. The third plot centers on Cymbeline's refusal to pay the annual fee ("tribute") demanded by the Roman ambassador, as the threat of war looms over the country. (That's not *so* complicated, right?)

THE SYMBOLS IN CYMBELINE

Of the many strange things that happen in *Cymbeline*, the strangest occurs in the play's final act. Posthumus, having been convinced of Imogen's infidelity, orders her killed; later he learns of her innocence, but mistakenly thinks his orders have been carried out. He had been traveling with the Roman army, but now switches sides and fights valiantly for Britain to defeat the Roman forces. Believing Imogen dead, however, he yearns for his own death, and puts on Roman garb to hasten his demise. Instead, he is taken prisoner. While he is in jail, something very peculiar happens.

Scene 4 opens with Posthumus being led into his prison cell. Welcoming the solitude, he collapses in slumber. He then has a dream, or perhaps a vision, involving the ghosts of four dead family members—relatives whom he never knew in life. The spirits are those of his mother,

---

* There are different spellings for this unusual name. It is "Imogen" in the 1623 folio, but some editors believe this to be a typo, and prefer "Innogen" (which would match the name of a ghost character in early editions of *Much Ado About Nothing*). Jachimo has alternate spellings, too (typically Iachomo and Giacomo). All excerpts here are from the recent Penguin edition, edited by John Pitcher, and I have adopted his spellings for the names of the characters.

father, and two brothers. As he lies in a daze, the ghosts move around him in a circle. (The stage direction says, "*They circle Posthumus round as he lies sleeping.*") Feeling Posthumus's anguish, they appeal to the Roman god Jupiter to come to his aid—and Jupiter obliges:

BROTHERS

Help, Jupiter, or we appeal, and from thy justice fly.

Jupiter descends in thunder and lightning, sitting upon an eagle. He throws a thunderbolt. The ghosts fall on their knees.

JUPITER

No more, you petty spirits of region low,
Offend our hearing. Hush! How dare you ghosts
Accuse the thunderer, whose bolt, you know,
Sky-planted, batters all rebelling coasts?

(5.4.62–66)

Jupiter continues to chastise the ghosts, and then gives them a book and instructs them to give it to Posthumus. Jupiter continues:

And so away. No farther will your din
Express impatience, lest you stir up mine.
Mount, eagle, to my palace crystalline.

He ascends into the heavens.

(5.4.81–83)

Shakespeare's plays cover a lot of ground, and employ many theatrical tricks—but as for gods descending from the heavens, this episode is unique; there is nothing else like it in the entire canon. Martin Butler calls the Jupiter scene the play's "spectacular high point," as it surely is. But the scene is also bizarre, unexpected, and extravagant—so much so that some have wondered if it represents Shakespeare's own work. Its authenticity, Roger Warren notes, "has often been questioned . . . on the grounds that it is a detachable episode and that the ghosts' speeches in particular are written in a style thought to be unworthy of Shakespeare." The consensus seems to be that it's as authentic as the rest of the play; still, the scene is so odd, and is so difficult to stage, that it is often cut in modern performances—*when* the play is performed, which is not very often.* Nonetheless, it was certainly acted on the Jacobean stage—no doubt pushing the skill of the backstage technical crews to the limit. (Because of its complex staging requirements, it may have been more readily performed at the fully enclosed Blackfriars Theatre rather than at the open-air Globe.) Butler suggests that the actor playing Jupiter would have been lowered in a chair from the ceiling (which was also, appropriately, "the heavens") using a crane-like mechanism operated via a winch hidden in the roof. Warren adds that "perhaps the head, wings, and claws of the eagle were a façade fixed to the front of the chair, so that Jupiter appeared to be *sitting upon an eagle*." The visual and auditory effects would have been more important than the acting: "Thunder" could be created by rolling a cannonball along a groove in the ceiling, while pyrotechnics (perhaps fireworks) delivered the necessary "lightning." (But the special effects aren't the only reason to imagine the play being staged at the Blackfriars; the intimacy of the "trunk scene" would lose much of its impact in the enormous space of the Globe.)

If anything in Shakespeare's late plays points to Galileo, this is it: Jupiter, so often invoked by characters in so many of the plays, never actually makes a personal appearance—until this point in *Cymbeline*. And of course Jupiter is not alone in the scene: Just below him, we see four ghosts moving in a circle. . . . Could the four ghosts represent the four moons of Jupiter, newly discovered by Galileo? Certainly the timeline seems to hold up: Although we don't know exactly when *Cymbeline* was written, the consensus is that it most likely dates from the summer or fall of 1610—in other words, it was written within the first few months (or at most half a year) after the publication of *The Starry Messenger.*

Remarkably, the idea of a connection between Galileo and Shakespeare's *Cymbeline* appears to have escaped scholarly attention for centuries—until roughly ten years ago, when three scholars, working independently, appear to have hit on the idea at about the same time. Scott Maisano, of UMass–Boston, discussed the *Cymbeline*–Galileo connection

* I feel privileged to have seen an uncut version at the Stratford Festival in Stratford, Ontario, in 2012. The godlike figure of Jupiter indeed descended on a giant eagle, just as the four-hundred-year-old stage direction calls for.

in a journal called *Configurations*, in the fall of 2004 (and in his PhD thesis, which was completed at about the same time), while John Pitcher, at Oxford, has given the idea its largest audience to date, addressing the matter in some detail in the new Penguin edition of the play, published in 2005. But they seem to have been beaten to the finish line (if only by a few months) by none other than Peter Usher, the astronomer we met in the previous chapter.

Usher's article, titled "Jupiter and *Cymbeline*," was published in the Spring 2003 edition of the *Shakespeare Newsletter.* Usher summarizes the bizarre happenings of the play's fifth act, noting the descent of Jupiter and the appearance of the ghosts: "These ghosts happen to be four in number, equal to the number of the Galilean moons." But that is just the beginning: Jupiter, as mentioned, gives a book to Posthumus (via the ghosts): "This tablet lay upon his breast, wherein / Our pleasure his full fortune doth confine" (Scene 4, 79–80). The book is not identified, though we later discover that it contains a message of hope for Posthumus, who will "end his miseries" while Britain will "flourish in peace and plenty" (112–13). However, Usher has an idea of what book we should imagine it to be: "The book placed on the bosom of Posthumus could represent Galileo's *Siderius Nuncius*, and the good fortune that is assigned could be praise for that part of the text that is genuinely new, the discovery of the Galilean moons."

"TO SEE THIS VAULTED ARCH"

For Usher, the ghosts-as-moons symbolism is just one element in a larger thesis. He goes on to interpret other lines from the play as suggestive of early Elizabethan telescopy, a subject we looked at briefly in the previous chapter. Later, in *Shakespeare and the Dawn of Modern Science*, he expands on the play's astronomical allusions, noting that "*Cymbeline* has mystical and cosmic overtones, and this chapter shows that it contains a subtext that chronicles the course of astronomical discovery in 1610." Once again, the English astronomer Leonard Digges and his son, the scientist Thomas Digges, are at the center of his argument. He speculates "that Posthumus's spirit is that of Thomas Digges, who by 1610 had graduated to a heavenly post and needed educating on new discoveries."

Of particular interest is a passage in act 1, scene 2, in which Jachimo meets Imogen for the first time. He has just arrived in England, and hands her a letter of introduction from her husband; she welcomes him. Seeing Imogen for the first time, Jachimo is struck by her beauty; winning the wager will be a pleasure. He concocts a story about how, back in Rome, Posthumus has forgotten she exists, and has been enjoying himself thoroughly (the implication is, with prostitutes). It's a lie, of course; Jachimo just wants to turn Imogen against her husband. He suggests her best course of revenge would be to sleep with *him* (naturally). The ploy fails; she gets mad; he pretends to have only been testing her, offers an apology, and stresses Posthumus's virtue—and for some reason she forgives him. But as he looks at her, he realizes the weakness of his own story:

> Thanks, fairest lady.
> What, are men mad? Hath Nature given them eyes
> To see this vaulted arch and the rich crop
> Of sea and land, which can distinguish 'twixt
> The fiery orbs above and the twinned stones
> Upon th'unnumbered beach, and can we not
> Partition make with spectacles so precious
> 'Twixt fair and foul?
>
> (1.6.30–37)

The passage seems to allude, at least in part, to the sights one might see in the heavens; at the very least, it has something to do with distinguishing different kinds of objects (including, it would seem, stars) from one another. But the context is crucial: The first line is spoken to Imogen; the remaining lines are clearly an aside, spoken only to the audience. He seems to be saying, *My story is unbelievable; why would Posthumus stoop so low, when his own wife is so beautiful?* After all, he reasons, the eye gives one the power to tell the stars apart, and even to distinguish one stone on the beach from another; can't Posthumus see the difference between his wife and a common whore? Usher passes over the sexual aspect of these lines, however, and focuses on the astronomical: The "vaulted arch" is surely the sky; the "fiery orbs above" must be the stars. Could the precious "spectacles" be a reference to a telescope-like device?

There is more: Usher sees a parallel between Galileo's observations of the night sky and Jachimo's observations in Imogen's bedroom. Jachimo emerges from the trunk in which he had been hiding, and begins to tally the various things he sees in her bedchamber. In a genuinely tricky passage, he refers to "ten thousand meaner movables." Jachimo has taken out a notepad, and begins to write down a description of all that he observes:

Such and such pictures; there the window; such
Th'adornment of her bed; the arras, figures,
Why, such and such; and the contents o'th'story.
Ah, but some natural notes about her body
Above some ten thousand meaner movables
Would testify, t'enrich mine inventory.

(2.2.25–30)

The footnotes in most editions suggest that the "ten thousand meaner movables" refers to small items of furniture. (Indeed, as a good dictionary will point out, "movables" can still mean "furniture" to this day, and the Oxford English Dictionary notes that Jonson had used the word in this sense in *Volpone* in 1607.) But Usher objects: Nobody has *that much* furniture in their bedroom. Instead, he sees it as a reference to the number of stars visible to the unaided eye (a number that, as mentioned in the previous chapter, also crops up in *Hamlet*).* This number, of course, is in the process of being rendered obsolete, as Galileo's telescope now reveals the existence of untold thousands of stars beyond those accessible to the unaided eye. Usher also points out that Jachimo emerges from a "trunk"—which was also one of the words used to describe a telescope-like device. (Usher points out that it can have a third meaning, too: It can refer to a person's midsection—for example, the headless body of the murdered Cloten, in act 4, scene 2.)

Of equal interest is Imogen's intriguing reference to an "astronomer" in act 3. She has just been handed a note from her husband, delivered by his servant, Pisano:

PISANO

Madam, here is a letter from my lord.

IMOGEN

Who, thy lord? That is my lord, Leonatus?
O learned indeed were that astronomer

---

* Roger Warren, in the Oxford edition, notes that "movables" can include small personal items as well as furniture (a claim supported by the OED). Still, ten thousand is a lot—if we are to take the number literally. Martin Butler, in the Cambridge edition, says we do not have to: "What counts is the evocation of a richly decorated bedchamber—and one broadly seventeenth-century in its furnishings" (Butler, *Cymbeline*, 120). Intriguingly, the OED also notes that "movables" had been used to denote the spheres in Ptolemaic astronomy, a fact that nobody seems to have linked to Jachimo's bedroom tally (and indeed ten thousand spheres would seem excessive in any planetary theory). See also my footnote on page 183.

> That knew the stars as I his characters—
> He'd lay the future open.
>
> (3.2.25–29)

In the standard interpretation of this scene, "astronomer" simply means "astrologer," and "characters" means "handwriting": *If only I could read the stars as easily as I recognize my husband's handwriting, I could know the future.* But Usher sees something else in these lines: He points to Thomas Digges's addition to his father's almanac, published in 1576, a book whose cover "is replete with zodiacal signs that could well be the 'characters' that Imogen's astronomer knew. . . . It is reasonable to suppose that 'the astronomer' refers to Thomas, and that (as posited) his subtextual representative on stage, Posthumus, has a store of information residing in memory comprised of contemporary celestial facts."

And, as mentioned in the previous chapter, Usher believes that the play contains references to sights that even Galileo never saw—such as the detailed structure of Saturn's rings. "*Cymbeline*," he concludes, "is a paean to the glories of the night sky revealed through telescopy."

A SCIENTIFIC ROMANCE

Scott Maisano and John Pitcher exercise a more restrained approach. They do not tackle the subject of early English telescope use, but instead focus on the appearance of Jupiter as a likely allusion to *The Starry Messenger.* Writing in *Configurations*, Maisano proceeds with caution:

> If it seems incongruous and unlikely, at first, for Shakespeare to have alluded to Galileo's startling scientific discovery at the conclusion of a play primarily set in Roman Britain, a millennium and a half before the invention of the telescope, it has seemed even more unlikely to many readers that Shakespeare would *not* have alluded to Galileo's discoveries, ever, in at least one of his plays.

Maisano cites the work of Marjorie Hope Nicholson, who, writing in the 1950s, finds no trace of the "new astronomy" in Shakespeare's works—even though he could hardly have been unaware of the latest discoveries:

> Shakespeare must have seen the new star of 1604, must have heard of Galileo's discoveries in 1610. . . . Yet his poetic imagination shows no response either to new stars or to other spectacular changes in the cosmic universe.

Maisano, like Usher, challenges that view. The appearance of Jupiter with the four ghosts, in the final act of *Cymbeline*, seems to have Galileo's fingerprints all over it. (As he put it when I spoke with him in Boston recently: "It seems an awfully big coincidence if it's *not* an allusion to Galileo.") Yes, the play is a "romance"—but a romance need not be a retreat from reality, Maisano explains. (Plus, the label "romance" was bestowed on the late plays only in Victorian times.) Maisano compares *Cymbeline* to a peculiar book written by Johannes Kepler the previous year. This was his *Somnium*—Latin for "Dream"—published in 1609. In the *Somnium*, the German scientist imagines what the Earth might look like from the moon, and how that view might change depending on the veracity of the Copernican model. This remarkable book has been called, among other things, the first work of science fiction, and yet it can also be seen as a vital work of science: It is here that Kepler first uses the word "gravity" in its modern sense. It could have been told in straight-ahead prose, but instead the entire story unfolds as a dream sequence. Kepler's vision may have been revolutionary, but his approach in communicating it in the *Somnium* was romantic, even antiquated. As Maisano puts it, "Kepler understood that in order to get at intellectual realities that defy our ordinary experiences . . . it is often necessary to resort to what looks on its surface like 'literary escapism.' " Similarly, Shakespeare's *Cymbeline*, composed just one year later, "might appear to be a backward-looking romance full of dog-eared devices from popular literature, but it is in reality a scientific romance." Incidentally, he agrees with Usher that the reference to a "learned . . . astronomer" does, in fact, point to a real-life astronomer—but while Usher sees it as referring to Thomas Digges, Maisano believes that "it is undoubtedly Galileo" who is being alluded to.

Another line in the play, from near the very end, catches Maisano's attention. One by one, the play's divergent plots come together, and the various loose ends are tied up: The misunderstandings are resolved, disguises removed, true identities revealed. King Cymbeline is overjoyed, but stunned. He asks, "Does the world go round?" (5.5.232). Maisano notes, "This is the only such utterance in Shakespeare's plays; and coincidentally, this precise question was part of intellectual discussion all across Europe in 1610."

And what of the book placed on Posthumus's breast? *The Starry Messenger* was perhaps the most provocative new book circulating at that time—but it was not the only one. As Maisano points out, scholars were also plugging away on another groundbreaking book, the King James version of the Bible, to be published the following year. The book mentioned in act 5 of *Cymbeline* may, he speculates, be an allusion to the new bible—an appropriate gift, perhaps, for characters in a pagan setting seeking to improve their fortunes. (Again the dates are a bit tricky; as Maisano notes, although *Cymbeline* likely dates from late 1610, the first known performance occurred the following year.) There is another link to connect these two books: The Copernican view, now supported by Galileo's observations, would soon call into question the interpretation of scripture (a subject Galileo would address in his *Letter to the Grand Duchess Christina* in 1615). This, of course, would eventually get Galileo into trouble; but already the tension could be felt. Shakespeare, Maisano writes, "calls our attention to how this new universe of unimaginable size fundamentally alters the human predicament." The playwright, he says, "seems to have set the two revolutions—Christian and Copernican—purposely and provocatively side by side."

Like Maisano, John Pitcher of Oxford sees *Cymbeline* as Shakespeare's attempt to come to grips with a changing world, a universe opened wide by the scientific discoveries of the day. The play involves fathers and kings and gods—throughout history, figures of authority—but all now finding their leadership challenged "by the evidence of modern experimental science." As Pitcher argues in the introduction to the Penguin edition of the play (2005), the Jupiter scene is almost certainly a reference to the discoveries newly announced by Galileo. Previous scholarly editions of the play (and there have been many) have, as far as I can tell, left Galileo out completely. Editors of course noticed the cosmological allusions in the play's dramatic climax, but seemed content to address it in Ptolemaic terms. When Jupiter returns to his "palace crystalline," for example, Martin Butler notes that "in Ptolemaic cosmology the 'crystalline heaven' was one of the universe's outermost spheres, next to the firmament."

For Pitcher, however, the nod to Galileo is more than a one-off allusion. The very essence of *Cymbeline*, he argues, involves the playwright's confrontation with a new worldview, a new way of thinking now sanctioned by the discoveries of Galileo and other promoters of the "new philosophy." Shakespeare is also being forced to give something up—to abandon ancient ways of thinking that, beginning in 1610, were no longer tenable:

> In that year, because of Galileo . . . the universe was finally proved to be not an enormous glass ball with the earth at its center but an expanding infinitude of galaxies, each packed with stars. It took a century or more for the old father, ruling in European courts and churches, to be unseated by this extraordinary scientific discovery, but everyone in the know realized its significance from the start, including Shakespeare.

This is a bold argument: Galileo looks through his telescope, the world suddenly changes, and Shakespeare knows it. The transformation began with Copernicus writing on the revolutions in the sky; soon, Pitcher is suggesting, there will be revolutions of a more dangerous kind, with political and religious orders turned on their heads. (He is perhaps getting ahead of himself with the reference to galaxies, whose nature was not understood in Shakespeare's time.*) But Pitcher's assertion is also the clearest statement yet from a mainstream Shakespeare scholar that *Shakespeare knew what was going on in science, and that this knowledge is reflected in his plays.*

Intriguingly, Shakespeare manages to allude to the new astronomy in a scene built almost entirely from elements of ancient mythology and Ptolemaic cosmology. Jupiter is not only a planet but a god; and when he appears he refers explicitly to his "palace crystalline." Moreover, the play is set not in Renaissance England but in ancient Britain. When Jupiter makes his appearance, Pitcher says, it is intended "as a deliberate and subtle twist in the game of old and new being played out constantly in *Cymbeline*." What Galileo has seen with his telescope is crucial, but what

* Galileo showed that *our* galaxy was composed of a multitude of stars; but he didn't know that it *was* a galaxy, or that other galaxies existed. While many "nebulae" were seen in the night sky, the fact that some of them are galaxies external to our own Milky Way, and that they each contain billions of stars, wasn't realized until the early decades of the twentieth century. (However, a few bold thinkers had guessed as much, more than 150 years earlier, the philosopher Immanuel Kant among them.)

he *doesn't* see may be just as relevant: There is no sign, for example, of the crystalline sphere through which Jupiter must pass in order to make his descent onto Shakespeare's stage. "If Galileo's telescope was correct," Pitcher writes, "the crystalline roof had been an illusion all along." In addition, Pitcher, like Maisano, sees something not-so-subtly Copernican in King Cymbeline's question, "Does the world go round?" The Earth's alleged movement through the heavens was, for an Elizabethan audience, every bit as disorienting as the discovery of new stars in the night sky. And, like Usher, he suspects the play also alludes to the new stars revealed by Galileo's telescope. The "stones on the beach / stars in the sky" comparison, in the "vaulted arch" speech from act 1, may be "possibly a way of saying that the stars too are uncountable." (Mind you, the comparison did not originate with Galileo; Pitcher points out that it has its roots in Genesis.)

A further note on *The Starry Messenger* is in order: While it was written in Latin, it would not have been a difficult read for educated Englishmen. Galileo's book, Pitcher writes, was "a scientific publication written in very simple Latin, unimpeded by courtly rhetoric, and illustrated with clear plates." I explored this point in more detail when I spoke with Pitcher in his office at St. John's College at Oxford.* "I think by the time Shakespeare had finished his grammar-school education, he would have had a reading fluency for Latin prose that matches the achievements of our undergraduates after, say, five terms," he told me. In school, "Shakespeare would have done very little else *but* learn Latin." Moreover, *The Starry Messenger* is not a difficult book. "The Latin is 'schoolboy Latin,'" Pitcher says. "The learned community, and the not-so-learned community, will be able to open up that book and know what it means." As mentioned, the book was distributed far and wide; it made such a splash that everyone who was anyone would have been discussing it. "I think it's the kind of stuff that would have been talked about in alehouses," Pitcher muses.

Here is a destination for your time machine: an alehouse in London—maybe the famous Mermaid, in Cheapside, which Ben Jonson and other playwrights are known to have frequented—circa April 1610. Mugs and

---

* When it comes to professors' offices—and I've seen quite a few—it is hard to compete with Oxford. The walls of Pitcher's quarters are painted eggshell blue, with white trim and lots of what Douglas Adams would have called "fiddly bits." The chairs and sofa, deep burgundy in color and thickly padded in the extreme, would not be out of place in M's office as portrayed in one of the early James Bond films.

plates clatter; a knight argues with a tailor; a silversmith's apprentice tries to chat up a barmaid; a vagabond looks for any unguarded foodstuffs or coin purses. A group of actors are making merry at a table in the corner. One of them, sitting at the end of the table, is an actor who is also a playwright. Two strangers walk in; the playwright doesn't recognize them. One of them has just returned from Italy, and is enthusiastically describing his adventures to his companion. He pulls out a small book; they start talking about its remarkable claims, and pointing to its crisp, copperplate engravings. . . . The playwright leans forward. *What was that about Jupiter again?*

Shakespeare, we can be sure, wouldn't have settled for just *hearing about* this remarkable book: Here was a provocative little pamphlet from Italy, describing sights never seen before in the heavens, *with pictures.* He would have wanted to see it with his own eyes. Some of those in Shakespeare's audience—not all, of course—would also have seen *The Starry Messenger*; many others would have at least heard of it. With up to a thousand people at each performance of *Cymbeline*, Pitcher speculates, you could easily have a couple of hundred who would understand a reference to Galileo's book.

Naturally, I was curious what Stephen Greenblatt, perhaps the best-known Shakespeare scholar in America today, might have to say about these interpretations of *Cymbeline*, and, more generally, the suggestion that Shakespeare's plays contain allusions to the "new philosophy." He's heard most of the theories at least in broad outline; he hadn't read Usher's work, however, so I summarized it as best I could. As the rain poured down in Harvard Square just beyond his window, he said that he wasn't quite willing to commit, explaining that, as a general rule, he is "somewhat allergic" to treating works of literature (not just Shakespeare) "as a kind of esoteric allegory." I asked him specifically about Jupiter's appearance in act 5 of *Cymbeline*: Might the scene have been intended as an allusion to the discoveries announced in *The Starry Messenger*? Greenblatt concedes that it's "a very strange moment in Shakespeare," and that it requires some sort of explanation. "I suppose it's conceivable," he said.

Looming over both *The Starry Messenger* and *Cymbeline* is the question of authority: who has it, who can challenge it, and where one might seek the truth. Galileo's discoveries question the supremacy of ancient teachings, and, by extension, those who supported and propagated those views; in *Cymbeline*, Shakespeare questions a whole array of once-unassailable authorities, from fathers to kings and beyond. Galileo's discoveries made another set of ancient beliefs equally obsolete. Paradigms that had managed to escape the slings and arrows of the past fifteen centuries now lay in ruins. "The authority of Jupiter, of the old king, of God—it's done with, finished," Pitcher says. "It's all back to human beings now."

10. "Treachers by spherical predominance . . ."

THE ALLURE OF ASTROLOGY

The parish register from Holy Trinity Church in Stratford—the site of Shakespeare's baptism, and his burial—does not make for compelling reading; normally, it is an endless tally of births, marriages, and deaths. Next to the date of July 11, 1564, however, the register contains these words: *Hic incipit pestis* ("Here begins the plague"). It is hard to imagine the terror that lurks behind these three little words, written three months after the playwright's birth. A tenth of the town's population was dead within six months. In London, the victims numbered in the tens of thousands. The plague was a recurring menace in Shakespeare's England, and no one knew when the next deadly visitation might come. When an outbreak occurred in 1593, the historian William Camden, a contemporary of Shakespeare, made a careful note of the circumstances. It was not the overcrowded streets or the lack of hygiene that captured his attention; rather, he noted that "Saturn was passing through the uttermost parts of Cancer and the beginning of Leo"—just as they had thirty years earlier, during another deadly outbreak.

The stars were never the sole explanation for human misfortune; disasters could also be interpreted as punishment by God for various moral transgressions (as countless pamphlets from this period show). Everyone agreed, however, that the movements of the stars and planets were a critical factor, and that their motion, along with the appearance of meteors or comets, could be read as portents of terrestrial events to come. One ignored such heavenly signs at one's peril. As Kirstin Olsen puts it, "People watched the sky with the same jumpy intensity of Wall Street analysts watching economic indicators; a bad omen could cause public confidence to plummet." Clearly the sun, moon, stars, and planets held power over people's lives. As Kent declares in *King Lear*, "It is the stars, / The stars above us govern our conditions. . . ."* And Kent is not alone in pondering heavenly influences. In the same play, Gloucester refers to "these late eclipses of the sun and moon," believed to be a reference to actual eclipses in September and October 1605. But note the rest of the line: These celestial events "portend no good to us" (1.2.91). Clearly there is little hope for disentangling astronomy from astrology: To Shakespeare—or at least to his audience—a profound link is suggested between celestial happenings and human affairs, a connection that would have made perfect sense to even the best-educated of Shakespeare's countrymen. Astrology was by far the most prevalent form of magic in Shakespeare's day; in fact, it simply reflected the prevailing wisdom of the time. As Olsen notes, anyone who denied the power of astrology would have been adopting "a fringe position." Man and nature, earth and sky, microcosm and macrocosm: It was all connected, and the profound influence of the stars and planets was not to be taken lightly.†

"THERE WAS A STAR DANCED"

Astrology loomed large in Shakespeare's world, but it was hardly new. The Babylonians had laid the groundwork three thousand years earlier; the system was further developed by the Greeks and Romans, and then by Arab astrologers in the Middle Ages. In England, astrology came to have two more or less distinct branches, known as "natural astrology" and "judicial astrology." Natural astrology was, in fact, something like straight-ahead astronomy; it focused on tracking and predicting the motions of the sun, moon, and planets. Judicial astrology was closer to what we think of today as just plain "astrology"—the attempt to link celestial happenings to earthly affairs, and to use astronomical knowledge to predict terrestrial happenings. (To avoid confusion, I will put natural astrology aside, and use the term "astrology" to refer solely to "judicial astrology.")

---

* The line is one of those tricky passages that occurs only in the 1608 quarto edition of *King Lear*, and not the 1623 folio edition. It is thus absent from the New Cambridge edition (ed. Halio), but can be found, for example, at 4.3.33–34 in the Arden *Complete Works*.

† Once again we must recall that an Elizabethan could view the heavens far better than we can, from our light-polluted skies. As you read this sentence, are you conscious of the phase of the moon? Do you know if Venus is currently in the evening sky, or in the morning sky? A villager in sixteenth-century England would have known. In Shakespeare's time, the nation was gradually becoming more urbanized, but artificial street lighting was still a century away. The night sky would have been a captivating sight on any cloudless night.

How, exactly, would the motions of the heavenly bodies affect human affairs? We must recall, first of all, the prevalence of the geocentric worldview: Most people believed that the Earth was the center of the universe, and that the stars revolved around the Earth. This by itself must have lent considerable weight to astrological beliefs. Also note that the sublunar world—the corruptible, changeable earth and its equally imperfect environs—was thought to be composed of the four elements: earth, air, fire, and water. These were in a state of constant flux, but their motion was thought to be governed by the pristine (and yet complex) motion of the heavenly spheres. This explains why any attempt to separate astrology from any other branch of "science" would have been a meaningless pursuit in Shakespeare's day. Astrological thinking, as Keith Thomas notes, "pervaded all aspects of scientific thought." It should not be thought of as an isolated discipline, but as "an essential aspect of the intellectual framework in which men were educated." Astrology, as J. A. Sharpe puts it, "had a fair claim to being the most systematic attempt to explain natural phenomena according to rigorous scientific laws then in existence." In other words, astrology, in Shakespeare's England, was seen as a scientific pursuit, and a rigorous one at that. Indeed, many of the scientists we've been looking at (again, noting that "scientist" is an anachronistic term) transitioned effortlessly between astrology and astronomy. Tycho Brahe, Johannes Kepler, Thomas Digges, and John Dee were all, to some extent, astrologers as well as astronomers. (Writing in 1570, Dee notes that "man's body, and all other elementall bodies, are altered, disposed, ordered, pleasured and displeasured, by the influentiall working of the Sunne, Mone and other Starres and Planets.") Dee consulted astrological charts to determine the best day for Elizabeth's coronation, and was called on to offer his views on the political significance of the comet of 1577. (Elizabeth herself asked that horoscopes be cast for her suitors, and used astrology to assess potential heirs.) As we've seen, the new star of 1572, and the Great Comet of 1577, were imagined to be fraught with astrological significance. Even Francis Bacon, now regarded as one of the key figures of the Scientific Revolution, seems to have embraced astrological thinking. (It has been argued that Copernicus and Galileo were exceptional in their *lack* of interest in astrology—although Galileo did cast horoscopes for his Medici patrons; presumably he had little choice in the matter.) But it was not just an educated man's hobby: The widespread popularity of astrology is reflected in the sales of almanacs, which were filled with astrological prognostications along with astronomical data; they outsold even the Bible. Astrology lay at the very foundation of humankind's attempt to understand the universe.

In an age when scientific explanations were in short supply, one can see astrology's appeal. After all, the sun and moon (if not the stars and planets) actually *do* play a vital role in regulating life on Earth. The sun, of course, provides warmth and light and, indirectly, is responsible for the wind and weather patterns, while the moon (together with the sun) controls the tides. The motion of the sun and moon, along with the lunar phases, was perfectly predictable, and farmers required an intimate knowledge of this cycle. A doctor would have been well aware that certain kinds of illness—say, a bronchial condition—would be more common in winter than in summer. But the connections were imagined to run much deeper. For example, the moon was believed to control not only the tides, but also the moisture in a person's body, including the humors thought to govern health and sickness (we will look at medicine more closely in Chapter 12); moreover, a person's personality, and even their actions at specific moments, were thought to be controlled, or at least swayed, by celestial influences.

To the astrologer, much depends on the positions of the planets at the time of a person's birth; the planets, like the moon, were thought to affect the person's natural humors, which in turn make one more likely, or less likely, to be influenced by various passions, and to be predisposed to either good or evil. The free-spirited Beatrice, in *Much Ado about Nothing*, notes that "there was a star danced, and under that star I was born" (2.2.316). It is true that, by force of will, one might overpower these predispositions; but as one Elizabethan astrologer put it, "the most part of men doe follow their affections, and there are but fewe that doe master and overrule them." It's not that the stars left one with no freedom; but it was only prudent to be fully aware of the various cosmic forces pushing and pulling on each individual as they navigated through life's decisions, big and small. As another practitioner put it, "An expert and prudent astrologer may through his cunning skill show us how to prevent the many evils proceeding from the influence of the stars." The skilled astrologer could also advise on the best time to perform certain activities, such as embarking on a long journey, choosing a wife, or having a baby. Here, for example, is a seventeenth-century tip for siring a male child: "If thou want'st an heir, or man-child to inherit thy land, observe a time when the masculine planets and signs ascent, and [are] in full power and force, then take thy female, and cast in thy seed, and thou shalt have a man-child."

We should not be surprised to learn that, in 1599, Shakespeare's company consulted an astrologer to decide on the best day to open the Globe Theatre. (They settled on June 12, which was the summer solstice and also a new moon.) Of course, astrologers needed to hedge their bets: They never offered certainty, only probabilities.* Purposeful ambiguity was the norm. A prediction might fail to come true, and yet one could not label the astrologer or the almanac as "wrong." (Indeed, as early as 1569 an English pamphlet had mocked the almanac makers by publishing three differing predictions from popular almanacs of the day.) Moreover, if you shell out enough predictions, some are bound to come true—as Montaigne observed in his *Essays*: "I know people who study their almanacs, annotate them and cite their authority as events take place. But almanacs say so much that they are bound to tell both truth and falsehood."

Astrology permeates the Shakespeare canon, with its endless references to celestial happenings and their earthly significance. The playwright understood, and exploited, the traditional symbolism associated with each of the heavenly bodies, linking the sun with masculinity and kingship; the moon with femininity, changeability, and of course madness ("lunacy"). The planets had their purported domains of influence, as did the twelve constellations of the zodiac in which they appeared. The characters in *Henry VI, Part 1* speak of "planetary mishaps" (1.1.22) and "adverse planets" (1.1.54); and in *The Winter's Tale*, Hermione bemoans:

> There's some ill planet reigns.
> I must be patient till the heavens look
> With an aspect more favourable.
>
> (2.1.105–7)

We have already looked at Helena's verbal sparring with Parolles in *All's Well That Ends Well*, in which she suggests that he was "born under a charitable star," but then goes on to insist that the star was Mars "when he was retrograde," an astrological as well as an astronomical reference. In *Richard III*, the king expresses a straightforward desire for a celestial "blessing" of his ambitious political maneuverings: "Be opposite, all

* While this sounds like a cop-out, we might note that, in many branches of science, probability (albeit educated probability) is still the norm: A doctor gives the chances that the patient will have a heart attack over the next ten years; the geologist estimates the odds of a major earthquake in the next century; an economist speaks of the risk of a recession. . . .

planets of good luck, / To my proceeding . . ." (4.4.402–3). And of course the entire plot of *Romeo and Juliet* is focused (as the prologue tells us) on the fate of the "star-crossed lovers." One senses that the playwright knew the basics of astrological practice nearly as well as the almanac writers.

"STARS WITH TRAINS OF FIRE AND DEWS OF BLOOD"

Astrology was intertwined with magical thinking in general, and with connections between humanity and nature in particular. In Shakespeare's world, these connections take center stage when important people are involved. In *Henry IV, Part 1*, the king chastises his son, Prince Hal, for behavior unbecoming of the heir to the throne. He says that in his prime, he was ". . . seldom seen, I could not stir / But, like a comet, I was wondered at; / That men would tell their children, 'This is he!'" (3.2.47–48). A similar sentiment can be found in Marlowe's *Tamburlaine the Great, Part 2*, where Orcanes, the king of Natolia, dismisses Tamburlaine as a "shepherd's issue, baseborn." Tamburlaine counters that although he was indeed of humble birth, "Heaven did afford gracious aspect / And joined those stars that shall be opposite / Even till the dissolution of the world . . ." (3.5.80–82). Kings and princes were, in a sense, heavenly beings. The stars could be expected to take notice when they were born, and when they died. Should they meet a premature end, even greater disruptions could be expected. In *Macbeth*, when the king is murdered, mayhem—both celestial and terrestrial—follows. Ross, one of the Scottish chiefs, asks an old man if

> Thou seest the heavens, as troubled with man's act,
> Threatens his bloody stage. By th'clock 'tis day
> And yet dark night strangles the traveling lamp.
> (2.4.5–7)

The old man agrees that "'Tis unnatural, / even like the deed that's done" (lines 10–11), and then the conversation turns from the celestial to the earthly, as he describes a bizarre sight that he's witnessed, involving a falcon and an owl. But Ross can top that: He observed that Duncan's horses "Turned wild in nature, broke their stalls, flung out . . ." (line 16) and finally—but only after prompting from the old man—the kicker: *The horses ate each other.* (You'd think Ross would have *started* the conversation with that item, wouldn't you?) "O horror, horror, horror," wails Macduff (2.3.56), comparing the murdered king's wounds to a breach in nature itself.

We have already noted Calpurnia's observation in *Julius Caesar*: "When beggars die there are no comets seen; / The heavens themselves blaze forth the death of princes" (2.2.30–31). Of course, Caesar himself meets an untimely end—an event recalled by Horatio, in *Hamlet*, who notes the upheavals in the earth and wonders in the sky that accompanied Caesar's murder:

> A little ere the mightiest Julius fell,
> The graves stood tenantless and the sheeted dead
> Did squeak and gibber in the Roman streets
> As stars with trains of fire and dews of blood,
> Disasters in the sun, and the moist star
> Upon whose influence Neptune's empire stands
> Was sick almost to doomsday with eclipse.
> (1.1.113–19)

Remember, Horatio is the wise "scholar" in *Hamlet*. But then, astrology purported to be an empirical, objective science; it was exactly the sort of thing a scholar *would* study. As one observer wrote in 1600, "Nowadays among the common people [one] is not judged any scholar at all, unless he can tell a man's horoscopes, cast out devils, or hath some skill in soothsaying." The more one studied astrology, the better one (supposedly) became at its practice. Astrology was, as Thomas notes, "probably the most ambitious attempt ever made to reduce the baffling diversity of human affairs to some sort of intelligible order"; it provided "a coherent and comprehensive system of thought."

GOD VS. THE STARS

Astrology was ubiquitous—but it was also controversial. Not because it was antiscience, of course; if anything, it was imagined to be *a part of* science. We've noted that there was no "war" between science and religion in early modern Europe; even so, some aspects of astrology did seem to pose a threat to Christianity. The central question was how much faith one ought to place in astrology's predictive power. As Paul Kocher notes, judicial astrology, which purported to govern all of the major events in one's life, "often looked suspiciously like a rival of religion contending for the emotional loyalty of mankind." After all, the Church of England nurtured the idea of God having a specific plan for every man, woman, and child in the land. The extreme version of this was the notion of "predestination"—the idea that everything that happens has already been determined by God, who had known the destiny of each living creature, even before the universe had come into being. By this line of reasoning, if it was God's will that something occur, then no course of action on the part of mortal human beings could prevent it coming to pass. The sentiment is suggested in *Julius Caesar*: Calpurnia warns her husband that he is in danger, but Caesar refuses to change his plans. Why bother? "What can be avoided / Whose end is purpose'd by the mighty gods?" (2.2.26–27). And more famously by Prince Hamlet, as he riffs on the Gospel according to Matthew (10:29):

> There is special providence in the fall of a sparrow. If it be now, 'tis not to come; if it be to come, it will not be now; if it be not now, yet it will come. The readiness is all. Since no man, of aught he leaves, knows aught, what is't to leave betimes? Let be.
>
> (*Hamlet* 5.2.215–20)

One approach was to compromise, and assert that God and the stars somehow worked together; that their powers were complementary. As Sir Walter Raleigh once said:

> If we cannot deny but that God hath given virtue to springs and fountains, to cold earth, to plants and stones, minerals and to the excremental parts of the basest living creatures, why should we rob the beautiful stars of their working powers? For, seeing they are many in number and of eminent beauty and magnitude, we may not think that in the treasury of his wisdom who is infinite there can be wanting, even for every star, a peculiar virtue and operation; as every herb, plant, fruit, flower, adorning the face of the earth hath the like.

While anyone might get caught up in the allure of astrology's (seeming) predictive power, a clergyman or theologian who did so was likely to draw condemnation from his superiors. (As Kocher notes, all five of the harshest polemics against astrology published in Elizabethan England were written by churchmen.) If the clergy were against astrology, who was in favor of it? Surprisingly (to the modern reader) it was the people we would now call scientists. Doctors, in particular, seemed to be its staunchest defenders. But it all depended on how far one presumed to go with one's astrological forecasting. The more specific the predictions, the greater the controversy, for the simple reason that such predictions seemed to pose the most direct challenge to God's own power over the lives of men and women. But of course it was exactly such specific predictions that people yearned for: No one cared all that much about general good or ill omens suggested by heavenly patterns; rather, they wanted to know what lay in store *for them*. When a woman came knocking on the astrologer's door, Kocher writes, she came

> not to hear about the stars as universal causes but to ask what lay in store for her, whether she was to inherit that pewterware from her aunt, find the ring she had lost, or marry the handsome stranger who had been eyeing her lately. The noble lord who summoned the astrologer to his castle to cast a horoscope for his new heir did not happen to be interested in a calculus of probabilities larded with "ifs," "buts," and "maybes." He demanded positive information about the boy's education, his marriage, and career.

While the published almanacs of the day tended to avoid making overly specific predictions, individual astrologers almost certainly went much further—especially if there was money to be made by telling clients what they wanted to hear. And if the client had experienced misfortune, there was comfort in knowing that the stars weren't aligned in one's favor: It's not that you're foolish, or lazy, or have poor judgment; the cosmic deck was merely stacked against you.

"THE EXCELLENT FOPPERY OF THE WORLD"

But there were always doubters: It was easy enough to see that the astrologers' predictions were wrong as often as they were right; and anyway the positions of the stars and planets were only known with limited precision. Besides, how many other factors might influence a person's destiny to an equal or larger degree? And even if an astrologer's predictions were occasionally borne out, might it not be a coincidence? Today, statistics professors (and science journalists) routinely caution that correlation does not imply causation—but a skeptic named William Perkins made the same point in 1585 when he wrote, "For in those thinges which happen together, the one is not the cause of the other." Chaucer, writing two centuries earlier, pokes fun at astrology (and, one might argue, astronomy) in *The Miller's Tale*, where an astrologer walks into a field to gaze at the stars, in the hope of seeing the future—and promptly falls into a well:

> So fared another clerk with astromy;
> He walked into the meadows for to pry
> Into the stars, to learn what should befall,
> Until into a clay-pit he did fall;
> He saw not that.

Shakespeare, it would seem, also found astrology deserving of some degree of mockery. His characters, at least, harbor doubts. It's worth focusing on *King Lear*, in which the key figures themselves debate the power of astrological thinking. The crucial scene comes in act 1, scene 2. As we've seen, Gloucester is concerned about the meaning of the eclipses recently observed in the heavens; he is afraid that they "portend no good to us." He goes on to list all of the disasters that such events may signal: "Love cools, friendship falls off, brothers divide. In cities, mutinies; in countries, discord; in palaces, treason; and the bond cracked 'twixt son and father" (1.2.94–96). For Gloucester, unusual sights in the heavens cannot be mere coincidence; they *must* be understood in relation to terrestrial happenings. They serve as signs, as forecasts, as warnings. Disharmony in the heavens will bring disharmony on earth, as surely as night follows day. As it turns out, of course, Gloucester has more to fear from his evil son, Edmond, than from the movements of the heavenly spheres. As soon as he has left the room, Edmond, in a remarkable speech directed only to the audience, dismisses his father's superstitious beliefs:

> This is the excellent foppery of the world, that when we are sick in fortune, often the surfeits of our own behaviour, we make guilty of our disasters the sun, the moon, and stars; as if we were villains on necessity, fools by heavenly compulsion, knaves, thieves, and treachers by spherical predominance, drunkards, liars, and adulterers, by an enforced obedience of planetary influence; and all that we are evil in, by a divine thrusting on. An admirable evasion of whoremaster man, to lay his goatish disposition on the charge of a star! My father compounded with my mother under the Dragon's tail, and my

> nativity was under *Ursa major*, so that it follows, I am rough and lecherous. I should have been that I am had the maidenliest star in the firmament twinkled on my bastardizing.
>
> (*King Lear* 1.2.104–16)

It's a wonderful passage, and the footnotes help us through some of the trickier phrases. In the New Cambridge edition, for example, we learn that "heavenly compulsion" means "astrological influence," and that "spherical predominance" means "planetary influence" (think of the heavenly spheres that carry the sun, moon, and planets). In other words, Edmond is a hardened skeptic, mocking his father's superstitious worldview in a fashion that would make Carl Sagan proud (or these days, Richard Dawkins). As David Bevington notes, Edmond's contempt for astrology "is likely to strike us as appealingly modern":

> He scorns the platitudes of his elders as without foundation. He is a true skeptic in the sense of interrogating received opinion, refusing to accept its timeworn notions without objective verification. In his view, such a verification would be impossible because the older ideas are worthless. They are myths, in his view, that human society invents to perpetuate privilege and hierarchy.

We find a similar view expressed more succinctly by Cassius, in *Julius Caesar*, who asserts that "The fault, dear Brutus, is not in the stars / But in ourselves, that we are underlings" (1.2.139–40). And again, from Helena, in *All's Well That Ends Well*:

> Our remedies oft in ourselves do lie,
> Which we ascribe to heaven; the fated sky
> Gives us free scope; only doth backward pull
> Our slow designs when we ourselves are dull.
>
> (1.1.216–19)

Shakespeare's characters understood astrology's appeal: It gets you off the hook. They also saw that, as often as not, the astrologers' predictions are flat-out wrong. Anyone who takes the trouble to examine their work seriously—as Shakespeare seems to have done—is led to an inescapable conclusion: The astrologer has no clothes. And if the astrologer is naked, so, too, is the magician, and perhaps anyone else claiming to have insights into the supernatural.

SHOW ME THE MAGIC

Consider the case of a great ruler who claims to have mystical powers; whose birth supposedly shook the heavens. We return to *Henry IV, Part 1*: As act 3 opens, King Henry is confronted by rebels on all sides—mainly Welsh and Scots, but also (to complicate matters) a smattering of Englishmen with loyalty issues. The rebel leader, Owen Glendower, boasts that his prowess on the battlefield is due to his magical powers; indeed, the forces of nature put on something of a spectacle at the moment of his birth. But young Harry Percy, known as Hotspur, isn't buying it. The two men, each with substantial egos, debate the matter at some length:

> GLENDOWER
>
> At my nativity
> The front of heaven was full of fiery shapes,
> Of burning cressets, and at my birth
> The frame and huge foundation of the earth
> Shaked like a coward.
>
> HOTSPUR
>
> Why, so it would have done at the same season, if your mother's cat had but kittened, though yourself had never been born.
>
> GLENDOWER
>
> I say the earth did shake when I was born.
>
> HOTSPUR
>
> And I say the earth was not of my mind,
> If you suppose as fearing you it shook.
>
> GLENDOWER
>
> The heavens were all on fire, the earth did tremble.

HOTSPUR

O, then the earth shook to see the heavens on fire,
And not in fear of your nativity.
Diseased nature oftentimes breaks forth
In strange eruptions; oft the teeming earth
Is with a kind of colic pinched and vexed
By the imprisoning of unruly wind
Within her womb, which, for enlargement striving,
Shakes the old beldam earth and topples down
Steeples and moss-grown towers. At your birth
Our grandam earth, having this distemperature,
In passion shook.

GLENDOWER

Cousin, of many men
I do not bear these crossings. Give me leave
To tell you once again that at my birth
The front of heaven was full of fiery shapes,
The goats ran from the mountains, and the herds
Were strangely clamorous to the frighted fields.
These signs have marked me extraordinary;
And all the courses of my life do show
I am not in the roll of common men.

(3.1.12–41)

No wonder Glendower is getting agitated: In case you missed it, Hotspur has just compared the Welshman's "magical" birth to a giant fart (". . . oft the teeming earth / Is with a kind of colic pinched and vexed / By the imprisoning of unruly wind" causing "strange eruptions . . ."). But Glendower won't back down, at which point Hotspur confesses that "there's no man speaks better Welsh"—a put-down, based on the reputation of Welshmen as boastful liars—and another rebel leader, Mortimer, cautions Hotspur not to provoke Glendower any further, lest he "make him mad." But as they prepare to retire to dinner, Glendower just can't let it go—and neither can Hotspur:

GLENDOWER

I can call spirits from the vasty deep.

HOTSPUR

> Why, so can I, or so can any man;
> But will they come when you do call for them?
>
> (3.1.51–53)

Hotspur is every bit as skeptical as Edmond. Indeed, he is probably even closer to being a twenty-first-century-style skeptic, along the lines of Neil deGrasse Tyson or Lawrence Krauss, than Gloucester's bastard son: Confronted with wild, outlandish tales of cataclysmic upheavals in the earth and sky, Edmond says, in effect, *I don't believe you. First of all, you're probably exaggerating; but even if something weird happened at the time of your birth, it's a coincidence. It can be explained through natural forces, and has nothing to do with you. You are just a man, as I am.* Of course, Edmond and Hotspur are "bad guys"; they are generally unsympathetic characters, and we cheer when they meet their comeuppance. But that fact doesn't necessarily mean that Shakespeare sympathized with astrology's supporters rather than its detractors. Indeed, we can see why Shakespeare chose to employ characters on both sides of the issue—those who embraced the idea of stellar influences on human society, and those who questioned it. As Thomas McAlindon puts it, Shakespeare endowed his characters with "cosmic imagination"; they speak "to and of the elements, the stars, the sun, the moon, and 'all the world.'" This achieves at least two goals: It makes their predicaments seem more intense, and it also makes them seem more relevant. It is, McAlindon says, "part of an endeavour to connect the tragic fate of the individual with the structure and dynamics of universal nature." In other words, Shakespeare has found a way of taking grand stories and making them even grander. His plays are, in quite a literal sense, universal.

THE FADING STARS

Astrology maintained a grip over the popular imagination for centuries, if not millennia—but that grip inevitably weakened, especially in the decades following Shakespeare's death. The solar eclipse of March 29, 1652, makes a useful case study. The date of the eclipse was known in advance, but not its consequences; as the day grew closer, people in England talked about little else. About one-quarter of the publications issued that spring were devoted to the eclipse and its implications. The day before the event, the Lord Mayor of London and a group of aldermen listened to a lecture on the eclipse. The diarist John Evelyn was skeptical. He observed the general air of panic in the capital, noting that the people were so alarmed that "hardly any would work, none stir out of their houses, so ridiculously were they abused by knavish and ignorant star-gazers." The rich fled; the poor gave away what few possessions they had, "casting themselves on their backs, and their eyes towards heaven and praying most passionately that Christ would let them see the sun again, and save them." Then came the eclipse—and, other than the sky becoming darker for a short time, nothing much happened. As another diarist noted, the ultimate effect of the eclipse had been to discredit the astrologers, who "lost their reputation exceedingly." Astrology's grip had loosened further by the time Jonathan Swift offered this indictment in 1708:

> For their observations and predictions, they are such as will equally suit any age or country in the world. "This month a certain great person will be threatened with death or sickness," This the newspapers will tell them; for there we find at the end of the year, that no month passes without the death of some person of note; and it would be hard if should be otherwise, when there are at least two thousand persons of note in this kingdom, many of them old, and the almanac-maker has the liberty of choosing the sickliest person of the year, where he may fix his prediction. . . . Then, "such a planet in such a house shows great machinations, plots, conspiracies that may in time be brought to light," after which, if we hear of any great discovery, the astrologer gets the honour; if not, his predictions still stand good.

Swift's indictment gets to the heart of the astrologer's craft: One's predictions could fall short, with little or no accountability. After all, no science was perfect; medical doctors hardly had a better track record for improving people's lives. It's true that different astrologers may come up with starkly different predictions—but then, this could happen with physicians, theologians, and lawyers, too. "The paradox," as Keith Thomas notes, "was that the mistakes of any one astrologer only served to buttress the status of the system as a whole, since the client's reaction was to turn to another practitioner to get better advice, while the astrologer himself went back over his calculations to see where he had slipped up."

There are any number of theories as to why astrology eventually lost its grip. Keith Thomas, who probed the various arguments in his book *Religion and the Decline of Magic* (1971), believes that the obvious answer is the most plausible one: As science came into its own, the weakness of astrological thinking was exposed. We have seen how the appearance of Tycho's star in 1572 showed that the heavens were changeable; if the stars were not "perfect," how could their influence be predicted? The problem was compounded in 1610, when Galileo announced his telescopic discoveries, including the existence of untold thousands of "new" stars. How could one speak of the influence of the stars with any confidence, when clearly *most* stars are invisible? More broadly, an awareness of the sheer vastness of the universe was beginning to sink in—and the idea that the stars held special messages for humankind seemed less and less plausible. As Thomas puts it, "The world could no longer be envisaged as a compact interlocking organism; it was now a mechanism of infinite dimensions, from which the old hierarchical subordination of earth to heavens had irretrievably disappeared." Astrology imagined itself as a science, but in terms of actually figuring out how the universe worked, it had come to be seen as a dead end. It would never quite disappear, but it had lost its intellectual cache.

The case of Halley's Comet is illustrative. For millennia, this celestial wanderer had been approaching the Earth, and then receding, and then approaching again—*but no one knew it was the same comet.* Each appearance was a fearful event. Finally, in the early years of the eighteenth century, astronomer Edmond Halley used Newton's laws to work out the exact trajectory of a comet that had appeared in 1682, and noted that its orbit had the same properties as comets seen in 1531 and in 1607. His conclusion: It was the same object, a comet that periodically makes its way into the inner solar system, at intervals of approximately seventy-five years. And although Halley wouldn't like to see it, it reappeared right on schedule, in 1758. Suddenly a comet, once a thing to be feared, was merely another of the solar system's wanderers, moving along a predictable path. In time, scientists would show that comets were composed of rock and ice—not unlike terrestrial rock and ice. A harbinger of doom had been reduced to a dirty snowball.

11. *"Fair is foul, and foul is fair . . ."*

MAGIC IN THE AGE OF SHAKESPEARE

Thunder and lightning. Enter three WITCHES.

Macbeth has always been my favorite of Shakespeare's plays—not just because it was the first one I saw performed (at the Barbican, in London, at about age ten), but because, from the very first scene, it grabs you and doesn't let go. For A. C. Bradley, it was a play in which "the action bursts into wild life"; A. R. Braunmuller describes its first moments as "perhaps the most striking opening scene in Shakespeare." No sooner have the witches finished their dance, observing that "Fair is foul, and foul is fair," than a blood-soaked soldier stumbles onto the stage, barely able to report to the king and his men the horrors he has seen on the battlefield. As one eighteenth-century critic observed:

> The weird sisters rise, and order is extinguished. The laws of nature give way, and leave nothing in our minds but wildness and horror. No pause is allowed for reflection: . . . daggers, murder, ghosts, and inchantment, shake and possess us wholly . . . we, the fools of amazement, are insensible to the shifting of place and the lapse of time, and till the curtain drops, never once wake to the truth of things, or recognize the laws of existence.

And we are enthralled by Macbeth himself—"a soul tortured by an agony which admits not a moment's repose," as Bradley puts it, "rushing in frenzy toward its doom." The play is by far the shortest of Shakespeare's tragedies, but the impression it leaves us with, as Bradley notes, is "not of brevity but of speed." It is, he says, "the most vehement, the most concentrated, perhaps we may say the most tremendous, of the tragedies." In other words, the play rocks. And the witches—who breathe life into the opening scene—are a big part of it. As Terry Eagleton asserts, the witches are in fact the play's heroines.*

In England, witchcraft was something close to a national obsession over a span of three hundred years, from the middle of the fifteenth century until the middle of the eighteenth. During that period, many people, including very intelligent and highly educated people, believed in the reality of witchcraft, and often took steps—usually with the law on their side—to persecute those imagined to be witches, often with tragic results. But such beliefs had ancient roots. Throughout the Middle Ages, there was a tradition of men and women who could supposedly tap into supernatural forces; who could cast or remove curses, tell one's fortune, find lost items, and provide trinkets with magical properties. I said "men or women," but the vast majority were, in fact, women. There would appear to be two explanations for this, one psychological and one social. Kathryn Edwards writes, "Long misogynistic traditions in late medieval and early modern society depicted women as more susceptible than men to corruption, demonic and otherwise"; women were, as J. A. Sharpe puts it, "less resistant to Satan's advances." But there were also economic and social factors; as Keith Thomas writes, "it was the women who were the most dependent members of the community, and thus the most vulnerable to accusation"—and it should be noted that it was not just any women, but elderly and helpless women, in particular, who were most often persecuted.

KNOW YOUR WITCHES

Pamphlets and handbooks offered pointers on recognizing witches in one's community: One such book warned of "all persons that have default of members naturally, as of foot, hand, eye, or other member; one that is crippled; and especially of a man that hath not a beard"; another warned against an "old woman with a wrinkled face, a furr'd brow, a hairy lip, a gobbler tooth, a squint eye, a squeaking voice, or a scolding tongue." Shakespeare had no difficulty exploiting these stereotypes:

* Aside from being short, *Macbeth* is also a fairly straightforward play; perhaps that's why it is so often taught in high school. Conversely, some Shakespeare aficionados see it as too simple a drama to deserve our admiration. I once heard a Harvard professor introduce *Macbeth* to his class by describing it—hopefully somewhat tongue in cheek—as "a not entirely contemptible play. . . ."

When confronted by the witches, in *Macbeth*, Banquo ponders their appearance, even questioning their gender:

> What are these,
> So withered and wild in their attire,
> That look not like th'inhabitants o'th'earth,
> And yet are on't?—Live you, or are you ought
> That man may question? You seem to understand me,
> By each at once her choppy finger laying
> Upon her skinny lips; you should be women,
> And yet your beards forbid me to interpret
> That you are so.
>
> (*Macbeth* 1.3.37–44)

There were ways in which a witch could be forced to reveal her identity. She could be interrogated or, if necessary, tortured. One might also look for a "devil's mark" on the body; as Olsen notes, one could also observe "whether the accused could say the Lord's Prayer without stammering."

We don't normally think of Lady Macbeth, comfortably ensconced in her castle, as having much in common with the "weird sisters" out on the heath. But as the play progresses, we see a deeply unsettling and vaguely demonic side to Macbeth's wife, and, as Braunmuller points out, it is quite possible that early audiences "might have understood Lady Macbeth as a witch, or as possessed by the devil"—and that's before she starts sleepwalking and muttering about the dark deeds she knows she is partly responsible for. Braunmuller draws our attention to Lady Macbeth's invocation of the "spirits / That tend on mortal thoughts" (1.5.38–39), and her request that these spirits

> Come to my woman's breasts
> And take my milk for gall, you murd'ring ministers,
> Wherever in your sightless substances
> You wait on nature's mischief.
>
> (1.5.44–48)

Lady Macbeth calls on spirits; ordinary witches engaged in "nature's mischief" also received help. A witch was assisted by her "familiar"—an animal companion (supposedly) controlled by the accused, typically a cat, dog, toad, or other common creature. The witch, it was said, had promised her soul to the devil, in exchange for this animal helper.

As with astrological forecasts, a witch's predictions find their strongest resonance when they happen to coincide with the recipient's desires: Note that the weird sisters' prophecy is treated with suspicion by Banquo, but is welcomed by the eager Macbeth. And yet, just as Hamlet is initially unsure of the ghost's identity, Macbeth hesitates—both on his first encounter with the witches, and again as his downfall looms:

> If chance will have me king, why chance may crown me
> Without my stir. . . .
>
> And be these juggling fiends no more believed
> Than palter with us in a double sense,
> That keep the word of promise to our ear
> And break it to our hope . . .
>
> (1.3.142–3 / 5.8.19–22)

On the stage, witchcraft was high entertainment; in real life, it was a crime to be prosecuted, and across much of Europe laws were enacted to counteract it. The total number of cases can't be known with certainty, but by one estimate, one hundred thousand people were charged with witchcraft, of whom about forty thousand were executed, with women forming about 80 percent of all cases. In England, records show that, from the middle of the sixteenth century to the early years of the eighteenth, about two thousand people were tried, of whom three hundred were executed.* (In Scotland, the per-capita numbers were higher; roughly as many people were tried, even though Scotland's population was one-quarter that of England.) In England, the first anti-witchcraft statutes were passed in 1542, and replaced by new legislation in 1563 and 1604. There were, of course, skeptics: Samuel Harsnett, the archbishop of York, wrote that only a man without "wit, understanding, or sence" could believe in the supposed power of witches; and Reginald Scot's *The*

* It is sobering to note that the period with the most intense preoccupation with witchcraft coincides almost exactly with the period labeled by modern historians as the Scientific Revolution. The reasons for this (if it is more than a coincidence) remain unclear.

Discovery of Witchcraft (1584), the first full-length treatise on the subject, was skeptical through and through.* Even so, the persecution continued, with perhaps five hundred hangings in England, before the laws were finally repealed in 1736. We might also note that the number of witchcraft cases reached its peak in the 1580s and 1590s—the very decades in which the young Shakespeare was beginning his career.

In hindsight, the tragedy of the persecution of alleged witches is all too clear. A witchcraft case usually began as a dispute between neighbors, triggering a complaint from one citizen against another. The typical charge involved *maleficium*—causing harm. The case of the "Chelmsford witches" in 1566 is typical. A woman named Agnes Brown accused another woman, Agnes Waterhouse, of sending her familiar—in this case, a cat—to interfere with Brown's work in the milk house. (Actually, the familiar was said to *have been* a cat, but was now taking the form of "a thing like a black Dog with a face like an ape, a short tail, a chain and a silver whistle . . . about his neck, and a pair of horns on his head.") Waterhouse was convicted and hanged—the first woman to be executed for witchcraft in England. Thirteen years later, the case took another life, when the woman who was said to have given the familiar to Waterhouse was hanged as well.

Witches were the ultimate scapegoat, as the case of an eleven-year-old Lancashire boy named Edmund Robinson illustrates. One day, when tending cattle, he returned home late, saying that he had been abducted by witches; for good measure, he named a number of local women. A judge was suspicious, and referred the case to Westminster, which sent a bishop to investigate; the boy and several of the accused women were brought to London for questioning. At length, young Edmund admitted that he made the whole thing up: He had been late getting the cattle home, and was afraid his mother would punish him.†

---

* Scot believed in the existence of witches, but denied that their "power" was supernatural. Yet his attempt to find a "scientific" explanation for witchcraft is, to the modern reader, quite troubling. He focuses on menopause, and the extreme psychological metamorphosis it supposedly triggers. Witches, "upon the stopping of their monethlie melancholike flux or issue of bloud," leave themselves prone to an increase in "melancholike imaginations," and these imaginations remain, "even when their senses are gone" (quoted in Braunmuller, p. 34).

† One can't help thinking that "blaming a witch" was to sixteenth-century England what "blaming a Jew" was to mid-twentieth-century Europe, and "blaming a black man" was (and perhaps is) to late-twentieth- and early-twenty-first-century America. The most famous such incident in recent times was the 1994 case of Susan Smith. The South Carolina woman told police that she had been carjacked by a black man who drove off with her children still inside the vehicle,

GOD VS. THE WITCHES

The connection between witchcraft and religion is worth exploring. On a psychological level, the appeal of witchcraft as an explanation for one's misfortune is clear enough. After all, to claim it was God's doing suggests that the creator of the universe went out of his way to punish *you*—hardly an appealing thought. Better to blame the lonely old woman who lives down the road. For the theologians, however, the very existence of witchcraft was troubling. Why would God allow witches to flourish in the first place? One contemporary writer attempted an answer. The Lord permitted witchcraft in order "to chasten sinful humankind; to punish sin directly; to punish humankind's ingratitude in not accepting revealed truth; to shake up the godly who were lapsing into sinfulness; [and] to test Christians to see if, under adversity, they would cleave to God or desert Him for the devil." The idea of witchcraft was, in a sense, an outgrowth of organized religion. It was, as Susan Brigden puts it, "just one part of the eternal, cosmic struggle between God and Satan, between good and evil, between salvation and damnation."

Churchmen were, of course, the most ardent opponents of witchcraft, and, given the religious turmoil of the time, it's no surprise that Catholics and Protestants treated the phenomenon of witchcraft quite differently. While the old religion may have called for an exorcism, the new faith rejected such practices "as a meaningless piece of popish superstition," as James Sharpe puts it; instead, prayer and fasting were the first line of defense. The most famous author to weigh in on the subject of witchcraft (and on demonology in general) was none other than King James VI of Scotland (later to be James I of England); his treatise, *Daemonologie,* was published in 1597. Six years earlier, a plot against the king was supposedly uncovered following the torture of several alleged witches.* It has long been suspected that *Macbeth*, first performed in 1606, was written specifically for the king's pleasure. It is "the Scottish play," honoring England's first Scottish king (and patron of the playwright's acting

---

triggering a nationwide manhunt; in fact, as she later confessed, she allowed the car to roll into a lake, intentionally drowning her two young sons. (See, for example, Katheryn Russell-Brown, *The Color of Crime* [1998].)

* Kirstin Olsen summarizes the plot against the king: It "chiefly involved trying to wreck a ship carrying James and his Danish bride by casting a spell with human body parts and a christened cat." As many as one hundred people were tried in connection with the alleged crime, and James supervised some of the torture sessions personally (Olsen, vol. 2, p. 676).

troupe)—but Shakespeare surely knew of James's fascination with the occult.

As the decades passed, cases of witchcraft came to be treated as curiosities. Pamphlets that documented the latest cases circulated widely, along with reports of monstrous births, earthquakes, fires, whales washed up on beaches, and the like. Eventually, the idea of witchcraft began to fade from the national consciousness. The last hanging was in 1685; the last conviction in 1712. Why it declined is perhaps easier to comprehend than why it was so prominent for so many centuries. As Kathryn Edwards writes, "the growing preoccupation with witchcraft and the danger it posed during this time has not been conclusively explained."

NATURAL AND UNNATURAL MAGIC

Witchcraft was just one kind of magic that confronted the citizens of early modern Europe. Just about everyone, from the university-educated to the rural poor, considered witches on a spectrum of the supernatural, alongside "cunning" or "wise" men and women, fortune-tellers, magicians, and sorcerers of various kinds. (Note that Shakespeare was especially prone to including soothsayers and fortune-tellers in those of his plays set in the ancient world: *Julius Caesar*, *Antony and Cleopatra*, *Cymbeline*.) As late as 1621, Robert Burton would write, in *The Anatomy of Melancholy*, "Sorcerers are too common; cunning men, wizards, and white witches, as they call them, in every village, which, if they be sought unto, will help almost all the infirmities of body and mind." As the quote suggests, these sorcerers and magicians offered competition to the doctors—which perhaps isn't a surprise, since doctors themselves had limited power to aid the sick. And yet, as Keith Thomas notes, even when a Renaissance magician specialized in "medicine," it was usually "only one branch of a very diverse repertoire." The magician was, typically, performing his magic at the request of a client with a specific problem. Of course, there were trustworthy magicians as well as charlatans, and the public was understandably wary.

Some kinds of magic were said to be readily learned, and could be used by anyone who had mastered the requisite skill—for example, techniques for recovering lost or stolen property. In the case of theft, the "sieve and shears" method was one way of rooting out the guilty party. The mechanics of this "test" seem to have been lost in the mists of time, though it likely went something like this: All of those involved would sit in a circle, with the sieve and the shears (presumably attached to each other) suspended at the center of the group; a verse from the Bible was probably recited. In the end, the sieve would point to the guilty person. Another peculiar superstition pertained to murder, the most serious of all crimes. If a murderer's guilt was in doubt, he might be asked to touch the victim's body; the theory was that the wounds of the deceased would bleed once again. (As Thomas notes, this practice was endorsed by scientists and judges well into the seventeenth century.)

Of course, quackery abounded. Often practitioners were exposed, and occasionally they were arrested. When that happened, as Thomas points out, it was not unusual for townspeople to rush to their defense—after all, who else could they turn to in times of crisis? The practice of cunning men and women was occasionally profitable; but if it brought prestige, that was likely enough. They flourished because they filled a need. A writer named William Perkins summed up the situation:

> Let a man's child, friend, or cattle be taken with some sore sickness, or strangely tormented with some rare and unknown disease, the first thing he doth is to bethink himself and inquire after some wise man or wise woman, and thither he sends and goes for help.

And if the sick party recovers, Perkins notes, "the conclusion of all is the usual acclimation: 'Oh, happy is the day that I ever met with such a man or woman to help me!' "

Needless to say, the Church opposed all kinds of lay magic; the official position was that only God, and perhaps the Devil, had the power to manipulate nature and harness supernatural forces. James himself asserted that magicians and witches served "both one Master, although in diverse fashions," and that both should be punished with death. Indeed, the links between religion and politics in Shakespeare's England were seamless; kings, after all, claimed to rule by divine right, and the Bible made it clear that rebellion and witchcraft are equally sinful. Magic was also seen as linked to paganism; after all, such traditions no doubt dated back many centuries, and many beliefs and rituals had pre-Christian origins. In 1554, a London bishop declared that "witches, conjurers, enchanters, and all such like, do work by the operation and aid of the Devil," and that "all such commit so high offence and treason to God, that there can be no greater." Of course the Church's *own* brand of magic, like that associated with Mass, or with the traditional healing power of the saints, was perfectly legitimate. The *source* of the magic was crucial: Magic was acceptable if it derived from God; and it was also acceptable if it came from nature, and was uncovered by careful study and investigation—this was "natural magic," a pursuit at least somewhat allied with what we now call "science." ("Natural magic" roughly parallels "natural astrology," which, as noted, was seen as a harmless pursuit akin to astronomy.) This is the question that looms over the dramatic final scene in *The Winter's Tale*, in which the statue of Hermione, the queen who had been dead for sixteen years, comes to life. As with the appearance of the ghost in *Hamlet*, the immediate question is whether the sight we are witnessing is heavenly or demonic. Paulina, who has overseen this seemingly miraculous return from the dead, insists she was not "assisted / By wicked powers" (5.3.90–91), a sentiment echoed by the king: "If this be magic," Leontes says, "let it be an art / Lawful as eating" (5.3.110–11).

Shakespeare's greatest magician is, of course, Prospero, protagonist of *The Tempest*. We have to assume that Prospero's magic is legitimate; for one thing, he talks of his "art" in contrast to the demonic power of his archenemy, Sycorax the witch. Sycorax uses her powers to trap the spirit Ariel in a tree for twelve years; only Prospero's magic is strong enough to break the spell and secure his release. Prospero seems to use his powers for good rather than for evil—and yet there are hints of something darker. As he explains to Miranda, it was a craft that required an intense and focused period of learning; he eventually became "rapt in secret studies" (1.2.77). Even so, he strikes many readers as more of a scientist than an alchemist. As Elizabeth Spiller notes, Prospero's art "can only be imagined to work for the same kinds of reasons that natural philosophers like Gilbert and Bacon understood their sciences to do so." He gives us "a history in small of the larger cultural transformation by which Aristotelian philosophy would become Baconian science." And while Prospero performs some rather impressive feats—creating and directing the storm that gives the play its title, for example—there is also something of the street magician, even the hustler, in his craft. Such performers were a common sight in Jacobean London, and Shakespeare's audience would have instantly recognized such a character on the stage. As Virginia Mason Vaughan and Alden T. Vaughan note in the Arden edition of *The Tempest*, the play's protagonist is "a combination of serious magician and carnival illusionist." We might also note the link between Prospero's magic and astrology: "I find my zenith doth depend upon / A most auspicious star . . ." (1.2.181–82). Of course, the playwright wields his own peculiar brand of magic: Attending a theatrical performance is, after all, accepting an invitation to be (benignly) deceived. No wonder that the character of Prospero, of all of Shakespeare's creations, is seen as a plausible reflection of the author himself.

"TELL ME WHO MADE THE WORLD"

Another quasi-legitimate "magic" was alchemy, the quest to turn one kind of substance into another (and especially, to turn cheap metals such as lead into gold). Again, the theory of the four elements is key; alchemists believed that by changing the balance of these elements, they could transform one kind of material into another. This could be achieved through burning, distilling, dissolving, sublimating, and melting, usually with the aim of purifying one of the ingredients. (The fact that the transmutation of metals was prohibited by law suggests that many people believed it could actually be done.) As with other kinds of magic, including astrology, there were countless quacks. Ben Jonson's satirical play *The Alchemist* (1610) serves as a kind of theatrical debunking of alchemy and those who practice it. Jonson makes endless fun of the main characters, a con man named Subtle and his sidekick, Face—as well as those gullible enough to fall for their trickery (the "gulls").

Who was Jonson's model for Subtle, the title character in *The Alchemist*? As we saw in Chapter 4, he may have been influenced by John Dee, the Elizabethan scientist-magician (and Dee is mentioned by name in the play); but another line of reasoning connects the play to Giordano Bruno's comedy *Candelaio*, published in 1582. After weighing the evidence of a Jonson–Bruno connection (including a tally of who may have known whom), Hilary Gatti concludes that it is "at least possible, if not probable, that Jonson had some knowledge of Bruno and perhaps of his works, even if only through conversations with those who had known him personally in London."

As with astrology, there would have been those who believed in the power of alchemy, and those who doubted—and some who would have harbored both views simultaneously, just as someone today may claim to dismiss astrology, but might check their horoscope in the newspaper, even if just for its entertainment value. As Gordon Campbell puts it, both Jonson and his audience would have regarded alchemy "as a combination of science and imposture." For Jonson, this duality creates the perfect dramatic and comedic vehicle.

Shakespeare seems to have had less use for alchemy than Jonson, though he mentions it on a handful of occasions. It comes up metaphorically in *Timon of Athens* ("You are an alchemist; make gold of that" [5.1.114]), for example; and in *King John*, where Philip, the king of France, says:

> To solemnize this day the glorious sun
> Stays in his course and plays the alchemist,
> Turning with splendour of his precious eye
> The meagre cloddy earth to glittering gold . . .
>
> (3.1.3–6)

But the greatest magician on the London stage was not one of Shakespeare's creations, nor was it one of Jonson's. Twenty years before Prospero cast his first spell, audiences were treated to Christopher Marlowe's masterly play *The Tragicall History of Doctor Faustus* (ca. 1592). Marlowe did not invent his doctor out of whole cloth: His play is based on age-old stories of learned men who sell their souls to the devil in return for knowledge. By the sixteenth century, these tales had become associated with the real-life figure of Johannes Faustus, a German astrologer who lived in the early part of the century. A fictionalized account was set down by an anonymous German writer in 1587, and an English translation had appeared in 1592, serving as Marlowe's immediate source.

Marlowe's Faustus was a magician and also a scholar; he studied at Wittenberg just like Hamlet and his friend Horatio. Faustus yearns to know the secrets of the universe; in keeping with the traditional story, he promises his immortal soul to the devil in exchange for worldly knowledge. He abandons the traditional fields of academic learning for black magic ("'Tis magic, magic that hath ravished me" [1.112]). As we saw in Chapter 1, one of the first things Faustus asks of Mephistopheles concerns cosmology: He inquires after the structure of the heavenly spheres. "Now would I have a book where I might see all characters and planets of the heavens, that I might know their motions and dispositions" (7.171–3). His questions become ever more dangerous: "Tell me who made the world . . . Sweet Mephistopheles, tell me" (7.66–8).

Faustus gradually gains both knowledge and power. He travels around Europe, casting spells, wreaking havoc in the royal courts, and playing tricks on the pope. As the end nears, the Devil comes to make good on

The Tragicall Hiſtoy of
the Life and Death
of Doctor Fauſtus.

With new Additions.

Written by *Ch. Mar.*

LONDON,
Printed for *Iohn Wright*, and are to be ſold at his ſhop without
Newgate, at the ſigne of the Bible. 1620.

Fig. 11.1 "Tell me who made the world . . . Sweet Mephistopheles, tell me." Both a scholar and a magician, the title character of Christopher Marlowe's *Doctor Faustus* yearns to understand the universe's workings. This is the frontispiece of a 1631 edition of the play. © British Library Board/Robana/Art Resource, NY

---

the bargain. Faustus is now filled with regret—and fear—and pleads for mercy. All of that learning, he laments, was a terrible mistake: "O, would that I had never seen Wittenberg, never read a book!" (14.19–20). It is to no avail (spoiler alert!): The Devil carries Faustus's soul off to hell.

Astrology, witchcraft, alchemy, magic . . . and science. It was all part of a package; all were thoroughly intertwined in the sixteenth century, and even into the early years of the seventeenth. Belief in fairies, demons, ghosts, and witches was common; like religion, these spirits were simply a part of everyday existence. What we now think of as "science" was only beginning to disentangle itself from magical thinking. We have heard about John Dee and his magic crystal, and the subtle mix of science and magic that informed the works of thinkers like Bruno and Gilbert. Indeed, Gilbert's magnetism seems almost tailor-made for mystical interpretations. As Keith Thomas notes, the very idea of magnetic forces "seemed to open the possibility of telepathy, magical healing, and action at a distance." (For example, if someone was injured by the use of a weapon, it made sense to apply the healing ointment not only to the wound, but also to the weapon; after all, if magnetic forces could affect planetary orbits, might not vital spirits readily traverse the short distance between weapon and wound? Even the Royal Society, in its early years, took an interest in such matters.)

Another pivotal figure who embodies both mysticism and the emerging scientific worldview is the German astronomer and mathematician Johannes Kepler (1571–1630). Today we remember Kepler as the man who completed what Copernicus had begun—the scientist who finally worked out the precise mathematical laws governing planetary motion. But there was another side to this Renaissance genius, a side that reveals his deep-seated connections to the thinking of past ages.

Born near Stuttgart and educated at Tübingen, Kepler studied theology with the expectation of becoming a Lutheran clergyman. Instead, he ended up teaching mathematics at a provincial school, where he read *De revolutionibus* and became interested in the mathematical underpinnings of astronomy. He would later work as Tycho Brahe's assistant in Prague, and eventually served as court mathematician to Emperor Rudolf II and his successors.

THE MUSIC OF THE SPHERES

But Kepler was more than just a mathematician: Like the ancient Pythagoreans before him, he was obsessed with numerology; he was sure that certain numbers had special properties.* Why, for example, were there six planets (counting the Earth), rather than five or seven or

* We like to imagine that we live in a more sophisticated age—but once again the truth is more muddled. As I was writing this chapter, newspapers were reporting that the town council in Richmond Hill, Ontario, has agreed to stop using the numeral "4" in street addresses. Residents had been complaining that the mere presence of a 4 in a house's address was lowering its resale value by tens of thousands of dollars. Richmond Hill is home to a large number of Asian immigrants, many of whom consider 4 to be unlucky because it sounds like the word for "death" in Cantonese and Mandarin. (And now you can deduce what the word "tetraphobia" means.)

some other number? The Creator must have had a reason for this state of affairs, and numerology presumably held the answer. When he was developing his model of the solar system, Kepler was careful to make it conform to his notions of mathematical beauty. He was enamored with the parallels between mathematics and music, and dreamed of translating the positions and motions of the heavenly bodies into a musical score. This was the "music of the spheres," another idea that goes back to the Pythagoreans. Shakespeare alludes to this ancient concept several times—for example, in the words of Olivia in *Twelfth Night*:

> O, By your leave, I pray you!
> I bade you never speak again of him;
> But would you take another suit,
> I had rather hear you to solicit that,
> Than music from the spheres.
>
> (3.1.107–11)

Kepler was particularly fascinated by another mathematical discovery of the ancient Greeks—the five "regular solids" of Euclidean geometry. (These three-dimensional figures have sides that are all of an identical polygonal shape. They are the tetrahedron, with four sides, each triangular; the cube, with six sides, each square; the octahedron, with eight sides, each triangular; the dodecahedron, with twelve sides, each pentagonal; and the icosahedron, with twenty sides, each triangular.) In his *Mysterium Cosmographicum* (*The Cosmographic Mystery*), published in 1596, he presented his first full-length defense of the Copernican model. But he also put forward a remarkable theory about the relative sizes of the planetary orbits. The insight is said to have occurred to him while teaching in Graz, Austria. He was thinking about the solar system, and the platonic solids, and came up with the notion that the orbits of the planets must have the same proportions as these perfect geometrical figures (see figure 11.2).

It was an ingenious, bold idea; unfortunately, it was wrong—as Kepler discovered when he checked his theory against the best available data on the sizes of the planetary orbits, and realized it didn't quite fit (although it was close). The first draft of the *Mysterium Cosmographicum* was simultaneously a work of science and a work of theology, with Kepler trying to reconcile the heliocentric theory with various passages from the Bible. He imagined the universe itself as analogous to God, with the sun corresponding to the Father, and the stellar sphere to the Son; the space separating them he imagined as the Holy Spirit. (The book was published only after his mentor, Michael Maestlin, urged him to simplify the argument, and to go easy on the theology.)

Kepler's mystical side tends to be downplayed these days, as we instead honor him for developing the laws of planetary motion, and for deducing the true shape of the planetary orbits (they're not circles, as the philosophers had for millennia imagined, but rather ellipses). These breakthroughs came only after he had moved to Prague; indeed, the great leap forward came only after Kepler gained access to Tycho's exquisite data on the positions of the planets (especially Mars), collected over many years. (Tycho and Kepler failed to see eye-to-eye on many things—recall that Tycho had rejected the Copernican view in favor of his own hybrid model—and Kepler got his hands on the vital data only after Tycho's death, and after much wrangling with his heirs.)

The resulting book was called the *Astronomia Nova* (*A New Astronomy*), published in 1609. The title suggests a certain amount of bravado, and perhaps it was justified. As mentioned, Kepler did not yet have the notion of gravity in the modern sense—but he realized that whatever force was holding the planets in their orbits decreased in strength in proportion to the distance.* Inspired by Gilbert's work on magnetism, Kepler was willing to accept the notion that it was indeed a magnetic force that kept the planets in their orbits. He pays tribute to both Gilbert and Tycho in his later astronomy textbook, *Epitome astronomiae Copernicanae* (*The Epitome of Copernican Astronomy*), published in three volumes beginning in 1615. He writes, "I erect the whole of astronomy on Copernicus's hypotheses about the universe, on Tycho Brahe's observations, and finally on the Englishman William Gilbert's science of magnetism." Perhaps the key point is not that Kepler believed that the force was magnetism, but that he was willing to consider that *a force*, a purely physical (if tangible) entity, governed the motion of the planets. For the first time, an astronomer was investigating not only the apparent motions of the heavenly bodies, but their presumed physical causes. We can see why

---

* And he *almost* nailed it: As Newton would show, the strength of the force decreases with the *square* of the distance (double the distance, and the strength of the force decreases by a factor of four). Still, as I. Bernard Cohen notes, the important thing is not that he got the exact formula wrong, "but rather that he should have conceived of a celestial force in the first place," and recognized that it must decrease with increasing distance (Cohen, *Revolution in Science*, p. 130).

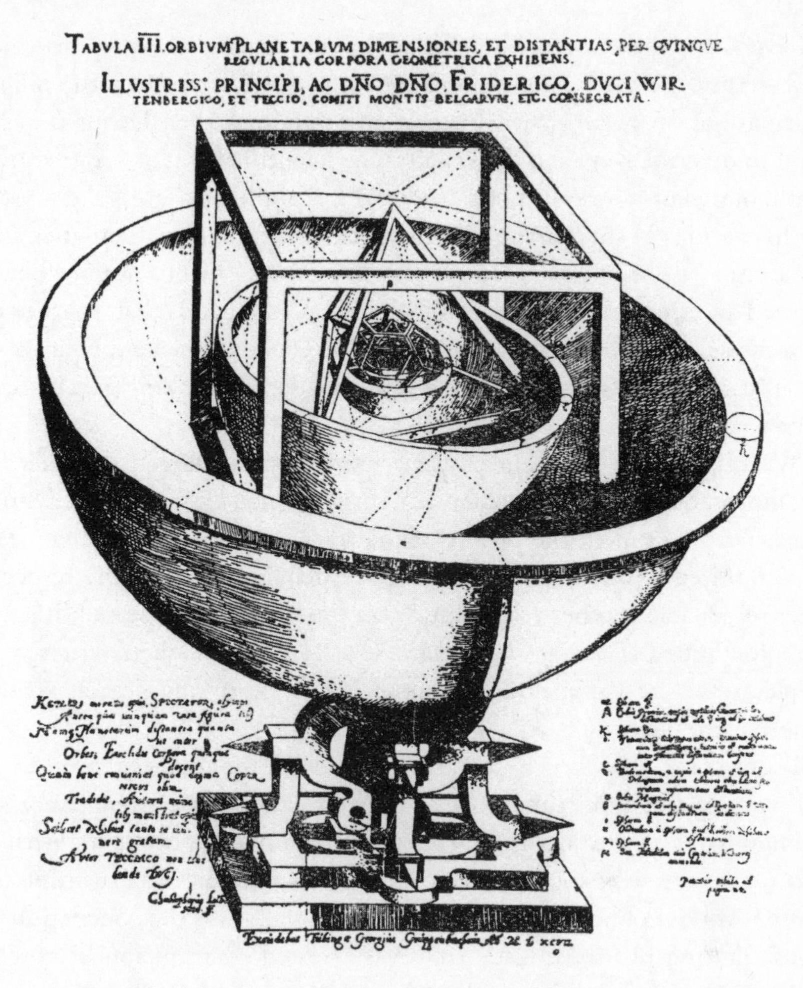

Fig. 11.2 In his *Mysterium Cosmigraphicum* (*The Cosmographic Mystery*), published in 1596, Johannes Kepler imagined the sizes of the planetary orbits as having the same relative proportions as the five "platonic solids" of Euclidean geometry. Bpk, Berlin/Art Resource, NY

Owen Gingerich has referred to Kepler as the first astrophysicist. His work, as historian I. Bernard Cohen puts it, "implied an end to the Aristotelian cosmos and readied the scientific stage for Newton"—even though, as Cohen stresses, many astronomers remained unconvinced of Kepler's theory, and, as usual, the paradigm changed only slowly.

And yet Kepler, in spite of being one of the greatest scientists of his age, produced a body of work in which we see the physical and the mystical fully integrated. The planets were held in their orbits by magnetic forces, but the rotation of the Earth (and the sun, for that matter) was best explained by an animistic or "soul principle." (For example, Kepler believed that the existence of sunspots demonstrated the presence of a soul within the sun.) His scientific writings were coupled with endless speculations on metaphysics, history, and religion. Allen Debus is quite right to describe Kepler as "a Renaissance scientific paradox—the superb mathematician whose inspiration derived from his belief in the mystical harmonies of the universe." His subtle mixture of mathematics and mysticism "is far removed from modern science," Debus notes, "but it formed an essential ingredient of its birth." Cohen adds that, in spite of Kepler's scientific genius, "we could easily assemble a whole volume of his writings that would show how unscientific his thinking and his science were."

We might note that Kepler was a practicing astrologer, and that he cast horoscopes for the German nobility. It's not clear, however, how much faith he put in the power of the stars to influence our lives. He once referred to astrology as the "foolish little daughter of the respectable, reasonable mother astronomy"; on the other hand, he published a pamphlet titled *The Sure Fundamentals of Astrology.* Cohen writes that Kepler was "the last major astronomer . . . to be in any degree a convinced astrologer."

Proof that Kepler was living at least partly in the medieval world can be seen in the unfortunate case of his mother, Katharina. Another woman had claimed that Katharina gave her a magic potion, and that it had made her sick; soon there were rumors that she was running an apothecary out of her home, and that her specialty was the concoction of mind-altering libations. Katharina was arrested on suspicion of witchcraft in 1615. Kepler did everything he could to help her, writing numerous letters on her behalf; but she was convicted and sent to jail. Kepler eventually left his family in Prague, traveling to join his mother at Württemberg, spending nearly a year by her side and probably saving her from torture and execution. She insisted throughout that she was innocent, saying that she would rather die than give a false confession. She was finally released—most likely because of a technicality (the prosecutors had failed to go through all the hoops required by law). One can readily imagine Katharina's fate had her famous son not been on hand to intercede.

For much of human history, magic was simply *everywhere.* As J. A. Sharpe puts it, "The overwhelming impression is that the majority of the population, certainly before 1700, to a greater or lesser extent accepted magical beliefs as part of their world view." Why did it hold such sway over people's lives and thoughts, for so many centuries? Part of the answer is that it was pragmatic: It offered, or at least appeared to offer, solutions to real problems, generally involving health or prosperity, in an era when few other solutions were at hand. Magician-scholars like Faustus were condemned for their overly curious nature—but today's scientists, too, are tireless questioners, and it sometimes seems that Marlowe's doctor is less like Prospero and more like an inquisitive graduate student in astrophysics or cosmology. As natural philosophy gained ground, certain kinds of magical thinking—"the bits that worked"—simply became absorbed into what would become "science." Indeed, as historian of science John Henry puts it, the reason that our view of magic has changed since Shakespeare's time "is precisely because the most fundamental aspects of that tradition have now been absorbed into the scientific worldview." Alchemy is the obvious example: As wrongheaded as the attempts to turn lead into gold may have been—at least in hindsight—there is no question that alchemy was the forerunner of modern chemistry. (At the very least, it provided the tools that chemists would come to rely on, from balances and beakers to filters and heat sources.)

As we have seen, magic and science were deeply intertwined, and it would take more than a century for chemistry and alchemy to permanently part ways. Still, when natural philosophy rose in stature toward the middle of the seventeenth century, it owed an enormous debt to the natural magic of an earlier age. Francis Bacon, like Kepler, was a figure poised between the age of magic and the age of science. He was clearly influenced by the magical tradition, but was crafty enough to try to separate the intellectual wheat from the pseudoscientific chaff. It wasn't easy. "The end of our foundation," he wrote, "is the knowledge of Causes, and secret motion of things; and the enlarging of the bounds of Human Empire to the effecting of all things possible." Are these the words of a scientist, a magician—or both?

12. *"A body yet distempered . . ."*

SHAKESPEARE AND MEDICINE

There are museums. And then there are museums built into the attics of old churches. I'm fond of the latter kind.

From street level, the brown-brick, Wren-esque Church of St. Thomas looks like any of a hundred other London churches, nestled among modern office buildings and flats, around the corner from London Bridge underground station in Southwark. To discover the building's significance, one must climb the narrow, spiral staircase located in the church's tower. A rope handrail assists the visitor up the thirty-two steps, and eventually one emerges from the dim stairway into the museum's brightly lit foyer. It soon becomes clear that St. Thomas's is not just any old church. It wasn't built to minister to the residents of the surrounding neighborhood, but rather to those affiliated with the hospital that once stood on the site. And while the church's present structure dates only from the late seventeenth century, the story of St. Thomas's Hospital goes back much further. It was named for Thomas Becket, who was murdered in 1170, and it may date back to "only" a few years after his death; but historians suspect that its roots likely go back to about 1100, which would make it the oldest hospital in the city, perhaps in all of England.

The hospital was already described as "ancient" in 1215, when a fire that started on London Bridge swept through the neighborhood. (When the flames finally died down, the Bishop of Winchester observed: "Behold at Southwark an ancient spital, built of old to entertain the poor, has been entirely reduced to cinders and ashes.") It was rebuilt after the fire, and would flourish for another six and a half centuries. The hospital was finally closed in 1862, when a new facility was opened a couple of miles upriver in Lambeth. Most of the old structure was demolished, and what was left was virtually forgotten—even as worshippers continued to make use of the church itself, the last surviving part of the original complex. The church's garret (meaning "attic") was rediscovered in 1956, and its historical value was finally recognized. "You don't normally get into the roof of a church," says Kevin Flude, the director of the Old Operating Theatre Museum and Herb Garret, as we gaze up at the weathered wooden beams. "It's really a remarkable and unusual building."

When historians poked around in the garret in the 1950s, they found the heads of dried opium plants in the rafters, which hinted at the building's remarkable history. The garret had been used for storing, and perhaps growing, medicinal herbs and plants; it would have been relatively dry, making it better suited for the purpose than other rooms in the church or the adjacent hospital. The luckiest find was the rediscovery of what had been a surgical operating theater. Located at the far end of the attic, it dates from 1822 and has now been lovingly refurbished. The word "theater" is appropriate: This was a place for learning, and the operating table was overlooked by a series of concentric wooden stalls, from which students could peer down at the operation. A sign at the front of the room reads MISERATIONE NON MERCEDE—"For Compassion, Not Gain." But compassion could only get the patient so far. One look at the surgical instruments, including an amputation saw, is enough to make any twenty-first-century visitor very glad *not* to be living in a previous age. What's frightening is not the collection of instruments itself, but the fact that, before the invention of anesthetics in the 1840s, surgery of any kind was akin to torture: *You were awake for everything.* Alcohol and various opiates could dull the pain slightly, but that was about it. Surgery was thus a last resort, with many patients simply choosing death over the inevitable, excruciating pain of an invasive procedure.

Other display cases in the museum are less horrifying. There are human skeletons and plaster skulls; rows of glass jars and metal tins that once contained various medicines; bowls of herbs, roots, and seeds; and arrays of glass beakers, pipettes, and syringes. A set of emerald-green bottles sparkles under the incandescent lighting; they have cryptic labels such as TR. HYOSCY, EXT. ERGOT. LIQ., and TINCT: HAMAMEL. The items on display are all from well after Shakespeare's time, although the hospital was certainly a fixture of the neighborhood in the playwright's day.

In fact, St. Thomas's wasn't far from the Globe Theatre or the bear-baiting halls, being located just a few hundred yards to the east, and was just a stone's throw from Southwark Cathedral. Not that the playwright would have set foot in this, or any other, hospital: They generally catered to the poor and downtrodden, and Shakespeare, a man of some means, could probably have afforded to have a physician call on him, if needed. "Middle-class people would tend to be treated in their own home," Flude explains. "They certainly wouldn't come to this hospital."

PHYSICIANS, SURGEONS, APOTHECARIES, AND MIDWIVES

Today there is little ambiguity to the term "doctor," other than distinguishing a medical doctor (with an MD) from a researcher or scientist (with a PhD)—and those in the latter group don't generally introduce themselves as doctors anyway, which helps minimize confusion. In Shakespeare's time, it was a bit more complicated. There were at least three distinct classes of medical practitioners, with physicians at the top, surgeons in the middle, and apothecaries and midwives at the bottom. Physicians were licensed, and were called "doctors" because they had studied at either Oxford or Cambridge. They had spent many years in school, learning from the Latin texts of Galen and Hippocrates. Physicians could diagnose ailments and prescribe treatments, but they did not perform surgery. A twelfth-century Church edict had prohibited doctors from shedding blood—thus anything that involved cutting a patient was left to another group of medical practitioners, the surgeons. Physicians made house calls, but their services were expensive; most ordinary people would never have consulted one. However, there were shortcuts for those on a tight budget. One could always send one's urine to the doctor for inspection, rather than ask for a personal consultation. Much could be gleaned, it was imagined, from the color and texture of the urine (and, if needed, the stool). Shakespeare has some fun with this practice, at Falstaff's expense, in *Henry IV, Part 2*:

FALSTAFF

Sirrah, you giant, what says the doctor to my water?*

---

* Why does Falstaff call the page a "giant"? The footnotes in the scholarly editions explain that he's making fun of the page, who is actually short.

PAGE

He said, sir, the water itself was a good healthy water, but, for the party that owed it, he might have more diseases than he knew for.

(1.2.1–5)

A similar reference in Marlowe's *Tamburlaine the Great, Part 2*, has Tamburlaine asking his doctor, "Tell me, what think you of my sickness now?" To which the doctor replies,

I viewed your urine, and the hypostasis
Thick and obscure, doth make your danger great;
Your veins are full of accidental heat,
Whereby the moisture of your blood is dried.

(5.3.82–85)

And the doctor continues in this vein (pardon the pun) for another fourteen lines, all of it more or less Galenic. Incidentally, although there is no reason to imagine that Falstaff attended university, he was certainly familiar with Galen, whom he mentions later in the same scene (one of five references to Galen in Shakespeare's writing). Even with a personal consultation, however, there was only so much the physician could do. Without the benefit of diagnostic tools such as X-rays, and treatments like antibiotics—or even an understanding of the circulatory system—the physician was limited to educated guesswork.

Surgeons were considered to be of a lower class than physicians. They did not attend university; rather, they trained as apprentices under a master, just like artists and craftsmen. As a profession, surgery was allied with barbering, since both involved cutting and the use of sharp metal tools; and they were licensed by the same body, the Barber-Surgeons Company. (When Shakespeare was living in London's Cripplegate neighborhood, he would have been just a couple of blocks away from Barber-Surgeons' Hall on Monkwell Street, headquarters for the Company of Barbers and Surgeons.*) Surgeons could stitch up wounds, remove "stones"

* The hall itself survived the Great Fire of 1666, but was destroyed by German bombs during the Second World War. The hall was rebuilt after the war, and, although the barbers and surgeons had parted ways in the mid-eighteenth century, the old name was retained, and a sign reading "Barber-Surgeons' Hall" still marks the front entrance, on Monkwell Square.

from internal organs, and, in the case of a head injury, relieve pressure by boring into the skull (a process known as trepanning). But opening up the body in this way was always a last resort. Of course, a diseased limb might have to be amputated—but without anaesthetic, it would have been a terrifying experience for the patient (and probably not much fun for anyone else within earshot).

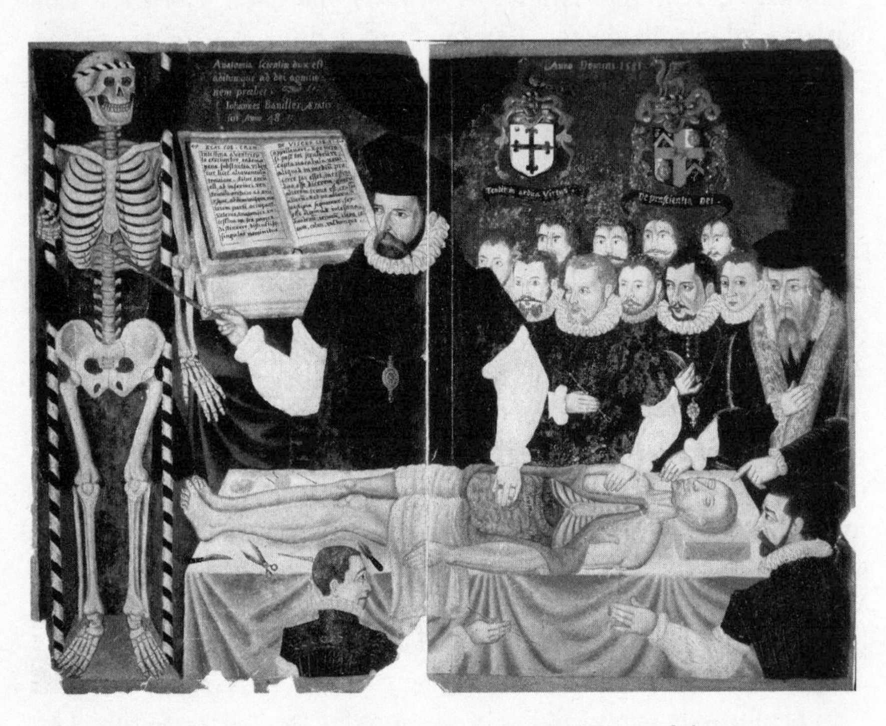

Fig. 12.1 This depiction of an anatomy lecture, including the dissection of a human corpse, dates from the early 1580s. (Note the prominence of the book, likely the writings of the second-century A.D. Greek physician Galen.) The Bridgeman Art Library, London

---

Finally there were the apothecaries and midwives, lower still in the social ranks than either physicians or surgeons. Like the surgeons, apothecaries learned their craft via apprenticeship. They could fill prescriptions given by a physician; and since they charged less than either physicians or surgeons, they were more readily accessible.

Shakespeare famously makes use of a rustic apothecary in *Romeo and Juliet*, a reclusive man who, in spite of being a hermit-like figure—his hovel is littered with "empty boxes"—manages to stock a remarkably diverse array of tonics and potions:

> I do remember an apothecary –
> And hereabouts 'a dwells—which late I noted
> In tatter'd weeds, with overwhelming brows,
> Culling of simples. Meagre were his looks,
> Sharp misery had worn him to the bones,
> And in his needy shop a tortoise hung,
> An alligator stuff'd, and other skins
> Of ill-shap'd fishes; and about his shelves
> A beggarly account of empty boxes,
> Green earthen pots, bladders, and musty seeds,
> Remnants of packthread and old cakes of roses
> Were thinly scatter'd to make up a show.
>
> (5.1.37–48)

The midwife was perhaps a more respected figure than the apothecary. She could be relied on to examine suspected witches, rape victims, and female prisoners. As Kirstin Olsen notes, however, if something went wrong with a delivery, it might well be the midwife who initiated rumors of a "monstrous birth." Along with the apothecaries and midwives, there were unlicensed "empirics" and alewives, and a vast array of amateur healers of all stripes, promising all manner of cures. Quackery was rampant. In all, these unlicensed medical men and women no doubt greatly outnumbered the licensed practitioners.

HUMORING GALEN

The Greek physician and philosopher known as Galen of Pergamon (A.D. 129–199), dead for fourteen centuries, was the leading medical authority in Shakespeare's England. His writings were perennially sought after; by one estimate, more than six hundred editions of his works were published between 1490 and 1598. Galen's theory focused on the four "humors" thought to govern the body: blood, phlegm, black bile, and yellow bile. These mirrored the Earth's own constituents, the elements earth, water, air, and fire, via the properties that they shared. Each humor, and each element, was associated with a particular combination of hot, cold, wet, and dry. Thus, blood was hot and wet; phlegm was cold and wet; black bile was cold and dry; and yellow bile was hot and dry. The key to good health was the maintenance of a proper balance or "temperature" among the four humors. "Distemperature"—of the body or of the nation—was a sure sign of more serious ills to come. In *Henry IV, Part 2*, the king complains that "rank diseases grew near the heart" of his kingdom; Warwick reassures him that

> It is but as a body yet distempered,
> Which to his former strength may be restored
> With good advice and little medicine.
>
> (3.1.40–42)

Galen also proposed three "souls": the rational soul, governed by the brain; the emotional, controlled by the heart; and the vegetative, controlled by the liver. It turns out that Galen greatly overestimated the importance of the liver. He thought it was the first organ to form in the fetus, and believed that it governed the entire circulatory system. He also thought that the liver wrapped around the stomach, warming it so as to aid in digestion. In fact, although the liver has this shape in certain animals, it does not take this form in humans; as Olsen points out, this is one of the mistakes that suggests that Galen examined only animals, not actual human corpses.

Galen's mistakes went unnoticed (or at least, uncorrected) for centuries. The first person to take Galen to task, in print, was the Flemish physician Andreas Vesalius (1514–1564). Born in Brussels, Vesalius studied in France and in Italy, and learned both from his classically trained professors and from his own investigations. His dissections of human cadavers at the University of Padua showed, among other things, that Galen was wrong about the shape and function of the liver. (Vesalius was also something of a showman: His dissections were open to the public, and drew crowds of curious onlookers.) He published his findings in a groundbreaking book, *De humani corporis fabrica* (*On the Fabric of the Human Body*), the first modern anatomy textbook, complete with detailed drawings based on the author's own work. (As historians like to point out, Vesalius's book was published in 1543, the same year that saw Copernicus's *De revolutionibus*—a good year indeed for science publishing!) Even so, Galen was so revered that his errors were only slowly coming to light.

Galen's biggest mistakes were physiological rather than anatomical.

For one thing, he failed to deduce the circulation of the blood, which would have to wait for the work of William Harvey in 1628 (a dozen years after Shakespeare's death). The tendency to animate all forms of matter didn't help. We have already noted this tendency in astronomy—for example, Kepler's insistence on planetary "souls." The human body was equally susceptible to such theorizing, and, as Olsen notes, by the sixteenth century Galen's followers imagined that each organ had not only a physiological function, but also a kind of personality. The spleen, for example, was thought to be filled with black bile, which was cold and dry; and so the organ was imagined to be the seat of irritability, impulsiveness, and quick-changing passions. Shakespeare mentions the spleen in this sense quite a few times—for example, Lady Percy's retort to Hotspur, in *Henry IV, Part 1*:

> Out, you mad-headed ape!
> A weasel hath not such a deal of spleen
> As you are tossed with.
>
> (2.3.74–76)

This also leads to some rather confusing passages; as Olsen points out, the modern reader may have trouble with the line from *Venus and Adonis* in which Shakespeare writes of Venus that "A thousand spleens bear her a thousand ways" (line 907). It just means that the goddess is in a highly changeable mood—but the phrasing, as Olsen playfully notes, calls to mind "an angry mob of tiny internal organs traipsing about with a bewildered goddess on their backs."

Yellow bile, believed to be hot and dry, was also known as choler. Too much choler was thought to make a person impatient and argumentative, and Shakespeare uses the word in exactly this sense—for example, in *Henry V*, where the king describes the Welsh captain Fluellen as "touched with choler, hot as gunpowder" (4.7.175). Similarly, words like "phlegmatic" (too much phlegm), "melancholic" (too much black bile), and "sanguine" (an excess of blood) crop up throughout the canon, describing not only medical conditions but also aspects of personality.

LET IT BLEED

A doctor's motto in Elizabethan England might well have been "When in doubt, bleed." Ailments were attributed to an imbalance in the humors, and if the problem seemed to be an excess of blood, the answer was obvious—let some of it out. Doctors urged even healthy people to be bled on a regular basis, to keep their humors in balance. A letter from a German student to his mother, dating from 1578, is illuminating: "I ask you to send me a blood-letting lancet," he wrote, noting that "the other students have their own special lancets" so that they don't need to go to the bathhouse and submit to the tools "used to bleed the peasants and everyone else."

Aside from the imagined medical benefits, there was a religious motivation for bloodletting: Some churchmen believed that by ridding himself of excess blood, a man could rid himself of sin. There was also an astrological motivation: Bleeding was recommended when the Earth itself was "growing," as it was imagined to be in springtime (the season associated with "cold" and "wet" weather), while bleeding in hot weather was discouraged. One relic from this blood-obsessed period can still often be seen, in the form of the red-and-white stripes on the barbershop pole: The white represented the foamy shaving cream; the red symbolized bloodletting.

Today the idea of intentionally causing a patient to bleed seems absurd, but there was a certain logic to it in Shakespeare's time. Blood was imagined to be the most refined form of food and drink, carried to all parts of the body through the veins. But it had to get "used up," so to speak, along the way—which is why, in *Coriolanus*, we hear that the "great toe" is "the worst in blood" (1.1.153). Excess blood could erupt in pimples or boils; it could also cause a fever. One solution was to stop eating (hence the expression "starve a fever"); but an even faster way to bring down the patient's temperature was to draw blood. Conversely, someone who seemed pale, or cold to the touch, might be told to eat more food (this producing more blood); red meat, in particular, might be recommended.

"They didn't have the insight into diseases that we have today," Flude says, as we sit at a desk nestled among the museum's exhibits. "They tended to confuse symptoms with causes. So, if you get a red face when you have a fever, they thought that was caused by blood. And blood *does* cause red faces. So they thought that was the cause of the fever." Either the patient's blood was bad, or there was too much of it; either way, the solution was to get rid of some of it. (They understood that the body could generate blood anew.) "So it was a quite simple cause-and-effect, as they saw it," says Flude. "And it kind of worked: When someone has a fever, and they're bled, they would likely go pale, and the symptom was relieved. It kind of makes sense. We know it *doesn't* make sense now, but at the time it made sense."

Doctors in Elizabethan England did what they could for their patients, but they lacked for real medical knowledge: They had little chance of deducing what was actually making a person sick. Luck almost certainly played as important a role as the work of the physician. Some patients were treated and got better—but perhaps they would have recovered just as quickly (or even more quickly) with no treatment. By the same token, some patients got steadily worse, and, in spite of treatment, they eventually died—but very likely they would have died even if no treatment had been given (and just the fact that *some* treatment was offered may well have brought comfort). Illnesses were not categorized in a rigorous way, and, worse, the symptoms themselves were often conflated with the disease. For example, fever is today recognized as a symptom common to many diseases; but in Shakespeare's day it was seen as an illness in itself. As Olsen notes, "No wonder that any one type of treatment, when applied to all fevers of whatever origin, usually failed to work." Diseases like tuberculosis, influenza, dysentery, smallpox, malaria, and syphilis were killers, while scurvy was particularly common among sailors. (The queen herself nearly died from smallpox in 1562, two years before Shakespeare's birth.) And of course there was the plague; because of its unpredictability, it was even more feared than the others.

THE PLAGUE'S THE THING

Even though Shakespeare was fortunate enough to be spared personally from the plague's terrifying effects, the dreaded disease would have been an integral part of daily life, as familiar as the weather. As mentioned, there were at least five outbreaks in Shakespeare's lifetime. Europe had in fact been subject to two distinct varieties of plague, bubonic and pneumonic (and we can't be sure which is being referred to in records from the time, which often use "plague" for both). The bubonic was the more common form; it took about six days to incubate, leading to nausea, fever, and swellings in the groin or armpit known as "buboes." Once the plague was contracted, the outlook was not good; the disease killed about six in ten of its victims. The pneumonic form attacked the lungs, and was even more deadly, killing virtually all of those who contracted it.

How the disease spread was a matter of conjecture. People blamed everything from bad air to inauspicious planetary alignments to (no surprise) foreigners; many also attributed the disease to the wrath of God.* We now know that the disease was spread by fleas that in turn had bitten infected rats. At the time, no one thought to blame the rats or the insects that hitched rides on them, perhaps because both were a ubiquitous presence in the towns and cities of Elizabethan England.† As Kristin Olsen notes, doctors were on the right track when they blamed animals, but they blamed the wrong animals, indicting dogs and cats instead of rats and fleas. The result was a government-ordered cull of stray dogs and cats—which in turn resulted in an increase in the rat population, and further outbreaks of the plague. The overall lack of sanitation and hygiene was the other major culprit. In cities and towns, waste was dumped onto the streets or into nearby ditches.

At least the contagious nature of the disease was recognized. It was believed to be spread—somehow—through the air, and people caught it from others who were infected. During outbreaks, foreign ships were kept at anchor, and travelers on foot were either turned back by guards posted well outside a town's limits, or forced to stay in makeshift hospitals for forty days—in Latin, *quarantina*—until they could be confirmed not to pose a risk. (It is from this practice, of course, that we have the word "quarantine.")

Houses in which someone had contracted the disease were sealed off, and family members were trapped inside with the afflicted person (hopefully aiding in the patient's recovery without catching the disease themselves). Overcrowding was recognized as a problem, and those who had the means fled to the countryside with each outbreak. Of course, nothing drew crowds like the theater, which, as mentioned, could attract as many as three thousand people for each performance. When plague deaths exceeded thirty per week, authorities in London closed the theaters in an effort to slow the spread of the disease. This of course had an

* The tendency to blame disease on foreigners can also be seen in the case of syphilis, another poorly understood malady. As John Hale notes, there was "much displacement of responsibility: To Italians it was either the Spanish Disease or—more popularly—the French Disease, to the French it was the Pox of Naples, to the Turks it was the Christian Disease" (Hale, p. 556).

† Also, the fleas would bite humans only if there were no more rats in the immediate area—which means that when an outbreak began, the rats may not have been particularly noticeable, making their role harder to infer.

enormous impact on Shakespeare and the actors in his company, who would have been forced to look for work in the country when the theaters shut down. Records show that the theaters were closed for some seventy-eight months between 1603 and 1613—closed, in other words, more than half the time. During an outbreak in 1609, Thomas Dekker noted that the playhouses stood with their "dores locked up, the Flagges . . . taken down"; in the surrounding neighborhoods one saw "houses lately infected, from which the affrighted dwellers are fled, in hope to live better in the Country."

Shakespeare makes only a handful of direct references to the plague, and usually it is at least somewhat metaphorical—for example, in *Timon of Athens*, where Timon is less than thrilled to hear that the Athenian ambassadors are at his doorstep: "I thank them; and would send them back the plague, / Could I but catch it" (5.1.137–38). None of Shakespeare's characters die from the plague; in fact, none of them even catch it. Perhaps the subject was simply too close to home. Nonetheless, more general references to disease, infection, and fevers are everywhere in the canon, and this surely reflected the concerns that ordinary people had with their health and well-being—neither of which was guaranteed to last you through to next Tuesday. "Mortality and anxiety," notes Peter Ackroyd, "were part of the air that the citizens breathed."

"THAT WAY MADNESS LIES"

There were, of course, mental ailments along with physical ones, and their effects were equally frightening. The causes of mental illness were not well understood; it was thought that madness could be triggered by emotional trauma, severe anxiety, and even unrequited love. There could also be physical causes, such as a fever or a bite from a mad dog. Or, as Shakespeare suggests in *Macbeth*, it could come from eating "the insane root" (1.3.82)—a reference, perhaps, to the root of the mandrake plant; similar to nightshade, it was known to have hallucinogenic properties. Even the the moon was thought to play a role, either due to its phase or (more rarely) due to its distance from Earth (hence "lunacy," from the Latin word for moon, *luna*). Shakespeare frequently alludes to the moon's influence on human affairs. In *Othello*, for example, the title character reacts to the news that "foul murders" have been committed:

> It is the very error of the moon:
> She comes more nearer earth than she was wont,
> And makes men mad.*
>
> (5.2.111–13)

(Not that Othello is one to talk; he murdered Desdemona barely a dozen lines earlier.)

There was little to be offered in the way of treatment. Some of those deemed "mad" were cared for by their families; others, as Olsen notes, "simply wandered from town to town, blamed for any increase in local crime." There were hospitals for the insane, but the level of care was appalling; patients could be beaten or even put on display as public entertainment. The most famous mental facility was London's Bethlehem Hospital, known by its nickname, Bedlam. In *King Lear*, Edgar aims to pass for a "Bedlam beggar" by smearing his face with dirt, messing up his hair, and speaking in riddles and nonsense rhymes. While Edgar is faking it, the king is terrified that he is losing his mind for real: "O let me not be mad," he moans, "not mad, sweet heaven! / Keep me in temper, I would not be mad" (1.5.37). To add insult to injury, madness was seen not only as a medical condition but as a character flaw. Whether Hamlet's madness is real or feigned (or both) is a matter for endless debate, but either way, his uncle is ashamed of the prince's behavior, chiding him for displaying "unmanly grief"; he tells Hamlet that his disposition reflects "a will most incorrect to heaven" (1.2.94–95).

The lack of any viable treatment for madness haunts act 5 of *Macbeth*, as Lady Macbeth's condition steadily deteriorates. Already in scene 1, the doctor declares that "This disease is beyond my practice" (5.1.49), and that "More needs she the divine than the physician" (line 64). By scene 3, the end is near, and Macbeth's frustration is palpable:

---

* Othello seems to be referring to the moon's distance rather than its phase. The moon's orbit around the Earth is an ellipse, causing its distance to vary (by about 12 percent) over the course of a month. This variation had been observed since ancient times, long before the true shape of its orbit had been deduced. The variation in size could be explained—although not very accurately—by means of the epicycles and deferents in the Ptolemaic system (see Chapter 1). It was Kepler who showed that planets (he did not mention the moon) do in fact move in elliptical orbits, beginning with his description of the motion of Mars in his *Astronomia Nova* of 1609 (less than six years after Shakespeare wrote *Othello*).

MACBETH

Cure her of that.
Canst thou not minister to a mind diseased,
Pluck from memory a rooted sorrow,
Raze out the written troubles of the brain,
And with some sweet oblivious antidote
Cleanse the stuffed bosom of that perilous stuff
Which weighs upon the heart?

DOCTOR

Therein the patient
Must minister to himself.

MACBETH

Throw physic to the dogs, I'll none of it.
(5.3.41–48)

Macbeth's doctor is just one of the physicians we meet in Shakespeare's plays. Medical men of various kinds appear frequently in the canon—more often, in fact, than workers of any other profession. We often hear their diagnoses and treatments; and they appear on stage not only in *Macbeth* but also in *King Lear* and *Two Noble Kinsmen*. We might take particular note of *All's Well That Ends Well*, in which the heroine, Helena, is the daughter of a famous doctor; she has learned many skills from her father, and uses her knowledge to save the king's life. But the medical talk does not sound forced in any of the plays; rather, it comes up naturally as the characters go about their business. It is, as Maurice Pope puts it, "unobtrusive."

Where did Shakespeare acquire his medical knowledge? It has occasionally been suggested—especially by people who question the plays' authorship—that only someone with medical training could have written so knowledgably about medicine. But that's taking it too far: Shakespeare didn't need to be a doctor in order to write about illnesses and treatments, just as he didn't need to be a nobleman to write about courtly intrigue, or to have visited Italy to write about Italian cities and customs. All he needed to do was to keep his eyes and ears open. As John Andrews points out, Shakespeare's knowledge of medicine, while seemingly thorough, wasn't particularly unusual for the time. "The number of medical references in Shakespeare's plays . . . doesn't necessarily indicate that

Shakespeare knew more about medicine than his contemporaries," Andrews writes. "Most Elizabethans were very concerned with their health and were thus familiar with basic medical theories."

A DOCTOR IN THE FAMILY

In the latter part of his career, Shakespeare had another route to medical know-how: His oldest daughter, Susanna, had married a successful doctor named John Hall in 1607. Perhaps this accounts for the respect that Shakespeare bestows on his medical men, especially in the later plays. There are, to be sure, numerous gibes—as in *Timon of Athens*, where Timon warns, "Trust not the physician; / His antidotes are poison" (4.3.433–34); and Lear's advice to "Kill thy physician, and thy fee bestow / Upon the foul disease" (*King Lear* 1.1.157–58). But overall, Shakespeare's doctors, and the medical profession in general, are shown in a positive light. And as Jonathan Bate has noted, Shakespeare's portrayal of doctors seems to take on an increasingly positive tone after his daughter's marriage. In the early plays we find comic figures like Pinch in *The Comedy of Errors* and Caius in *The Merry Wives of Windsor*—but in the later works we encounter "several dignified, sympathetically portrayed medical men."

Hall had studied at Queens' College in Cambridge, and began to practice medicine in Stratford around 1600. We don't know how often Shakespeare spoke with his son-in-law; as far as we can tell, the Halls lived in Stratford, while Shakespeare, by this point, was spending most of his time in London. Still, it's hard to imagine that Hall's work *never* came up on those occasions when the two men spoke, over the years. As it happens, we know more about John Hall than we do about most physicians of the period, because he left behind detailed notes about his cases (later published with the title *Select Observations on English Bodies of Eminent Persons in Desperate Diseases*).

Remarkably, Hall has even left us with the details of a treatment he applied to Susanna herself:

> Mrs *Hall of Stratford*, my Wife, being miserably tormented with the Colic, was cured as followeth. R. *Diaphen. Diacatholic.* Ana 3i. *Pul. Holand* 3ii. *Ol. Rute* 3i. *Lact. Q.s.f. Clyt.* This injected gave her two Stools, yet the pain continued, being but little mitigated; therefore I appointed to inject a Pint of Sack made hot. This presently brought forth a great deal of Wind, and freed her from all Pain. To her stomach was applied a Plaister

> *de Labd. Crat. Cum Caran. & ol. Macis.* With one of these Clysters I delivered the Earl of *Northampton* from a grievous Colic.

Dr. Hall is describing an enema ("clyster"), and even if the casual reader isn't quite sure what all those ingredients are, it provides a fascinating glimpse into the lives of Shakespeare's daughter and her doctor husband. As Bate notes, Dr. Hall likely procured his various ingredients from Philip Rogers, the apothecary who ran a shop on High Street in Stratford (and who was once sued by Shakespeare over an unpaid debt); other medicinal plants and herbs could have come from his own garden at New Place. Visitors to the Halls' home can still see the garden—now purely decorative in function. As Bate points out, it has become difficult to sense "the intimate relationship between plants and medicine in Shakespeare's time."

Hall's notes also reveal interesting correlations with the records at St. Thomas's Hospital. For example, Kevin Flude has noted that in 1612 the hospital paid nine pence for a pigeon to "lay at the feet of a patient." This may sound like straight-ahead quackery, but Dr. Hall himself prescribed the very same remedy for himself in 1632: "Then was a pigeon cut open alive, and applied to my feet, to draw down the vapours; for I was often afflicted with a light Delirium." The pigeon-at-the-feet treatment "seems fairly weird," Flude admits. "And yet that was a very well-educated, mainstream doctor—Dr. John Hall—who prescribed that. So the difference between a mountebank and a proper doctor is difficult to tell at this particular stage."

On other occasions, Flude has found, Hall prescribed treatments involving such ingredients as hart's horn, shavings of ivory, spiderwebs, dried windpipes of roosters, and a cosmetic known as virgin's milk.* More benign herbal remedies were more typically prescribed, and some of the more common "cures"—including wormwood, plantains, and egg whites, for example—can be found both in the records of St. Thomas's Hospital and throughout Shakespeare's plays. The witches' brew in *Macbeth*, containing such appetizing ingredients as "eye of newt, and toe of frog / Wool of bat, and tongue of dog" (4.1.13–14), sounds rather alarming today; but in Shakespeare's time, Flude notes, it was "not so very far from the reality."

---

* The Merriam-Webster dictionary says that it consists "either of the tincture of benzoin or some balsam or of lead subacetate precipitated by addition of water."

13. *"Drawn with a team of little atomi . . ."*

LIVING IN THE MATERIAL WORLD

O cursed, cursed slave!," moans Othello, as he comes to grip with the fact that he has just murdered his wife:

> Whip me, ye devils,
> From the possession of this heavenly sight,
> Blow me about in winds, roast me in sulphur,
> Wash me in steep-down gulfs of liquid fire—
> (5.2.275–79)

Throughout Shakespeare's works, there is remarkably little talk of heavenly rewards and hellish punishments—but here we at least have a vivid description of the imagined fate that awaits the sinner after death. (Though such treatment is normally a thing to be feared, in this case Othello yearns for it in the aftermath of his vile act.) But what if there are no angels and harps to reward the virtuous, and no pools of sulfur to torment the wicked? In fact, a forward-thinking Greek philosopher had imagined this very scenario some sixteen centuries earlier. In Shakespeare's time, that philosopher's message—lost for more than a millennium—was slowly finding its way back into the cultural life of Renaissance Europe.

The Roman poet Lucretius (99–55 B.C.) left us only one extant piece of writing—but what a piece it is. Spanning some 7,400 lines of rhyming Latin hexameter, his poem *De rerum natura* (*On the Nature of Things*) presents a radical description of the natural world. It is reductionist, materialist, and virtually godless. For Lucretius, the universe and all the wondrous diversity within it came about not because of God (or the gods), but because of the unthinking, random jostling of atoms, tiny particles that "Fly all around in countless different ways . . . Perpetually driven by an everlasting motion." Lucretius didn't invent the atomic theory, which had in fact been circulating for some five hundred years. It is generally credited to the Greek thinker Leucippus and his pupil, Democritus, and it was later expounded on by the philosopher Epicurus, who attracted many followers.* The school of thought Epicurus gave rise to, built around materialism and the legitimacy of valuing pleasure for its own sake, became known as Epicureanism—and we might note that variations of the word "epicure" crop up at least four times in Shakespeare's plays. Epicureanism, while not quite synonymous with atheism, was the next-worst thing; its practitioners, as Benjamin Bertram writes, did not necessarily reject God outright, but "they were dangerously close to doing so and were risking eternal damnation." While various writers made note of the atomic theory, it was Lucretius who, more than any other ancient thinker, explored its philosophical implications—and who gave the theory its most eloquent expression. The world, Lucretius says, is self-made; it is brought into existence by nature herself,

> Of her own, by chance, by the rush and collision of atoms,
> Jumbled any which way, in the dark, to no result,
> But at last tossed into combinations which
> Became the origin of mighty things,
> Of the earth and the sea and the sky and all that live.

Lucretius goes on to describe the physical properties of atoms, and how the motion of these atoms is able to account for the vast array of phenomena we see in the natural world. He asserts that earthquakes, volcanoes, and lightning—which have struck fear into so many, and which had often been seen as requiring a supernatural explanation—have physical causes. To be sure, certain key ideas are missing; he did not quite have the notion of evolution and natural selection, for example—although Lucretius did suspect, as other ancient writers had, that species that once flourished have since become extinct. Nonetheless, *On the Nature of Things* has a strikingly modern feel.

---

* For a brief history of atomic theory, I (humbly) direct the reader to my first book, *Universe on a T-Shirt: The Quest for the Theory of Everything.*

THE BEST SCIENCE POEM EVER WRITTEN

Lucretius's most recent champion has been Harvard scholar Stephen Greenblatt, who examines Lucretius's influence in his Pulitzer Prize–winning *Swerve* (2011). When he read *On the Nature of Things*, Greenblatt couldn't help thinking of a trio of more recent thinkers: "So much that is in Einstein or Freud or Darwin or Marx was there," he told *Harvard Magazine*. "I was flabbergasted." In *The Swerve*, he describes the ancient poet's worldview:

> There is no master plan, no divine architect, no intelligent design. All things, including the species to which you belong, have evolved over vast stretches of time. The evolution is random, though in the case of living organisms it involves a principle of natural selection. That is, species that are suited to survive and to reproduce successfully endure, at least for a time; those that are not so well suited die off quickly. But nothing—from our own species to the planet on which we live to the sun that lights our days—lasts forever. Only the atoms are immortal.

This is a remarkably modern way of seeing things (even though, as Greenblatt stresses, the path from Lucretius to our own culture is neither straight nor smooth). Nonetheless, there is at least a hint of twenty-first-century skepticism in Lucretius's epic poem, in which we are confronted with a universe that is neither for, nor about, humans. Here, for example, is Lucretius's take on "intelligent design":

> For certainty not by design or mind's keen grasp
> Did primal atoms place themselves in order,
> Nor did they make contracts, you may be sure,
> As to what movements each of them should make.
> But many primal atoms in many ways
> Throughout the universe from infinity
> Have changed positions, clashing among themselves,
> Tried every motion, every combination,
> And so at length they fall into that pattern
> On which this world of ours has been created.

Lucretius's vision provides, among other things, a new and much less frightening view of death. There can be no heaven or hell, he reasons, because the soul is mortal. He and his followers regarded death as part of life; they dismissed notions of an afterlife as superstitions. He even urges his readers to keep their minds "clean of the taint of vile religion." Even so, Lucretius and his followers were not, in the modern sense, atheists: They did, in fact, have a panoply of gods. And as Gavin Hyman writes, "the notion, intrinsic to the modern understanding of atheism, of immanence—of the world existing quite free of any sort of transcendent realm—would have been almost unintelligible to them." But such gods as they *did* believe in kept their hands off of humans and human affairs. If they were not entirely absent, the gods were at least indifferent.

There is no need to recount the gradual rediscovery of Lucretius in the late Middle Ages and the Renaissance, a story that Greenblatt covers quite thoroughly in *The Swerve*. It is enough to note that the ideas that Lucretius explores with so much energy in his poem were just beginning to resurface in the time of Shakespeare. The message of the poem was still too radical to be openly embraced—it was too close to all-out atheism—but nonetheless, as Greenblatt notes, "Lucretian thoughts percolated and surfaced wherever the Renaissance imagination was at its most alive and intense."

Some thirty Latin editions of *On the Nature of Things* were published between 1473 and 1600, from thick scholarly versions to cheap pocket editions. We don't know if Shakespeare got his hands on one of these editions, although we *do* know that Ben Jonson did; his own pocket-sized copy, its pages ink-stained, corroded, and brittle (it is literally falling apart), can be seen in the Houghton Library at Harvard. But, as we will see, Shakespeare at least knew *of* Lucretius, thanks to Montaigne, whose work we will examine in a moment.

THE THEORY THAT WOULDN'T DIE

And what of Lucretius's bold theory of atoms in motion? References to atomic theory are rare in English writing through to the middle of the sixteenth century, but then, as Ada Palmer has noted, they become increasingly common—presumably inspired by a renewed awareness of Lucretius's poem—beginning in the 1560s (as it happens, the decade of Shakespeare's birth). We can at least say that Shakespeare had some understanding of what Lucretius's atoms were about. In act 1 of *Romeo and Juliet*, the young lover is talking to his friend Mercutio about dreams and dreaming. Mercutio replies that Romeo must have been visited by Queen Mab, referring to a tiny fairylike creature possibly originating in

Celtic mythology. She causes her "victims" to dream, by entering their brains through their noses while they sleep:

> She is the fairies' midwife, and she comes
> In shape no bigger than an agate stone
> On the forefinger of an alderman,
> Drawn with a team of little atomi
> Over men's noses as they lie asleep.
>
> (1.4.55–59)

This little speech is nothing if not vivid, and we are left wondering how small Queen Mab must be, and her little coach, and the even tinier "atomi" that pull it. Shakespeare mentions "atomi" on a couple of other occasions, as when Celia declares, in *As You Like It*, that "It is easier to count atomies as to resolve the propositions of a lover" (3.2.229–30). This of course is poetry, not physics—but then, we can say the same for Lucretius's *On the Nature of Things*. Of course, one doesn't need to postulate a theory of atoms in random motion to acknowledge the haphazard nature of life's journey. As Florizel admits in *The Winter's Tale*, ". . . we profess / Ourselves to be the slaves of chance," always at the mercy "Of every wind that blows" (4.4.536–37).

"QUE SAIS-JE?"

If Lucretius was the great skeptical thinker of the first century B.C., Montaigne filled that role in the sixteenth century. Michel de Montaigne (1533–1592) was brave enough to doubt much of the accepted dogma in his time. He questioned the authority of religious and political leaders; he questioned the wisdom of the ancient philosophers; he questioned mankind's privileged status in the cosmic hierarchy. He even questioned the power of reason to make sense of the world. In his private study, he had the words of the skeptic philosopher Sextus Empiricus, "All that is certain is that nothing is certain," painted along one of the wooden beams; another bore the phrase "I suspend judgment." And for good measure he had a medal struck with what has come to be thought of as his motto: *Que sais-je?*—"What do I know?" For Montaigne, nothing was to be taken for granted.

Montaigne, whom we met briefly in the introduction, was born into a wealthy family that owned property not far from Bordeaux, in the southwest of France. He was something of a child prodigy, and his father, recognizing his son's talents, arranged for the boy to be spoken to only in Latin. It seems to have worked, as the boy is said to have mastered Latin even before he was fluent in French. He studied law, advised kings and princes, and served two terms as mayor of Bordeaux—but it is as a man of letters that we remember him.

Montaigne spent much of the final twenty years of his life in self-imposed exile within the tower of his chateau, where he amassed a personal library of more than a thousand books. His greatest influences were the skeptical philosophers of the ancient world; along with Sextus Empiricus, he was attracted to the ideas of the Greek thinker Pyrrho of Elis (ca. 360–275 B.C). Pyrrho and his followers believed that human beings were in no position to judge questions for which there was no clear answer. Indeed, they weren't sure if anything at all could be known with certainty. Pyrrho's own writings have been lost, but an account written by Sextus, nearly five centuries later, survived. New copies of that text began to circulate in Europe in Montaigne's time, and he eagerly devoured its contents. Sometimes he agreed with the conclusions of the ancient writers, and sometimes he didn't—he read everything with a critical eye. And then he started writing. And writing. And writing. The result is his sprawling *Essays*, spanning 107 chapters and filling three books. They were published over a twenty-two-year period, beginning in the 1570s. In the *Essays*, Montaigne has given us his thoughts on life, the universe, and everything.

Montaigne, writing in plain French, sets down exactly what is on his mind—and because of this clarity and honesty, his words sound as fresh today as they did four and a half centuries ago. His foremost goal was to know the mind itself—a task he described as "a thorny undertaking," one that asks us "to penetrate the opaque depths of its innermost folds." Indeed, he more or less invents "stream of consciousness" writing:

> I turn my gaze inward, I fix it there and keep it busy. Everyone looks in front of him; as for me, I look inside of me; I have no business but with myself; I continually observe myself, I take stock of myself, I taste myself . . . I roll about in myself.

Montaigne was able to project his imagination outside of his own immediate world. He knew that, in a sense, everything is relative: Everything depends on one's point of view. He saw that what was sacred to members of one culture was blasphemy to members of another. As a young man he had traveled widely, and became acutely conscious of what we would now call "cultural relativism." Neighboring countries, even neighboring regions, had different customs, laws, and beliefs. "What kind of Good can it be," he asked, "which was honoured yesterday but not today and which becomes a crime when you cross a river! What kind of truth can be limited by a range of mountains, becoming a lie for the world on the other side."

To question culture and customs leads one inevitably to question religion and religious practice. Montaigne understood that people come to their religious beliefs through a series of accidents: of birth, of location, by exposure to particular teachers, and so on. We defend our particular brand of faith—but we should not be surprised that our neighbors defend theirs just as vigorously. He quotes approvingly from the Roman poet Juvenal: "The fury of the mob is aroused since everyone hates his neighbours' gods, convinced that the gods he adores are the only true ones." Montaigne had witnessed the horrors of religious warfare with his own eyes, as well as the needless suffering brought on by the witch-hunt craze. As he once reflected, "It is taking one's conjectures rather seriously to roast someone alive for them." And although Montaigne was fascinated with (perhaps obsessed by) death, he seems to have had little interest in what, if anything, came after it.

Montaigne questioned *almost* everything. He refused to embrace the godless world described by Lucretius. However much he may have tried, he could not bring himself to doubt the existence of a creator. (He at one point calls atheism a "monstrous thing," and later speaks of "the dreadful, horrible darkness of irreligion.") He remained a practicing (and presumably believing) Catholic to his dying day. But he saw no conflict between his faith and his skepticism: He accepted the teachings of his religion not by any process of reasoning, but simply *as faith* (a perspective known as *fideism*, from the Latin word for faith, *fides*). Without such faith, he believed, there was no way to anchor one's life. Yet he allowed for a separation of private and public life. Whatever inner thoughts one harbored, there was no need to let it interfere with one's place in the community. "The wise man ought to retire into himself, and allow himself to judge freely of everything," he wrote, "but outwardly he ought completely to follow the established order." As a result of this duality, James Jacob notes, Montaigne's skepticism "had revolutionary intellectual consequences, while scarcely producing so much as a social ripple." In spite of Montaigne's avowed devotion to his Catholic faith, many people came to see the *Essays* as an irreligious, even dangerous work. A century after its publication, the Vatican, which had seen no problems with the book initially, decided not to take any chances, and placed the *Essays* on its Index of prohibited works.

MONTAIGNE, MAN OF SCIENCE?

Montaigne, like Lucretius, was prepared to question one of the most cherished notions of all—the idea that the universe was made for our benefit. Montaigne's take on such pompous self-importance—which presages not only Sagan, but Feynman, Weinberg, and a host of late-twentieth-century writers on physics and cosmology—is worth examining (we looked at it briefly in Chapter 2, but here I will quote the relevant passage in its entirety). Who, Montaigne asks,

> has convinced [himself] that it is for his convenience, his service, that, for so many centuries, there has been established and maintained the awesome motion of the vault of heaven, the everlasting light of those tapers coursing so proudly overhead or the dread surging of the boundless sea? Is it possible to imagine anything more laughable than that this pitiful, wretched creature—who is not even master of himself, but exposed to shocks on every side—should call himself Master and Emperor of a universe, the smallest particle of which he has no means of knowing, let alone swaying!

There is a stark humility here, and something very much like the assertion of the modern scientist that the universe "isn't about us." But was Montaigne a proto-scientist? His biographers tread carefully on this question. M. A. Screech, translator for the massive Penguin edition, says that later Enlightenment thinkers came to see Montaigne as an early proponent of "atheistic naturalism," while Hugo Friedrich notes Montaigne's tireless, inquisitive approach to the study of mankind, which he characterizes as "anthropological curiosity." But Friedrich adds a cautionary note: Montaigne's way of reasoning "bypasses the mathematical, physical, and technical sciences," and it would be an "erroneous conclusion" to label Montaigne an early scientist. The problem, I think, is that Montaigne sounds remarkably modern at times, and yet conservative and traditional at others—as one might expect, perhaps, from a figure who lived so close to what we imagine as the temporal divide separating the old from the new. Consider his stance on medicine: He seems to have embraced Galen's theory of the humors—but he also spoke enthusiastically of the recent work of the physician Paracelsus, who emphasized the role of observation over adherence to ancient texts. At one point Montaigne examined the contents of a goat's stomach and, on finding stones in it, concluded that goat's blood was probably useless as a cure for such ailments, as had popularly been imagined. In this early endorsement of experimentation and observation, writes R. A. Sayce, "we can see how his scepticism and empiricism are leading him towards the standards of the scientific age." His approach "points to the scientific method, and it seems likely that he exercised a direct influence on its development through Descartes and Pascal and perhaps even Bacon."

At the very least, Montaigne's way of thinking often has the *flavor* of science. One of my favorite examples of his hard-nosed skepticism concerns the fate of sailors lost at sea. (Like so many of his stories, it has classical origins; this one he borrowed from a Greek poet named Diagoras of Melos, also known as Diagoras the Atheist, who lived in the fifth century B.C.) Montaigne writes:

> [Diagoras] was shown many vows and votive portraits from those who have survived shipwreck and was then asked, "You, there, who think that the gods are indifferent to human affairs, what have you to say about so many men saved by their grace?"—"It is like this," he replied; "there are no portraits here of those who stayed and drowned—and they are more numerous!"

When I read this passage, I hear something akin to the modern skepticism of Lawrence Krauss or Stephen Hawking. Montaigne is a sober and skeptical debunker, and perfectly able to outthink the philosophers.

Consider his evaluation of the senses: We have our eyes and ears to take in the world around us—but while our senses serve as our window on the world, they also necessarily restrict our view: "We have formed a truth by the consultation and concurrence of our five senses," Montaigne writes, "but perhaps we need the agreement of eight or ten senses, and their contribution, to perceive it certainly and in its essence." Moreover, the senses can mislead us; what we see or hear can often lead us astray. One of his examples—again borrowed from one of the ancient writers—involves the theater, and is thus one that Shakespeare could certainly relate to: If colored glass is used to cover the torches that illuminate the stage, the audience can be fooled into thinking that the people and objects in front of them are blue or red or some other hue quite different from their natural color. But even when no tricks are being played on us, we can never be quite sure of what we're seeing. As Montaigne noted, different people do not necessarily perceive things in exactly the same way: Some people can hear or see better than others; as well, our senses deteriorate as we age. He was also careful not to confuse the sensory impression of an object with the object itself. "So whoever judges from appearances," he wrote, "judges from something quite different from the object itself." Snow, he said, may "seem white to us," but how do we know "that it is truly so in essence?" And once one recognizes this basic problem, "all the knowledge in the world is inevitably swept away." He even wondered what the world might seem like to a dog or to other animals, and tried to put himself inside their minds. He famously asked, "When I play with my cat, how do I know that she is not passing time with me, rather than I with her?"

Even if he wasn't a "scientist," Montaigne took a deep interest in what scientific thinkers—both ancient and contemporary—had to say about the structure of the world. He read Lucretius's epic poem, and his own heavily annotated copy of *De rerum natura* can now be seen in the library of Eton College. As Stephen Greenblatt notes, Montaigne's *Essays* contain more than a hundred quotations from Lucretius. And so, whether he found it convincing or not, Montaigne at least had some awareness of the atomic theory outlined at length by Lucretius. He also took an interest in the thinkers of his own century. As mentioned briefly in the Introduction, Montaigne was aware of Copernicus's heliocentric theory. Throughout history, everyone believed that the heavens moved, he states, until certain ancient thinkers proposed the idea that the heavens in fact remain fixed, and that

> it was the earth that moved, by the oblique circle of the *Zodiake*, turning about her axel. . . . And in our daies *Copernicus* hath so well grounded this doctrine, that hee doth very orderly fit it to all Astrological consequences. What shall we reape by it, but only that we neede not care, which of the two it be? And who knows whether a thousand years hence a third opinion will rise, which happily shall overthrow these two præcedent?

Note that Montaigne—quoted here in John Florio's translation of the *Essays*—is very far from embracing Copernicanism; indeed, what he seems to be questioning is not so much the authority of the ancients, but the authority of philosophers in general.* Even so, as Friedrich notes, Montaigne "was one of the first in France who became aware of Copernicus and took him seriously." Sayce goes further: In his brief discussion of cosmology, "Montaigne shows . . . his awareness of the greatest scientific discovery of the sixteenth century and an accurate appreciation of its significance as then understood." This, together with his reference to ideas about the Earth's motion, suggest "that he took the theories of Copernicus more seriously than the brief account [in the *Essays*] might lead us to suppose." And it's not just that he mentions Copernicanism and its competitors; it's what he says about the nature of science. The twentieth-century philosopher Karl Popper would have been proud of Montaigne's assertion that all knowledge is temporary; that a new theory "a thousand years hence" may overthrow our best theories of today. For Montaigne, everything was open to debate; all knowledge could be questioned. He was well aware of the "new" territories that French and Spanish and Dutch and English explorers were discovering, and the surprises they encountered in these far-off lands. Perhaps the cosmos holds some surprises for us too? "Is it not more likely," he asks, "that this huge body which we call the universe is very different from what we think?" Montaigne's particular brand of skepticism "threw everything into doubt, even itself," writes Sarah Bakewell, "and thus raised a huge question mark at the heart of European philosophy."

Montaigne's Essays were first translated into English in 1603 by John Florio (whom we met in Chapter 7), and they sold well. What was it about Montaigne's writing that appealed to the English mind? It wasn't necessarily his philosophy. Rather, as Bakewell explains, it was his style, and his utter lack of pretension:

> Montaigne's preference for details over abstractions appealed to them; so did his distrust of scholars, his preference for moderation and comfort, and his desire for privacy. . . . On

* It seems strange that scholars have had so little to say about Montaigne's discussion of Copernicus. M. A. Screech's hefty 1991 translation, running to nearly thirteen hundred pages, does not even have a listing for Copernicus in its index, and one has to hunt for a brief mention of the subject in the introduction.

> the other hand, the English also had a taste for travel and exoticism, as did Montaigne. He could show unexpected bursts of radicalism in the very midst of quiet conservatism: so could they. Much of the time he was happier watching his cat play by the fireside—and so were the English.

Florio, as Bakewell puts it, brought out the "hidden Englishman" in Montaigne. Born in London of an Italian father and an English mother, he traveled widely as a young man, and seems to have had a knack for languages. Shakespeare, as we have seen, may have known Florio, and was perhaps one of the first English readers of the *Essays*—and they had a profound impact on his work. (Because Montaigne's influence turns up as early as *Hamlet*, which predates Florio's published version of the *Essays* by a few years, it is thought that the playwright had access to earlier copies of Florio's translation, which had likely been circulating in manuscript.)

SHAKESPEARE AND MONTAIGNE

Scholars have long remarked on the similarities between numerous passages from Shakespeare's plays and the *Essays* of Montaigne. Dozens of examples have been found over the years; one will suffice to give the flavor of Shakespeare's debt to the French writer. Consider Shakespeare's *The Tempest* and Montaigne's essay "On the Cannibals." Montaigne was fascinated by reports from the New World, and devoured the various accounts from explorers who had returned from the Americas. Some of them, like the French explorer Nicolas Durand de Villegaignon, brought native captives back to France. Montaigne met some of these unfortunate men and women on a visit to Rouen—in this case, a group of Tupinambá people, from what is now Brazil. Here is Montaigne's description of the South American natives (as translated by Florio):

> It's a nation . . . that hath no kinde of traffike, no knowledge of Letters, no intelligence of numbers, no name of magistrate, nor of politike superioritie; no use of service, of riches or of poverty; no contracts, no dividences, no occupation but idle; no respect of kindred, but common, no apparell but naturall, no manuring of lands, no use of wine, corne, or mettle. The very words that import lying, falsehood, treason, dissimulation, covetousness, envie, detraction, and pardon, were never heard of amongst them.

And here, from Shakespeare's *The Tempest*, is Gonzalo's explanation of how he would turn Prospero's island into an earthly paradise, if only he had the chance to rule it as governor:

> I'th' commonwealth I would by contraries
> Execute all things, for no kind of traffic
> Would I admit; no name of magistrate;
> Letters should not be known; riches, poverty
> And use of service, none; contract, succession,
> Bourn, bound of land, tilth, vineyard—none;
> No use of metal, corn, or wine or oil;
> No occupation, all men idle, all . . .
>
> (2.1.148–55)

The parallel between these two passages was first noted in the nineteenth century, and since then scholars have found a multitude of such borrowings.

Of course, we can't, in a technical sense, *prove* that Shakespeare read Montaigne; after all, as Greenblatt said during our interview, "no one was around with a video camera" to catch him in the act—"but Montaigne's fingerprints are in many, many Shakespeare plays, after a certain point in his career." And it's not just the words themselves; in the case of the cannibals essay, as Virginia Mason Vaughan and Alden T. Vaughan note in the Arden edition of the play, Shakespeare employs a "rhetorical strategy of exploring different, often opposite, perspectives, never settling on a definitive view"—an approach that echoes that of Montaigne. Incidentally, the British Museum has a copy of Florio's translation of Montaigne's *Essays* that contains a signature, on its inside front cover, that may, perhaps, be that of William Shakespeare.* Knowing that Shakespeare read Montaigne, we are now in a position to restate a point made briefly in the Introduction: Shakespeare must at least *have known of* Copernicus's heliocentric theory, thanks to Montaigne's *Essays*, even if he had not come across it by another route.

---

* Like so much else that may illuminate the playwright's life, the signature is the subject of significant dispute. Fake Shakespeare signatures abound, and the one on the museum's copy of Montaigne had been declared a forgery decades ago. However, the debate reopened when Nicholas Knight compared the signature with that found on another copy of Montaigne, housed at the Folger Shakespeare Library in Washington; Knight believes both are authentic.

Why was Shakespeare so enamored with Montaigne? The two men, in spite of living on opposite sides of the English Channel and swearing loyalty to different kings and practicing different religions, are something like kindred spirits. Both men, as Bakewell puts it, have been "held up as *modern* writers, capturing that distinctive sense of being unsure where you belong, who you are, and what you are expected to do." Both men devoted their lives to investigating the human condition; both were insatiably curious; and neither was afraid to doubt.

In *King Lear*, too, we find Shakespeare borrowing extensively from Montaigne. As Jay Halio notes, it is in the *Essays* that Shakespeare finds unrighteous judges (4.5.146–48), blind men who can see (4.5.144–45), and dogs that could be "obeyed in office" (4.5.151); more generally, Shakespeare "seems indebted to the French essayist not only for phrases and ideas but for the skeptical attitudes that pervade the play." Scholars have found at least twenty-three passages in *Lear* that borrow directly from the *Essays*, and, as Millicent Bell notes, the play contains more than a hundred words that Shakespeare had not previously used, but that occur in Florio's translation of the *Essays*.

SHAKESPEARE'S SKEPTICAL VILLAINS

Let's take a closer look at the questioning and uncertainty that haunts *King Lear*. We have seen that Edmond is a skeptic—Bell describes him as a "skeptic philosopher"—but he can also be seen, perhaps, as something of a scientist. He is analytical and doggedly single-minded in pursuit of his goals, and is rational through and through. Let us look for a moment at Edmond's character, as sketched by the noted twentieth-century Shakespeare scholar A. C. Bradley:

> Edmond is an adventurer pure and simple . . . he regards men and women, with their virtues and vices, together with the bonds of kinship, friendship, or allegiance, merely as hindrances or helps to his end. They are for him divested of all quality except their relation to this end; as indifferent as mathematical quantities or mere physical agents.

> *A credulous father and a brother noble,*
> *. . . I see the business,*

he says, as if he were talking of *x* and *y*.

> *This seems a fair deserving, and must draw me*
> *That which my father loses; no less than all:*
> *The younger rises when the old doth fall:*

he meditates, as if he were considering a problem in mechanics.

"Mathematical quantities"? "*x* and *y*"? "A problem in mechanics"? If these aren't the hallmarks of a scientific thinker, I don't know what is; all that is missing is the white lab coat and black-rimmed glasses.*

Stephen Greenblatt, when we spoke in his Harvard office, focused on Edmond's relationship to nature—and the fact that Edmond sees no reason to look *beyond* nature: "I think of Edmond . . . as articulating what we could call a 'naturalistic' position: that the world is what it is by its *nature*; that people are what they are by their nature, not by astrological signs, not by divine impulsion, but because of how they are put together. Edmond has a very strong version of this naturalism. He thinks it is 'biological'—that's not the term that he uses, but that's effectively what he's saying."

Modern biology, of course, was not even in its infancy; modern science—what the "new philosophy" would evolve into—was only just being imagined by the likes of Bacon and Galileo. (Bacon's *The Advancement of Learning*, as we've noted, dates from the same year—give or take a few months—as *King Lear*.) Even so, as Jonathan Bate writes, "Edmond is the embodiment of the 'new man' who emerged in tandem with the 'new philosophy.'" A new way of thinking was just coming into being, but Edmond was already there as its first ambassador.

And Edmond is not alone in displaying this urge to scrutinize, dissect, and quantify. We have already seen how Hotspur, like Edmond, dismisses astrology as superstition (Chapter 10); but the two men also share a no-nonsense, analytical approach to their problems. In *Henry IV, Part 1*—in the same scene as the astrological dispute, in fact—we find Hotspur

* Bradley's analyses haven't survived the last hundred years unscathed: Today's scholars believe he erroneously applied early twentieth-century concepts to turn-of-the-seventeenth-century works. But then, what will scholars a hundred years from now think of today's literary criticism?

discussing with Glendower and Mortimer how their newly won lands ought to be divided. Glendower produces a map, and shows the proposed threefold division: "The Archdeacon hath divided it / Into three limits very equally" (3.1.69–70). The Welsh lord describes who gets which parcel of land, and is about to turn to other matters—when Hotspur jumps in. "Methinks my moiety [share], north from Burton here, / In quantity equals not one of yours. / See how this river comes me cranking in, / And cuts me off from the best of all my land. . . ." But, without skipping a beat, he proposes a solution: "I'll have the current in this place dammed up, / And here the smug and silver Trent shall run / In a new channel, fair and evenly" (lines 93–100). If Shakespeare's skeptical villains could be scientists, they could perhaps also be engineers.

I'm not sure why it is Shakespeare's villains, rather than his heroes, that display such a "scientific" orientation. Of course, we mustn't fall into the trap of thinking that Shakespeare would put only statements that he agreed with into the mouths of his heroes, and only those that he disagreed with into the mouths of his villains. On the other hand, Edmond and Hotspur are not *entirely* unlikable: We may despise their actions, yet admire the sharpness of their minds. As Harold Bloom notes, we may be forgiven for finding Edmond "dangerously attractive."

Lear himself toys with philosophical questions. In act 3 we find the king and his companions enduring the ravages of a storm on the open heath. Gloucester and Kent implore the king to seek shelter from the wind and rain. But Lear has more pressing concerns: "First let me talk with this philosopher," Lear says, turning to "Poor Tom" (actually Edgar in disguise). He asks, "What is the cause of thunder?" (3.4.138–39). Of course, Lear may be quite mad by this point—but he made similar inquiries even when he was in full possession of his wits. In act 1 he engages with his Fool in a similar vein:

FOOL

The reason why the seven stars are no more than seven is a pretty reason.

LEAR

Because they are not eight.

FOOL

Yes, indeed, thou wouldst make a good fool.
(1.5.28–29)

The Fool, it seems, has a knack for such riddles; a few lines earlier he asks, "Canst tell how an oyster make his shell?" (1.5.21). Lear doesn't know the answer—but the key here seems to be in the questions themselves. The stars, thunder, and oysters are all *supposed* to be God's handiwork; to seek physical explanations is to challenge not only the established faith, but faith itself. As William Elton writes in *King Lear and the Gods* (1968), "Lear's resort to natural, rather than divine, causation is a measure of his developing scepticism. . . . This appeal to second causes rather than to first, to nature rather than to God, was a mark of the new materialist doubt." He even seeks a materialist explanation for his daughter's behavior: Knowing that his once-beloved middle child has turned against him, he ponders a reductionist explanation: "Then let them anatomize Regan; see what breeds about her heart," he demands. "Is there any cause in nature that makes these hard hearts?" (3.6.33–34).

The betrayal by Regan was already fait accompli by this point. In act 2, Lear had implored her to be the sort of daughter who would care for him in his old age—"Dear daughter, I confess that I am old . . . on my knees I beg / That you'll vouchsafe me raiment [clothing], bed, and food." But to no avail; Regan calls his pathetic request "unsightly" (2.4.146–48). In the New Cambridge edition, Jay Halio remarks, "Lear aptly summarizes Gonerill's and Regan's Darwinian outlook, in which survival of the fittest rules and the elderly are superfluous." We glimpse something of this attitude in Edgar, too. Edgar is just as skeptical as his brother, though, unlike Edmond, he tends toward compassion rather than manipulative self-advancement. Even so, as David Bevington notes, Edgar "is like his brother Edmond in his matter-of-fact understanding of nature. He recognizes that competition for survival is a fact of existence." Atomism, atheism, reductionism, Darwinism: *King Lear* is a believer's worst nightmare. It is Shakespeare's darkest play. It may also be his greatest.

14. *"As flies to wanton boys . . ."*

THE DISAPPEARING GODS

H*amlet* versus *King Lear*: The question of which is the greater achievement is an age-old debate, and as with any vexing contest—Mac vs. PC, Coke vs. Pepsi, Betty vs. Veronica—passions run high. Beginning in the early years of the twentieth century, however, the tide seems to have turned in *Lear*'s favor. A. C. Bradley, writing in 1909, describes Shakespeare's *Lear* as "the tragedy in which he exhibits most fully his multitudinous powers"; if we were to lose all of the plays except for one, he argues, *Lear* would be the one to salvage. In fact, there is an entire book on the question of which is the greater of the two plays—R. A. Foakes's aptly titled *Hamlet versus Lear* (1993). The author suspects "that for the immediate future *King Lear* will continue to be regarded as the central achievement of Shakespeare, if only because it speaks to us more largely than the other tragedies to the anxieties and problems of the modern world."

Anxieties indeed. It is *King Lear*, not *Hamlet*, that somehow reflects the malaise that has loomed over the planet since the horrors of the World Wars. Trench warfare, the Holocaust, nuclear weapons, environmental destruction, terrorism—all have left their mark on the psyche of the twenty-first century, and all contribute to a kind of helplessness of the sort that hangs over *Lear*. The play was likely written in 1605, and was probably staged for the first time that year, although the first documented performance is from Christmas 1606, when it was performed at court in front of King James.

King Lear is surely the bleakest of Shakespeare's plays. It's not just that everybody dies—*Hamlet* had already been there and done that; and *Titus Andronicus* tops *Lear* for sheer blood and guts—but in *Lear*, the collateral damage is incalculably higher. It seems as though justice and morality and meaning are poised to perish along with the main characters. By the time Lear comes on stage with Cordelia's lifeless body, the audience must believe they've hit rock-bottom. Samuel Johnson couldn't bear the play's ending, saying it violated our "natural ideas of justice." John Dover Wilson writes that in *Lear* "horror is piled upon horror and pity on pity," making it "the greatest monument of human misery and despair in the literature of the world." Throughout the play, observes Thomas McAlindon, Shakespeare "has been asking how much misery life can inflict on human beings, and how much they can endure. Now he is wondering how far tragedy can go, and how much an audience can take." Where other dramatists paused or turned back, the author of *King Lear* pushed on. Shakespeare, notes McAlindon, "resolutely brings us to the edge of the abyss and beyond." Perhaps the playwright pushed too far: In 1681, Nahum Tate came up with an alternate version of the play—a softened rewrite in which Cordelia survives. This was the preferred version for the next 150 years, until the early decades of the nineteenth century, when Shakespeare's version again seemed palatable.*

BELIEF IN A JUST WORLD

Human beings either need or crave a handful of essentials—air, food, water, sex. But we yearn for justice, too. Seeing a person wrongly made to suffer fills us with anger; hearing that someone is getting away with wrongdoing fills us with outrage. Miscarriages of justice leave us steaming mad. Experiments have shown that even very young children have a sense of justice: They get cranky when a misbehaving puppet is rewarded rather than punished, and, if the child is put in charge of handing out rewards, she'll reward a "good" puppet rather than a "bad" puppet. Roughly put: We're hard-wired to want good deeds to be rewarded, and to see bad deeds punished. By the time we reach adulthood, the desire for justice is an integral part of who we are.

In fact, we desire justice so much that it shapes our view of the world. Sometimes we end up perceiving justice even when it isn't there. We *read*

---

* We might note that the previous *Lear* story—Shakespeare's immediate source—was also a happier one. A version dating from 1594, titled *The True Chronicle History of King Leir, and his three daughters, Gonerill, Ragan, and Cordelia*, had been a success on the London stage; in this version, both the king and Cordelia live, and the two evil sisters have their comeuppance.

it into nature, imagining that the universe itself has some kind of moral aspect. Psychologists believe there's a cognitive process at work in our brains that often causes us to imagine that the world itself is inherently just—that people "get what they deserve." Our language is filled with everyday expressions that reflect this desire to imagine that such "cosmic justice" has indeed been meted out: "He got his just deserts"; "What goes around comes around"; even "karma." Psychologists have a name for this way of seeing things: It's called the "just-world theory," or sometimes "belief in a just world" (BJW). Social psychology has produced a vast literature on the subject of just-world beliefs over the last few decades, beginning with the seminal work of Melvin Lerner in the 1960s. The idea, as one psychologist puts it, is that "good things tend to happen to good people and bad things to bad people despite the fact that this is patently not the case." Exactly why we evolved this way of thinking remains a subject of investigation, but psychologists suspect that, like many other "cognitive biases" that have been identified over the years, it confers a survival advantage. Our best guess is that it reinforces the idea that we're in control of our lives; that our actions, and those of others, have predictable consequences. As a recent review article puts it, "the BJW seems to provide psychological buffers against the harsh realities of the world as well as personal control over one's own destiny."

Unfortunately, bad things very often *do* happen to good people. The "slings and arrows of outrageous fortune" that Prince Hamlet spoke of rarely discriminate in their choice of targets. Lerner cautioned that belief in a just world is an "invention"; his book on the subject was titled *The Belief in a Just World: A Fundamental Delusion* (1980). But its effects are very real, and often quite repugnant, including a tendency to blame those who are suffering, or who have been victimized, for their fate. This can apply to individuals (e.g., blaming a rape victim because of the way she dressed); to groups (e.g., the poor must be lazy); and even to natural phenomena (e.g., arguing, as Pat Robertson did, that Hurricane Katrina was divine punishment for America's tolerance for abortion). Note that the first two cases require a human instigator, while the third is "an act of God." In every case, however, innocent people suffer; and of course innocent people *shouldn't* suffer in a just world. And therefore some of us will bend over backward to find a reason, any reason, why the victims of misfortune *deserve* their fate—and why those who have prospered deserve their good fortune. Belief in a just world has another negative consequence: It can make us doubt scientific evidence. A study by psychologists at the University of California–Berkeley, for example, has shown that some people resist the idea of anthropogenic climate change, even in the face of substantial and growing evidence, because it conflicts with their notion of a just world. As Matthew Feinberg and Robb Willer put it, acknowledging global warming "threatens deeply held beliefs that the world is just, orderly, and stable."

Shakespeare never got around to teaching psychology at Berkeley—but it would seem that he had some sense of these ideas four centuries ago. At the very least, he recognized that the idea of an inherently just world stretches credulity. He either knew, or suspected, that the universe does not have a moral aspect; that things *just happen*. Sometimes, as disturbing as it may be to witness, good things happen to bad people, and, even worse, bad things happen to good people.

Nowhere is the idea of cosmic justice explored more thoroughly than in *King Lear.* The word "nature" is used more often in *Lear* than in any of the other dramas, and the play itself can be seen, as David Bevington asserts, as "a battleground over which rival concepts of nature are being fought." As Thomas McAlindon puts it, the play is an investigation of "the nature of nature." It asks what sort of universe we inhabit. The choice is between one that is "essentially moral, dictating altruism, community, limit, and reason" versus a newer view in which "nature is an amoral system which encourages egoism and the unscrupulous use of force and cunning to achieve one's desires." And the answer is not comforting. Samuel Johnson asked if Lear was "a play in which the wicked prosper," and it certainly looks that way. The wicked often go unpunished, and the good reap few rewards. No one who has seen the play will forget the blinding of Gloucester, surely one of the most cruel (even sickening) scenes in Renaissance drama. But note what one of the servants says upon witnessing the horror inflicted by Cornwall on his victim: "I'll never care what wickedness I do / If this man come to good" (3.7.98–99).* The servants, as Bevington notes, "pose disturbing questions about the threat of universal disorder that must surely result if crimes are unpunished by the gods." The gods similarly fail to punish Iago, in *Othello*, so the Venetian authorities have to make up for the gap in cosmic justice: "If there be any cunning cruelty / That can torment

* This passage occurs only in the quarto text, and not the folio (and so it can be found, for example, in the Arden *Complete Works*, but not in the New Cambridge edition).

him much and hold him long, / It shall be his" (5.2.332–4). This sounds more like vengeance than justice, but it is needed, Stephen Greenblatt writes, as "a gesture, however inadequate, toward repairing the damaged moral order." The gods seem similarly uninterested in the protagonist in *Macbeth*, but at least in that play the title character meets his match, and we have the satisfaction of seeing Macduff come on stage carrying Macbeth's decapitated head. But in *Lear*, the gods, if they exist at all, seem to be looking the other way. In Shakespeare's time, churchmen often condemned the theater as immoral, but *Lear* presents something far worse than immorality. It suggests a universe that is neither just nor unjust, but rather one in which justice, unless we take steps to establish it ourselves, is simply absent. We are confronted with a universe that is terrifyingly *amoral*.

Edmond, the bastard son of the Earl of Gloucester, is one of Shakespeare's great villains. What is alarming about his role in *Lear* is that he prospers: He is resolutely evil—he is more than willing to harm others to achieve his own selfish goals—and he gets away with it. Bevington writes:

> What is truly frightening about *King Lear* is that the battle over "nature" seems to run in Edmond's favor to such an extraordinary degree and for so long a time. His creed of self-reliance gives him, as he readily perceives, a tactical advantage over those who credulously submit to the moral restrictions of the social order. Holding the view that moral codes are simply part of the mythology by which the power structure enforces its grip on society, Edmond sees no reason not to lie, cheat, or otherwise overwhelm those who stand between him and the goals of his limitless ambition. Confident that there are no gods to reward or punish, and no afterlife in which to suffer eternal pain, Edmond proceeds with relentless energy and tactical brilliance.

Edmond is, along with Aaron, Iago, and Richard III, one of a set of characters that provide Shakespeare with "the stage panache of the unapologetic villain," as Jonathan Bate puts it. But Edmond is more than a brilliant, calculating, manipulative villain. He is also a skeptic. We have already seen (in Chapter 10) his refusal to buy into astrology, rejecting the superstitions embraced by his father, but he is also—as Bevington suggests in the above passage—perfectly willing to reject the notion of life after death, and, indeed, to reject the gods themselves.

A BRIEF HISTORY OF ATHEISM

Just as "science," in the sense we use the word today, didn't quite exist in Shakespeare's day, atheism, too, was absent in its modern, Dawkins-like form. There had been millennia of debate on the extent of the involvement of God (or the gods) in human affairs—but the idea of the complete nonexistence of God is, as Gavin Hyman writes in *A Short History of Atheism* (2010), "an intrinsically modern disposition"; a way of seeing things whose birth was "roughly contemporaneous with the birth of modernity itself."

The word "atheism" begins to crop up in English writing in the sixteenth century, almost always as a put-down; the term was used as derogatory label, bestowed on anyone imagined to hold heretical views of one kind or another. Even so, the seeds of unbelief had been planted, and Hyman points to the years from 1540 to 1630 as a period in which "the notion of a worldview that was entirely outside a theistic framework was . . . gradually becoming conceivable." As it happens, Shakespeare's life falls wholly within this transitional period; and, just as his works hint at the beginnings of science, so, too, do they hint at the possibility of unbelief.

Once can easily perceive the danger that irreligion presented to the established faith; and, given the very real connections between religion and politics, it is hardly surprising that Parliament eventually passed laws against atheism. The first of these was enacted in 1667, calling for anyone "who denies or derides the essence, persons, or attributes of God the Father, Son or Holy Ghost given in the Scriptures . . ." to be jailed (at least until the payment of a fifty shilling fine). A similar piece of legislation, from 1678, requires that if any person over the age of sixteen "not being visibly and apparently distracted out of his wits by sickness or natural infirmity, or not a mere natural fool, void of common sense, shall . . . by word or writing deny that there is a God . . . [that person] shall be committed to prison." For the authorities to have sensed so much smoke, there must have been at least the occasional fire. As Benjamin Bertram writes, atheism "must have flourished in the sixteenth and seventeenth centuries."

We have already looked at the atomic theory of the ancient Greeks, championed by Lucretius in his epic poem—and it is not hard to see why it could be taken as an affront to the established faith. As Greenblatt notes, wherever Lucretius's poem surfaced, "the implications—for morality, politics, ethics, and theology—were deeply upsetting." The reasons were straightforward enough: Atomism challenges the idea of divine providence; it eliminates the need for a prime mover (another challenge to the divine); and it does away with the idea of an afterlife. And because of the deeply intertwined relationship between religion and politics, believing in atoms was a hairsbreadth away from treason. (Well into the seventeenth century, young Jesuits at the University of Pisa were required to recite a prayer denouncing Lucretius's atomic theory. The prayer concludes, "Atoms produce nothing; therefore atoms are nothing." It was, as Greenblatt notes, an attempt "to exorcise atomism" and to declare the universe to be God's handiwork.)

Lucretius imagined life to end with death, as the atoms that make up one's body disperse back into the chaos from whence they came—and by Shakespeare's time he was not alone. In his book *De animi immortalitate* (*On the Immortality of the Soul*), published in 1545, the Italian philosopher Girolamo Cardano wondered "whether human souls are eternal and divine or whether they perish with the body"—and goes on to list dozens of arguments for and against the immortality of the soul. The debate quickly took root in England. In 1549, the reformer Hugh Latimer warned against the "great many in England who say there is no soul, that think it is not eternal, but like a dog's soul; that think there is neither heaven nor hell."

No heaven or hell—and thus, one might ask, what place for the gods? To the extent that the atomists bothered with the gods at all, they imagined them to be utterly disinterested in human affairs. But even without the atomic theory, one might question the plausibility of divine providence—as a few bold thinkers were prepared to do. Did God in fact have a plan for each living creature, past, present, and future—the "special providence in the fall of a sparrow" that Hamlet imagined? As early as 1550, a theologian named Roger Hutchinson warned against those who would deny God's direct providence over his creation:

> Others grant God to be the maker of all things: but they suppose that, as the shipwright, when he hath made the ship, leaveth it to the mariners, and meddleth no more therewith; and as the carpenter leaveth the house that he hath made; even so God, after he formed all things, left all

> his creatures to their own governance, or to the governance of the stars. . . .

And then there were the peculiar (and highly unorthodox) views of Giordano Bruno, whose writings on science and philosophy we looked at in Chapter 4. Bruno wasn't an atheist; as we've seen, his science and his theology were deeply intertwined. Perhaps the best word to describe his worldview is *pantheism*: For Bruno, God and the universe are as one. Christianity, in this view, is little more than a delusion; Christ was not divine but, as Jennifer Michael Hecht puts it, "merely an unusually skillful magician"; he also dismissed heaven and hell, the Virgin Birth, and the Resurrection. As mentioned, Shakespeare is unlikely to have met Bruno himself. However, Shakespeare was certainly friendly with England's most famous alleged atheist of the time, the playwright Christopher Marlowe. Just over a dozen lines into Marlowe's *The Jew of Malta*, the Italian political thinker Niccolò Machiavelli (anglicized to "Machevil") declares, "I count religion but a childish toy . . ." (prologue, line 14). *Doctor Faustus*, Marlowe's most important play, was even more dangerous. "This was no simple morality play," notes Susan Brigden, "but a work terrifying in its intensity and daring which hinted at a dangerous questioning." Faustus declares, "I think hell's a fable" (5.129)—and the playwright may well have agreed.

THE CURIOUS CASE OF CHRISTOPHER MARLOWE

Matters are complicated, however, by the fact that Marlowe wasn't *just* an atheist—he was also a government spy; while traveling in France, he monitored the activities of English Catholics living in exile. He was also openly gay in an age when homosexuality was punishable by death—*and* was daring enough to portray, in *Edward the Second*, the doomed love between the young king and his "sweet favourite," Piers Gaveston. Marlowe, in other words, lived quite far from the respectable mainstream of Elizabethan life. In Corpus Christi College in Cambridge, there is a portrait said to be of Marlowe; it bears the mysterious motto *Quod me nutrit me destruit*—"That which nourishes me destroys me."

Accusations of Marlowe's atheism stem from several sources, beginning with testimony from another famous playwright, Thomas Kyd. When a fragment of a heretical tract was found in Kyd's living quarters, he said it belonged to Marlowe, with whom he had once shared the rooms. But the most damning testimony delivered to the Privy Council came from a man named Richard Baines (who was, just to make things even more convoluted, *also* a spy). In addition to condemning Marlowe, Baines's testimony is notable for referencing astronomer Thomas Harriot, whose work we looked at in Chapter 5. Baines tallies Marlowe's heretical views regarding specific passages in the Bible, and adds that the playwright believed "that Moses was but a Jugler, & that one Heriot being Sir W Raleighs man Can do more than he." As we've seen, Harriot was dogged by accusations of atheism, and so, too, was Sir Walter Raleigh. (Compared with Marlowe, however, there is little evidence that Raleigh was much of an atheist; what little of his writing that has survived, says George Buckley, contains "no evidence of religious incredulity.") And then there were the harsh words of Thomas Beard, a Puritan churchman. In a book called *The Theatre of God's Judgement* (1597), Beard outlines the array of punishments that await various kinds of sinners—and he wasn't afraid to name names. Most of his victims are Italian or French, but one Englishman is singled out (even if he mangles the name somewhat):

> Not inferior to any of the former in Atheisme and impiety . . . was one of our own nation, of fresh and late memorie, called Marlin, by profession a scholler, brought vp from his youth in the Vniuersitie of Cambridge, but by practice a Play-maker . . . [who] denied God, and his sonne Christ, and not only blasphemed the Trinitie, but also (it is credibly reported) wrote bookes against it. . . .

Not surprisingly, Marlowe's death is seen as somewhat suspicious: Just twelve days after a warrant had been issued for his arrest on suspicion of heresy, the playwright was fatally stabbed—just above the right eye—during a brawl in a Deptford bar. (Did the Crown take out a "hit" on a particularly irksome troublemaker?) It all has the flavor of a 1960s-era Cold War spy movie, with secret lives, dangerous documents, and double agents. As George Buckley puts it, Marlowe "was evidently playing some kind of very deep game"; and no doubt the charges of atheism were linked to the ongoing political machinations. A man named Richard Cholmeley said he had been "converted" by the playwright, claiming that "Marlowe is able to show more sound reasons for atheism than any divine in England is able to give to prove divinity," and that Marlowe once read an atheist lecture "to Sir Walter Raleigh and others."

Most likely, however, Marlowe's atheism was not so much rooted in his philosophy, but was rather, as Buckley puts it, "a temper of mind that expressed itself in life and action." Being a man of the theater didn't help: The stage had always been linked to sinfulness and debauchery. The theater, as Bertram puts it, served as a symbol "of a society insufficiently committed to God"; its actors and playwrights were imagined to have "no proper place in the social order; they blasphemed God, engaged in homosexuality, and followed Machiavelli."

Incidentally, suspicion of irreligion was even more dangerous on the other side of the English Channel. In 1623, a French poet and playwright named Théophile de Viau was accused of atheism, tortured, and sentenced to death, though, thanks to his connections, the sentence was commuted to banishment. Like Marlowe, de Viau was also suspected of homosexuality. As A. C. Grayling has noted, this is not entirely a coincidence: The word "atheism" was an all-purpose label for unacceptable beliefs and practices, and homosexuality was "taken to be expressive of atheism, or identical with it." Things did not go so well for an Italian philosopher named Lucilio Vanini. Like Bruno, he was a priest with radically unorthodox views; and, like Marlowe and de Viau, he was suspected of being a homosexual. Vanini was arrested in Toulouse in 1618 on charges of atheism, and, after a lengthy trial, convicted. Death was too good for him, the authorities believed—so they cut off his tongue and strangled him before burning his remains.

GODLESS SHAKESPEARE?

In the case of Shakespeare, we have no direct evidence, as there are no accusatory letters, no diatribes warning of his disbelief—or, indeed, of any sort of threat to the established order. (How very *dull* his life was, compared with Marlowe's!) And so we turn, with caution, to his dramatic works. The case for Shakespeare's lack of belief has been argued most recently by Eric Mallin in his book *Godless Shakespeare* (2007). Mallin begins by examining a remarkable scene in *Measure for Measure*, in which the hapless Claudio is in prison, awaiting execution. His sister Isabel, in training to be a nun, pays him a visit. At this point, Claudio has an idea: Maybe if Isabel were to sleep with the duke, Angelo, she could secure his release. She (quite reasonably) refuses. And then, as Mallin notes, we have an extraordinary speech on the nature of death. Claudio says:

> . . . to die, and we go we know not where;
> To lie in cold obstruction, and to rot;
> This sensible warm motion to become
> A kneaded cold; and the delighted spirit
> To bathe in fiery floods, or to reside
> In thrilling region of thick ribbed ice
> To be imprison'd in the viewless winds
> And blown with restless violence round about
> The pendant world . . .
> . . .'tis too horrible!
>
> (3.1.115–27)

For Mallin, the issue is not Claudio's fear, but its effect on Isabel, whose faith seems truly shaken by what she is hearing. The picture being offered, Mallin writes, is one of "religion as terrified sadism, the product of faith's deep, frustrated inadequacy to meliorate the darkness, or to cope with the complexities of selves who are touched by desire, the law, loneliness, despair." What happened to Isabel's faith? What we see, Mallin says, here and throughout the canon, is that "spiritual convictions crumble under pressure."

Shakespeare goes even further in *Titus Andronicus*, by presenting the audience with the only self-avowed nonbeliever in the canon, the Moorish villain Aaron. When Aaron is taken prisoner, he tries to bargain with his captor, Lucius. But Lucius asks, What good is a vow from a nonbeliever? Aaron, however, has a snappy comeback: Those who *do* believe, he says, are often fools and liars; yet we imagine their oaths to be worth something. (Note how quick-witted Shakespeare's villains are!)

Except, Aaron isn't *just* a villain. He is also a master manipulator, as Mallin, who teaches at the University of Texas–Austin, told me in an interview: "Aaron arranges things—he arranges plots, he sets the stage for his deeds, he has props that he uses." In other words, Aaron is also "one of Shakespeare's early models for his own work. Aaron is a 'playwright.' The parallel to Shakespeare is really quite compelling."

What led Shakespeare in this direction? One possibility, Mallin speculates, is that he was following Marlowe's lead—or perhaps trying to one-up his colleague. Consider the plot of *Doctor Faustus*: The doctor makes a pact with the Devil, and God doesn't seem to care. "What never really appears in the play is God's intervention; what never appears is

God's goodness," Mallin says. "This is a very upsetting possibility that Marlowe introduces, and that Shakespeare plays on, particularly in his tragedies."

And then, of course, there is *King Lear.* In this most somber of Shakespeare's plays, the gods are often called upon—by the king and Gloucester and others—but they do not respond. As Jay Halio observes, "their presence is nowhere found or felt"; for Greenblatt, the gods "are conspicuously, devastatingly silent." In their absence, justice cannot be guaranteed; indeed, it becomes fragile in the extreme. Lear, in desperation, hopes that events will "show the heavens more just" (3.4.36), but it is a lost cause. The play ends, as William Elton puts it, "with the death of the good at the hands of the evil." In one of the play's most famous—and darkest—lines, Gloucester laments, "As flies to wanton boys are we to th'gods / They kill us for their sport" (4.1.36–37). (A line, incidentally, that closely echoes a passage from Montaigne, who wrote, in Florio's translation, "The gods perdie doe reckon and racket us men as their tennis-balles.")

In *King Lear and the Gods,* Elton presents a kind of checklist of what makes a "Renaissance skeptic"—denying divine providence; denying the immortality of the soul; placing mankind among the beasts; denying God's role as creator of the universe; attributing to nature what is properly the work of God—and then shows that Lear, over the course of the play, develops into precisely such a skeptic. It is a gradual process, but it is relentless: "Lear's disillusionment, once begun, sweeps all before it, toppling the analogical edifices of God and man, divine and human justice." The play, writes Thomas McAlindon, "must have at least evoked for most Christians the dark night of the soul when faith seems groundless even to the most devout believer." We live, we die, and, it would seem, that's the end of it. Whether we led good lives or bad, the universe does not seem to care. In the play "there is no firm hint of an afterlife where flights of angels sing the afflicted to their rest, or where the wicked meet with a punishment commensurate with the evil they have done," McAlindon writes. "Human beings are here left utterly alone with nature, their own and the world's." Or, as Mallin put it in our interview, *King Lear* is "essentially a godless document"; it describes a world "emptied of divinity."

We have already spoken of Shakespeare's urge to outdo his colleague

Marlowe—to take greater risks, to shock, to subvert. But as Harold Bloom suggests, the character of Edmond—whom Bloom describes as "a pagan atheist and libertine naturalist"—may also have been inspired by Marlowe himself: "Marlowe the man, or rather Shakespeare's memory of him, may be the clue to Edmond's strange glamour, the charismatic qualities that make it so difficult not to like him." However it came about, the result was bold in the extreme: *King Lear*, writes Elton, is "fraught with danger, both politically and artistically."

The canon offers other hints of a godless Shakespeare: Hamlet's obsessive contemplation of death and decay, with no mention of an afterlife; Helena's assertion in *All's Well That Ends Well* that "Our remedies oft in ourselves do lie / Which we ascribe to heaven" (1.1.216–17); Macbeth's assertion that life is "a tale, told by an idiot, signifying nothing" (5.5.26–27). None of this proves that Shakespeare was an atheist, Mallin acknowledges—but it at least shows that he could *imagine* a godless world. And what better place to exercise that imagination than the London stage—the one place where one could dethrone a king, ridicule a nobleman, compare a prince to a beggar, and ignore the divine; the one place where one might be subversive and yet avoid the gallows.

MUCH ADO ABOUT NOTHINGNESS

The idea of an "atheist Shakespeare" seems to have taken root in the early years of the twentieth century, by coincidence—or perhaps not—the same time when *King Lear* was first imagined to surpass *Hamlet* in greatness. As George Santayana has written, the playwright was faced with a stark choice:

> For Shakespeare, in the matter of religion, the choice lay between Christianity and nothing. He chose nothing. . . . The cosmos eludes him; he does not seem to feel the need of framing that idea. He depicts human life in all its richness and variety, but leaves that life without a setting, and consequently without a meaning.

"Nothing," of course, is one of the great themes in *Lear*; in the very first scene, we hear it four times. The wicked sisters, Regan and Goneril, shower their father with extravagant declarations of devotion. Lear then asks his third daughter, Cordelia, what she can say to top her sisters' claims:

CORDELIA

Nothing, my lord.

LEAR

Nothing?

CORDELIA

Nothing.

LEAR

Nothing will come from nothing, speak again.

(1.1.82–85)

Shakespeare is just setting the stage; the mayhem and darkness are yet to unfold. Did the playwright "choose nothing"? Eric Mallin doesn't go quite that far. But he says that *King Lear* does lack "an image of a benevolent cosmos, of a benevolent deity." This may be partly due to a lack of belief on the part of its author, Mallin says—but it could also be because the supernatural is not Shakespeare's first concern. "He is interested in the social, in the worldly, in the sexual, in the linguistic," Mallin says. "He's interested in what happens on this planet. What matters is existence; what matters is what we do while we're here. And that strikes me as pretty modern."

The philosopher Colin McGinn, author of *Shakespeare's Philosophy* (2006), considers the question of labeling Shakespeare an atheist, but prefers the term "naturalist." His moral thinking is "entirely secular," McGinn writes. "He is simply saying, *This is the way things are, like it or not.*" When I met with McGinn in his Miami apartment, we explored these ideas a bit further—including the notion of "cosmic justice," which seems to be conspicuously absent in *King Lear.* For Shakespeare, "justice is entirely man-made," McGinn says. "And that may explain why there's so much interest in *law* in Shakespeare, and the way you have to use the law in order to *get* justice—because you won't get justice outside of human constructions or human inventions. You can't rely on nature to mete out justice in the right way." In *King Lear* in particular we find "a very progressive, radical position," one that receives relatively little attention until the existentialism movement of the late nineteenth century.

"People always use this phrase 'Things happen for a reason,'" McGinn says. "But they don't. Sometimes, things happen for no reason at all. I think that's part of his whole worldview [in *King Lear*]. There's a strong vein of pessimism, I think, in Shakespeare. It's a very bleak view of the meaninglessness of everything."

Shakespeare often emphasizes the role of happenstance—those "slings and arrows of outrageous fortune"—and it was still on his mind when he came to write *The Winter's Tale* in 1609. The play has, as Stephen Orgel puts it, a "disturbingly amoral" flavor. At the start of act 4, "Time," as the Chorus, declares, "I that please some, try all; both joy and terror / Of good and bad, that makes and unfolds error . . ." (4.1.1–2). Good or bad, it is all the same to the passage of time.

The atheist-Shakespeare theories may be gaining currency, but they can also be seen as the latest chapter in the never-ending story of Shakespeare's religious beliefs—a subject of boundless inquiry and speculation. He has been called everything from a closet Catholic to an apologist for the Protestant state religion; the truth, one suspects, is murkier. We have already noted the religious turmoil that marred the English psyche in the first half of the sixteenth century; Shakespeare, whatever he believed, was all too aware of the anguish brought on by religious quarreling. As Stephen Greenblatt writes, Shakespeare often "seems at once Catholic, Protestant, and deeply sceptical of both." While his father may have teetered between being a committed Catholic and its Protestant opposite, "William Shakespeare was on his way to being neither." Even so, Greenblatt cautions that the playwright's private beliefs are "wholly inaccessible."

We can't definitively label Shakespeare an atheist, just as we can't call him a scientist—even if we suspect we are seeing hints of such a worldview. All we can say is that he lived at a pivotal time in English history; a time when long-held beliefs were up for debate; a time of competing ideas and clashing values; a time of doubt and confusion. As mentioned, Jonathan Bate sees Shakespeare's mind as poised between "rational thinking and visceral instinct, faith and scepticism." McGinn, too, sees Shakespeare as inhabiting a world in transition. "Shakespeare is prescientific but he's post-magical," he said. "So I think of him as kind of prescientific naturalist. . . . Whether Shakespeare was himself an atheist, I just don't think you're able to say. But I wouldn't at all be surprised if he was. I really wouldn't be surprised."

"A GREAT PIECE OF CLOCK-WORK"

The revival of the ancient theory of atomism was one force that could have pushed a well-read Elizabethan toward atheism—or at least, toward a less religious worldview. But there were other intellectual developments that could have had a similar effect. One of the most profound changes involved the way philosophers came to imagine the kind of world we inhabited. The medieval worldview had been highly *animistic*—treating objects as though they had soul-like properties (from the Latin word for soul, *anima*. We have seen, for example, how the body's organs were imagined to have their own personalities, and how both Kepler and Gilbert had imagined the cause of planetary motion to have involved "souls." But in 1605, as Steven Shapin notes, Kepler began to have second thoughts. In his new work, Kepler would occupy himself "with the investigation of the physical causes. My aim in this is to show that the machine of the universe is not similar to a divine animated being, but similar to a clock." The chemist Robert Boyle would express a similar idea half a century later, writing that the natural world was, "as it were, a great piece of clock-work."* In the physical sciences, the clockwork metaphor is most closely associated with Isaac Newton, whose theory of universal gravitation would finally provide the mathematical underpinning of the model of the solar system described by Copernicus and Kepler. This required, among other things, a new conception of God: Rather than imagining a God who constantly interacted with his creation, guiding events from day to day, it was enough to picture God as a sort of divine architect—a cosmic watchmaker. Once set in motion, the universe could look after itself; God had made the universe and the laws that govern it, and that was enough. (Intriguingly, this was not Newton's own vision: A devout if unorthodox believer, he insisted that God was still an integral part of the physical world, and was required to tweak the system from time to time.)

A "clockwork universe" is all fine and well—unless, or until, it wears down. This is Gloucester's great fear in *King Lear*: "O ruined piece of nature! This great world / Shall so wear out to naught" (4.5.130–31). By the nineteenth century, physicists would coin a scientific name for this

---

* Although he didn't explicitly make use of the clock metaphor, the English philosopher Thomas Hobbes expressed similar doubts about the plausibility of animism. In *Leviathan* (1651), he makes fun of those who treated inanimate objects as though they could act with purpose, or had goals. Who can believe, he asks, that "stones and metals had a desire, or could discern the place they would be at, as man does" (quoted in Shapin, p. 30).

final act of the cosmic drama: It's called the "heat death of the universe," and it's not going to be pretty. Fortunately, it's billions of years off—that provides *some* comfort, one imagines. With the body, there is not quite so much time: Your best bet is to enjoy it before entropy takes its toll. But the picture of the human body as a machine comes to us not from Newton but from Descartes, who did more than any other thinker to usher in the mechanical age.

René Descartes (1596–1650), after he had come to grips with his own existence, argued that plants and animals functioned through processes analogous to the workings of a machine. There is, he wrote, "no difference between the machine built by artisans and the diverse bodies that nature alone composes," other than the range of their sizes. In either case, the principles of mechanics hold the key, for "it is not less natural for a clock, made of the requisite number of wheels, to indicate the hours, than for a tree which has sprung from this or that seed, to produce a particular fruit." In hindsight, this striving for mechanistic explanations may seem like a natural extension of the atomism of the ancient Greeks; in both cases, we have a focus on reductionist, materialist, explanations for most—perhaps all—natural phenomena. Nonetheless, Descartes himself rejected atomism; he imagined matter to be infinitely divisible, so there could not be any "smallest" component. Yet many of his contemporaries, notably his countryman Pierre Gassendi, were pushing for a revival of this ancient idea, reincarnated as "corpuscularism." Whether the world was comprised of atoms, corpuscles, or something else not yet envisioned, it was clear that its various parts pushed and pulled on each other: To explain some particular natural phenomenon, the key was to seek out the underlying mechanism.*

The body, of course, is a living thing, but it need not be thought of as inherently organic. Descartes conceived of the body as "nothing but a statue or machine made of earth." He describes the various functions that our bodies perform unconsciously—digestion, the circulation of blood, respiration, the workings of nerves and muscles—and declares that they occur "in the same way as the movement of a watch is produced merely by the strength of the spring and the configuration of its wheels." He draws an analogy with the automata in the royal gardens of Saint-Germain in Paris, which featured water-driven mechanical

* For a useful discussion of the rise of the mechanistic worldview, see Steven Shapin's *The Scientific Revolution* (1996), pp. 30–46. See also Richard DeWitt's *Worldviews* (2004), pp. 178–82.

creatures that could produce music and even "speak"—all of these functions "depending on the various arrangements of the pipes through which the water is conducted." In the case of animals, Descartes was willing to go all the way: There was nothing, he argued, that could distinguish a flesh-and-blood animal from a highly sophisticated mechanical simulation. For humans, however, he stopped short of allowing a purely mechanical description; he believed that human beings had souls that gave them the ability to reason, to converse, and to perform a multitude of tasks compared with the limited abilities of animals.

Sophisticated mechanical devices like the automata in the royal gardens were new, but the clock itself was well established, with the first cathedral clocks dating from the late thirteenth century. The clock, as Shapin notes, was "an exemplar of uniformity and regularity." If philosophers sought an analogy for something that exhibited order—say, the universe—the clock was the obvious choice. While the innards of a clock may be complex, they are not mysterious: They are, in principle, completely knowable (and were certainly known to the craftsmen and mechanics who built them). As Shapin puts it, the clock metaphor serves as "a vehicle for 'taking the wonder out' of our understanding of nature."

I hesitate to say that Shakespeare saw this coming—it is all too easy to imagine that Shakespeare saw *everything* coming—but he does seem to be enamored with machines, and with clocks and timekeeping in particular. In the canon, he uses "clock" 85 times; "minute" 63 times, and "hour" a whopping 462 times—an average of more than a dozen times in each play. As Scott Maisano points out, some of Shakespeare's metaphors seem to hint at the new mechanical philosophy: Coriolanus is called an "engine"; Bolingbroke is described as "Jack of the Clock"; and Hamlet, who describes himself as a "machine," famously asks if Guildenstern imagines him to be an "instrument" that can be "played upon" as one plays a flute. In *The Tempest*, we find a remarkably personal application of the clock metaphor, as Sebastian and Antonio await Gonzalo's next pompous utterance: "Look, he's winding up the watch of his wit; by and by it will strike—" (2.1.14–15). In *Richard II*, we find the deposed king languishing in prison, and ruminating on the nature of time; in a remarkable passage, he uses the word "time" seven times in the space of eight lines—and then imagines himself as a timepiece: "Time made me his numb'ring clock" (5.5.50). Imaginary clocks can be also found in the Forest of Arden in *As You Like It*. It's nothing but trees as far as the eye can see, but Shakespeare can't seem to get clocks out of his head:

ROSALIND

I pray you, what is't o'clock?

ORLANDO

You should ask me what time o' day; there's no clock in the forest.

ROSALIND

Then there is no true love in the forest, else sighing every minute and groaning every hour would detect the lazy foot of Time, as well as a clock.

(3.2.295–300)

This exchange is followed by Rosalind's famous reflection on the relativity (so to speak) of time's apparent passage—"Time travels in divers paces with divers persons . . ."—an exposition that would not be so far out of place in a twenty-first-century psychology textbook.

One of Shakespeare's last plays, *The Winter's Tale*, is of particular interest. In the climactic final scene, Queen Hermione, believed to have been dead for sixteen years, is presented to us, and to King Leontes, as a statue—which then springs to life, as the king and his court watch in amazement. "Statues," as we usually think of them, cannot do this—but automata, of the kind that inspired Descartes, and which could already be found in royal courts and gardens across much of Europe by Shakespeare's time, can indeed seem very much alive. As Maisano points out, Shakespeare may well have had such machines in mind when he wrote *The Winter's Tale*. Leontes, of course, is awestruck; he hopes that what he has seen is a case of "lawful" magic, and not something darker. But wouldn't a sophisticated automaton seem magical to the uninitiated? As Arthur C. Clarke famously said, "Any sufficiently advanced technology is indistinguishable from magic." No wonder Maisano sees *The Winter's Tale* as proto–science fiction—and *The Tempest*, with its magician-on-an-island, even more so. These last few plays have long been considered "backward-looking and profoundly nostalgic," Maisano writes, but they can better be viewed as "forward-looking speculations." Mary Shelley's *Frankenstein*, first published in 1818, is usually seen as the first work of literary science fiction—but Shakespeare, argues Maisano, was already inventing the genre more than two centuries earlier. And he's not the only one who sees Shakespeare in this light. In *Shakespeare the Thinker* (2007), Anthony Nuttall points to Prospero's relationship with Ariel, described in *The Tempest* as an "airy spirit": Does such a spirit have feelings and emotions, or is he (it?) like the unfeeling Vulcans on TV's *Star Trek*? Nuttall believes that, in *The Tempest*, "Shakespeare is inventing science fiction."

Just a year before *The Tempest* made its debut on the London stage, Galileo aimed his telescope at the night sky. What he observed in "heaven" was not particularly godly; in fact, the mountains and valleys on the moon seemed remarkably like those on the Earth. He did not make it impossible to believe in heaven—but for many people, he certainly made it harder.* A half century later, Blaise Pascal, the French scientist and philosopher, was humbled by the scale of the universe described by the astronomers. Pascal was born seven years after Shakespeare's death; as it happens, this was also the year that the First Folio saw the light of day. He was a Christian as well as something like an existentialist (before there were existentialists), and wrestled with the enormity of the cosmos, the smallness of mankind, and the contingency of human affairs. And, like Lucretius and Montaigne before him, he had a way with words:

> When I consider the short duration of my life, swallowed up in the eternity before and after, the little space which I fill and even can see, engulfed in the infinite immensity of spaces of which I am ignorant and which know me not, I am frightened and am astonished at being here rather than there; for there is no reason why here rather than there, why now rather than then. Who has put me here? By whose wonder and direction have this place and time been allotted to me? . . . The eternal silence of these infinite spaces frightens me.

---

* In his 1940 play *The Life of Galileo*, Bertolt Brecht has the astronomer himself make this point. When he observes the moons of Jupiter, Brecht's Galileo declares, "Today is 10 January 1610. Today mankind can write in its diary: Got rid of heaven" (Brecht, p. 24).

Three centuries later, Steven Weinberg—physicist, cosmologist, and atheist—was more blunt. He didn't ask "Who has put me here," because he already knew the terrible truth: His existence, like the existence of humankind, was an accident. The universe is vast, and cold, and indifferent to our joys and our sorrows. In his book *The First Three Minutes* (1977), he famously concluded, "The more the universe seems comprehensible, the more it also seems pointless."

It was the scientists who worked out the equations, but it was the artists and writers who showed us what they meant. Whether Shakespeare would have agreed with Weinberg that the universe was "pointless," we will never know, although my suspicion is that he would not: Whatever one may discover (or fail to discover) in the depths of space, here on Earth there are places to go, friends to cherish, lovers to woo, and the occasional regicide to avenge. But Shakespeare, writing four hundred years ago, was painfully aware of the possibility that that is *all* there is.

Conclusion "They say miracles are past . . ."

It's fashionable these days to deny that there was such a thing as a "Scientific Revolution." In fact, today's history books shy away from labeling *anything* as a revolution, lest we mistakenly conclude that "before X, things were like *this*; after X, things were like *that*." Best to avoid any mention of "turning points" or "pivotal moments." And yet by the middle of the seventeenth century we find a world very different from what it had been in the middle of the sixteenth. For one thing, the debate over Copernicanism was essentially over; the heliocentric model had triumphed. When Kepler wrote his *Epitome of Copernican Astronomy*, published between 1618 and 1621, he described a science that, as Paula Findlen puts it, "no longer had its origins in antiquity but had begun in 1543," the year of Copernicus's *De revolutionibus*. For more than a thousand years, writers on astronomy had quoted from Aristotle and Ptolemy. Now they quoted from Copernicus, Tycho, Gilbert, and Galileo.

But the change encompassed much more than just cosmology. Magical thinking was on the decline, mechanistic thinking was on the rise, science was becoming grounded in mathematics, and the first dedicated "scientific societies" would soon be flourishing in London and Paris. It is instructive to look at the year 1600: As it happens, the year was something of a transitional year for Shakespeare; it is the year of *Hamlet*, the work that marks the end of the playwright's learning curve, and his coming of age as the leading dramatist of his day. It was also a big year (or at least, the start of a big decade) for science. In his book *It Started with Copernicus*, Howard Margolis lists some of the discoveries that came to light in, or close to, 1600: The distinction between electricity and magnetism; the law of free fall; the law of inertia; the idea of the Earth as a

Fig. 15.1 Science as a voyage of discovery: The frontispiece of Francis Bacon's *The Great Instauration* (1620). Bacon would come to be seen as one of the key figures of the Scientific Revolution. Image Select/Art Resource, NY

magnet; the theory of lenses; the laws of planetary motion; Galileo's telescopic discoveries; the law of hydrostatic pressure; the law of the swing of the pendulum. That is quite a list. Revolution or not, something big does seem to have been going on.*

One notable development is that people began to place greater value

* Of course, not everyone denies the Scientific Revolution. Historian of science Richard Westfall, who died in 1996, stood by the claim that the world did indeed change profoundly during that period: "A once Christian culture has become a scientific one. The focus of the change, the hinge on which it turned, was the Scientific Revolution of the sixteenth and seventeenth centuries" (Westfall, p. 43).

on the new, the innovative, the original. In the introduction to his book *Most Worthy Discourses* (1580), the French craftsman Bernard Palissy wrote that "the theories of many philosophers, even the most ancient and famous ones, are erroneous in many points," and that "you will learn more about natural history from the facts contained in this book than you would learn in fifty years devoted to the study of the theories of the ancient philosophers." By 1620, an Italian writer named Alessandro Tassoni could boast of all the things discovered in his time (or in the recent past) that would have made the ancient Greeks and Romans drool with envy: the printing press, clocks that strike, the compass, the nautical chart, the telescope. All of these "surpass by far any Latin and Greek inventions that were discovered in the whole of their so-much-celebrated course of years."

This embracing of "the new" is reflected in the frontispiece of Francis Bacon's *The Great Instauration* (1620), part of a massive but unfinished project in which Bacon sets out a plan by which mankind might come to understand, even conquer, the natural world, and describes the "new philosophy" that will be required. The frontispiece depicts a grand sailing ship passing through the Pillars of Hercules (the Strait of Gibraltar, often thought to symbolize the boundary of the known world) and out into the Atlantic Ocean (figure 15.1). For Bacon, the world of science was a great undiscovered country, and more could be learned by setting sail into these uncharted waters than by reading all of the writings of the ancient philosophers. Steven Shapin calls the image "one of the most vivid iconographical statements of new optimism about the possibilities and the extent of scientific knowledge."

We have explored in some detail the question of what, if anything, Shakespeare made of the new philosophy, and of the new picture of the cosmos in particular. The traditional answer is "very little"; the playwright, it is imagined, was either unaware, or at best marginally aware, of these developments. I cited a few examples of this point of view in the Introduction; I repeat a few of them here, and offer some additional examples. Historian Dorothy Stimson, writing in the early decades of the twentieth century, put it bluntly: Various passages in the canon "indicate that Shakespeare accepted fully the Ptolemaic conception of a central, immovable earth." Marie Boas Hall, writing in the 1960s, says that Shakespeare "was no amateur of mathematical astronomy"—that much is certainly true!—". . . still less was he aware of the revolutionary ideas being tentatively developed by his contemporaries." Thomas McAlindon, writing in 1991, admits that Shakespeare was deeply concerned with cosmological matters, but writes that there is "no sign of [the Copernican] revolution" in the plays. Even Leslie Hotson—who, as we've seen, went further than anyone to show the connections between Shakespeare and the Digges family—writes that "among Shakespeare's myriad minds, there was not the mind ready to kindle the truth of Digges's vast vision." And David Levy, who has written extensively on Shakespeare's fascination with the night sky, has downplayed the playwright's awareness of the new astronomy. "Even if Shakespeare had believed in the new cosmology," he writes, "it would not have served his purpose well, for the old system, with its emphasis on the Earth and mankind at the center of the universe, is more sound for the purpose of drama."

But the tide may finally be turning. As we've seen, a growing number of scholars are examining the question of what Shakespeare knew, and when he knew it. Some examples warrant a second mention: We have Jonathan Bate, for example, writing that Ulysses's famous speech in *Troilus and Cressida* "may hint at the new heliocentric astronomy." James Shapiro says that "Ptolemaic science . . . as Shakespeare knew, was already discredited by the Copernican revolution." And we have seen the support that both John Pitcher and Scott Maisano have given to the notion that Shakespeare knew of Galileo's telescopic discoveries in time for them to have influenced his last few plays—a view wholeheartedly endorsed by astronomer Peter Usher, whose work—contentious as it may be—has sparked a renewed interest in the question of "Shakespeare and Science."

"ALL COHERENCE GONE"

Other writers and poets followed scientific developments more closely than Shakespeare, and in some cases their discomfort with the new philosophy is palpable. The poetry of John Donne (1572–1631) is full of celestial imagery, with innumerable references to the sun, moon, and stars. The context is almost always Ptolemaic, but his *First Anniversary* (1611) and *Of the Progress of the Soule* (1612) are, as Margaret Byard puts it, "laments for a dying world and a dying cosmology":

> And freely men confesse, that this world's spent,
> When in the Planets, and the Firmament

They seeke so many new; they see that this
Is crumbled out againe to'his Atomis.
'Tis all in pieces, all cohaerence gone . . .

In this new cosmology, nothing was certain; nothing could be depended upon. In the sky, Donne writes, there are "New stars, and old do vanish from our eyes. . . ." Donne was "an intelligent student of science," writes Francis Johnson, fully capable of assessing the merits of the old and new pictures of the cosmos. "The discoveries of Galileo and Kepler were scarcely published," Johnson notes, "before they were used by Donne as material for his writing"—and, in fact, he mentions both astronomers by name. As Byard points out, Donne had connections to the astronomer Thomas Harriot through mutual friends and associates—and it seems he was taken aback by what the astronomers had discovered. In one of his most moving passages, Donne writes that the heavens were once divine and unknowable; now, thanks to the astronomers, they are losing some of their mystery. Like the new lands across the ocean, they have been conquered:

For of Meridians, and Parallels,
Man hath weav'd out a net, and this net throwne
Upon the Heavens, and now they are his owne.

Perhaps we can sympathize with those poets who found the medieval system more to their liking. In *The Orchestra* (1595), poet John Davies writes:

Only the Earth doth stand forever still:
Her rocks remove not, nor her mountains meet;
(Although some wits enrich'd with learning's skill
Say heaven stands firm and that the Earth doth fleet
And swiftly turneth underneath their feet)
Yet, though the Earth is ever steadfast seen,
On her broad breast hath dancing ever been.

Writing more than seventy years later, even John Milton seems to hold a certain reverence for the old worldview—even though his detailed comparison between the old and new models of the universe in *Paradise Lost* shows that he understood the science quite clearly. Astronomical references are everywhere in the epic poem, and they reflect a level of precision that can come only from a masterly study of the night sky. As Thomas Orchard puts it, Milton's knowledge of the heavens indicates "a proficient and intimate acquaintance with this science." He clearly understood the tilt of the Earth's axis, which gives rise to the seasons; he knew when the sun passed through each part of the zodiac; he understood how the planets move against the backdrop of fixed stars (including the retrograde motion of the outer planets); he even understood the precession of the equinoxes.* Some fourteen constellations are mentioned by name, and they're always where they're supposed to be. *And* he visited Galileo in Florence. In book 1 we find a description of the moon, "Whose orb / Through optic glass the Tuscan artist views / At evening from the top of Fesole" (I.287–89), referring to the hills outside Florence. He later refers to the scientist's "Glazed optic tube" (III.590), and, lest there be any doubt, in book 5 he gives us the scientist's name:

> As when by night the glass
> Of Galileo, less assured observes
> Imagined lands and regions in the moon:
>
> (V.261–63)

Milton didn't *like* the Copernican system—but that doesn't mean he rejected it. Orchard believes that he "doubtless recognized the superiority of the system," but found the Ptolemaic system more agreeable, and—more to the point—better suited to his poetic purposes: Milton had a tale to tell, and the Ptolemaic system provided the needed scaffolding.

What Milton could not do—what no poet, by that time, could do—was to ignore the new picture of the cosmos. As Margaret Byard puts it, "By the end of the 17th century the poet could no longer write with quite the same belief in himself as a prophet and seer into the nature of things; intuition had increasingly to give way to the revolution in scientific knowledge that was to follow." The world had changed, and there could be no turning back.

By the end of the seventeenth century, the case for Aristotle and Ptolemy was hopeless. The idea of experimental science—of trusting repeatable

* *Precession of the equinoxes* refers to a slow, circular motion traced out by the Earth's axis. Just as the axis of a spinning top slowly rotates, the Earth's axis traces out a circle (actually a pair of cones, joined at their vertices). The period for this precessional motion is about twenty-six thousand years. The discovery of this motion is usually credited to the ancient Greek astronomer Hipparchus in the second century B.C.

observations and experiments over ancient authority—had won out. The universe was larger, perhaps infinite, and the question of other worlds, perhaps similar to Earth, was quite imaginable. New ways of understanding mankind and the natural world were taking root. These ideas spread slowly but steadily, like ripples on the surface of a pond. Old beliefs, even the most cherished ones, were in jeopardy. The Bible was still the most popular book in the world—it still is, in fact—but it was becoming less and less plausible to treat it as a science textbook. Many of its stories—ancient floods, virgin births, the raising of the dead—came to be seen as metaphorical rather than literal. Even in the closing years of the sixteenth century, the writing may have been on the wall. The French philosopher Jean Bodin, writing in the 1590s, applied a bit of physics to the Last Judgment, questioning whether it was physically possible for everyone on Earth to be resurrected in a single day. He estimates that it is 74,697,000 miles to the fixed stars (a rather precise figure for an estimate!), and even if people could be whisked up to heaven at a rate of fifty miles per day (a pretty good speed in the days of the horse and carriage), it would take some eighty thousand years to make the journey. Such a calculation obviously highlights the perils of mixing science and faith—and is roughly the equivalent of the playful twenty-first-century newspaper articles that appear each year around Christmas, calculating how fast Santa's sleigh has to move in order to deliver toys to all the good boys and girls of planet Earth in a single night. Of course, God could always pull off a miracle, in which case the precise distance to the stars is irrelevant—but as Lafew declares in *All's Well That Ends Well*, the days of God's tinkering with the natural world may be over: "They say miracles are past; and we have our philosophical persons to make modern and familiar, things supernatural and causeless" (2.3.1–3).

WHERE SHAKESPEARE FITS IN

How, then, are we to think of Shakespeare? We have explored the remarkable transformation that was unfolding in Europe as a result of what we now call the Scientific Revolution—but how relevant were these developments for the mind of the playwright from Stratford? In the preface to the First Folio, Ben Jonson famously declares that "He was not of an age, but for all time"—but of course Shakespeare *did* inhabit a specific time, and there is no shame in saying so. "I think the moment that Shakespeare was born *into*—1564—was terrifically important for him," Stephen Greenblatt said during our interview. Many features of the intellectual landscape found at that time would disappear within a generation, including the animistic, soul-drenched worldview that, at the time of Shakespeare's birth, still held enormous appeal. In was "a late-medieval worldview in which the universe is a kind of 'magical lyre,' in which you pluck one string and all other strings start resonating," Greenblatt says. In such a universe, "the fate of human beings seems to be at the very center of the whole project of nature."

But then came Copernicus, Tycho, Bruno, Harriot, and—just in time for Shakespeare's last few plays—Galileo. Even though they lived a thousand miles apart, Greenblatt sees the English playwright and the Italian scientist as kindred spirits, or perhaps mirror images, of a sort. "In the case of Galileo, we have a scientist of stupendous power and intelligence who also has a startlingly literary sensibility. In Shakespeare you have an artist of stupendous and incredible power who has an oddly interesting scientific sensibility." This does not make Shakespeare a scientist—but, says Greenblatt, the playwright "is actually surprisingly alert to and interested in what we could call the 'scientific naturalism' of his time." Not as interested, perhaps, as some other writers who would shortly follow—Donne or Milton, for example—but not because he didn't understand it, or found it boring. As Scott Maisano said during our interview, Shakespeare "is neither ignorant nor indifferent to the new science." He doesn't necessarily write *about* science, the way Donne does; instead "he is writing *with* it." The new science, Maisano says, can be thought of as serving as Shakespeare's "setting"—but it is more than that. It is also "an integral, essential part of his story."

Our journey has been a perilous one, with many traps: The trap of imagining Shakespeare to be "ahead of his time" (the dreaded sin of Bardolatry); the trap of confusing Shakespeare with his dramatic creations; the trap of seeing the turn of the seventeenth century through twenty-first-century eyes (sadly, the only eyes I have); the trap of equating science and atheism; and, perhaps most shameful of all, the trap of bestowing on science a privileged status relative to other ways of engaging with the world—the sin of "scientism." The surest way to offend is to suggest that science is somehow superior to religion; but hinting that it rises above the arts is equally problematic. A few years ago, a journal called the *South Central Review* published a special issue on "Shakespeare and Science," and this was a primary concern. One of the contributors warned against "the privileging of scientism as an authorizing mode," which could lead "to problems of anachronism and disciplinary superposition." Well, we don't want *that*. But caution is in order: We often see science as being unique in having the power to actually change society, with art, music, and literature, as vital as they are, merely coming along for the ride, providing an occasional diversion. As George Levine has put it, scholarship has been preoccupied, until recently, "with the way scientific ideas shaped literary ones. The traffic was all one way. . . ." Such an approach "implicitly affirmed the intellectual authority of science over literature." Even worse, when the traffic *does* flow the other way, it is often "cute": "We've named twenty-four of Uranus's moons after Shakespearean characters. Isn't that charming? Now, back to calculating their orbital parameters. . . ."

Of course, science would never have been in danger of acquiring a more privileged status than the humanities if the two disciplines hadn't diverged in the first place—a rift whose negative consequences were famously decried by C. P. Snow in his essay *The Two Cultures*, more than fifty years ago. But the two cultures were not so far apart in Shakespeare's time, and all branches of "the arts" were involved in shaping our picture of the world. And Shakespeare himself shaped it more than anyone. One could argue, as Harold Bloom has, that it was Shakespeare who made us fully human. The playwright probed more deeply into human nature than any psychologist. Again, this does not mean that Shakespeare was a "scientist"—but he was certainly an acute observer of human nature, and had a mind every bit as disciplined as that of a scientist. Perhaps a better word is "explorer," the term used by John Dover Wilson. A play like *King Lear* is very much an exploration, Dover argued, one "more dearly won and far more significant than that of a Shackleton or an Einstein." But it's not a contest, and there is no need to imagine art and science in competition. I prefer to think of it as a partnership. Science has given us a new world, and Shakespeare illuminates our place within it.

Notes

PROLOGUE

Although a work of fiction, the Prologue "could have happened": At least two Englishmen had sighted the new star by November 19, and word would have spread quickly. The moon was indeed full on that date, and its location in the sky—and that of the planets, Cassiopeia, and the supernova—is accurate. Hunt and Jenkins are the names of schoolteachers known to have taught in Stratford in Shakespeare's time. I have taken the liberty of imagining that Shakespeare's father, John, knew a decent amount of Latin; he may not have. (And the weather that evening is a matter of conjecture.)

INTRODUCTION

1 **"The poet's eye . . ."** *A Midsummer Night's Dream* (5.1.12–14).
6 **"The new philosophy . . ."** quoted in Bate, p. 60.
6 **"a scientific force . . ."** Boas Hall, p. 143.
6 **"small Latin and less Greek"** www.bartleby.com/40/163.htm.
7 **". . . shows little awareness or interest . . ."** Cartwright, p. 35.
7 **". . . took almost no interest in science"** Burns, p. 171. (Thanks to Scott Maisano for bringing this reference to my attention.)
7 **"no sign of [the Copernican] revolution"** McAlindon, p. 4.
8 **a bright new star lit up the night sky . . .** See Olson et al., "The Stars of Hamlet," for a good overview of the 1572 appearance of "Tycho's star."
8 **. . . a remarkably eventful period in terms of celestial drama** For a list of the astronomical events that took place in Shakespeare's lifetime, see Levy, *The Stars in Early Modern English Literature*, p. vii.
9 **"a Ptolemaic conception"** Bevington, *Troilus and Cressida*, p. 162.
9–10 **"may hint at the new heliocentric astronomy"** Bate, p. 62.

10 **". . . already discredited by the Copernican revolution"** Shapiro, *1599*, p. 259.

11 **"cosmic imagination"** McAlindon, p. 4.

11 **885,000 words** www.folger.edu/template.cfm?cid=862.

1. A BRIEF HISTORY OF COSMOLOGY

13 **"Arise, fair sun . . ."** *Romeo and Juliet* (2.2.4).

13 **known as "the heavens"** See, for example, Greenblatt, *Will in the World*, p. 183; Kermode, p. 109.

14 **Shakespeare was "aware of the passage of time . . ."** Ackroyd, p. 264.

14 **that the Globe was constructed in alignment with . . .** Ackroyd, p. 374.

23 **As . . . Sacrobosco noted** This was the thirteenth-century French scholar Johannes de Sacrobosco. See Meadows, pp. 4–5.

24 **"If the Lord Almighty . . ."** quoted in I. Bernard Cohen, *The Birth of a New Physics*, p. 33.

26 **"presides over the whole scheme . . ."** Bate, p. 64.

26 **"I, thy Arthur, am . . ."** These are lines 65–70 of the masque, quoted in Fowler, p. 98. Known as *The Speeches at Prince Henry's Barriers*, it was sometimes called *The Lady of the Lake*. It was written in honor of the Prince of Wales, son of King James I, and performed in 1610.

26 **Astraea, the "star maiden"** Fowler, pp. 99–100.

28 **"filled with purpose . . ."** Principe, p. 21.

29 **"The skillful ordering of the universe . . ."** quoted in Heninger, p. 8.

29 **"We must lay before our eyes . . ."** quoted in Heninger, p. 11.

30 **". . . mystery, wonder, and promise"** Principe, p. 38.

30 **"left most features . . ."** Dear, p. 28.

30 **"a rich tapestry of interwoven ideas . . ."** Principe, p. 4.

31–2 **"Oh worthy temple . . ."** quoted in Kocher, p. 155.

32 **"there was never any good astronomer . . ."** quoted in Kocher, p. 156.

32 **"was more often cited . . ."** Kocher, p. 256.

32 **"an inherently religious activity"** quoted in Numbers, p. 105.

32 **"Never was theology demoted . . ."** Principe, p. 36.

33 **". . . the historical situation"** Principe, p. 37.

2. NICOLAUS COPERNICUS, THE RELUCTANT REFORMER

34 **"He that is giddy thinks the world turns round"** *The Taming of the Shrew* (5.2.20).

35 **"What sort of person was Copernicus?"** Gingerich, *The Book Nobody Read*, p. 29.

35 **"The ancients had the advantage . . ."** quoted in Sobel, p. 27.

36 **"He discusses the swift course . . ."** quoted in Sobel, p. 15.

36 **. . . Copernicus apparently did not** Rosen, *Copernicus and the Scientific Revolution*, pp. 110–11.

36 **"neither sufficiently absolute . . ."** quoted in Kragh, p. 49.

36 **"All the spheres revolve . . ."** quoted in Findlen, p. 655.

37 **only a handful of manuscript copies** Gingerich, *The Book Nobody Read*, p. 31.

37 **". . . would compose a monster, not a man"** Danielson, p. 106.
38 **"follow the wisdom of nature . . ."** quoted in Kragh, p. 49.
38 **"What appear to us as motions . . ."** quoted in Sobel, p. 20.
38 **"Remember that in the sixteenth century . . ."** Gribbin, p. 12.
39 **"I think it is a lot easier to accept . . ."** Danielson, p. 116.
40 **"The Aristotelian uniqueness . . ."** I. Bernard Cohen, *The Birth of a New Physics*, pp. 50–51.
40 **"Had the Christian theologians . . ."** Johnson, *Astronomical Thought*, p. 94.
40 **Pope Clement VII's personal secretary . . .** Principe, p. 49.
42 **"that the world is eternal . . ."** quoted in Kragh, p. 42.
42 **"So far as hypotheses are concerned . . ."** quoted in Rosen, *Copernicus and the Scientific Revolution*, p. 196.
43 **"Holy Father, I can guess . . ."** Danielson, p. 104.
43 **"claim to be judges of astronomy . . ."** quoted in Rosen, *Copernicus and the Scientific Revolution*, p. 185.
43 **"Truly indeed does the sun . . ."** Danielson, p. 117.
43 **"For who would place this lamp . . ."** quoted in Heninger, p. 47.
44 **down from about eighty to thirty-four** Johnson, *Astronomical Thought*, p. 102; see also Appendix I in Gingerich, *The Book Nobody Read*.
44 **"Copernicus did no more than a bit of tinkering . . ."** Heninger, pp. 46, 48.
44 **"a kinship of geometrical methods . . ."** Cohen, *Revolution in Science*, p. 120.
45 **"was presented explicitly . . ."** Dear, p. 33.
45 **". . . of an intellectual radical"** Heninger, p. 48.
45 **". . . an invention of later historians."** Cohen, *Revolution in Science*, p. 106.
45 **"his theory was widely read . . ."** DeWitt, p. 135.
45 **as Arthur Koestler once described it** See Gingerich, *The Book Nobody Read*, p. vii.
45 **"So far as our senses can tell . . ."** Danielson, p. 112.
45 **"a staggering distance . . ."** Principe, p. 50.
46 **by a factor of four hundred thousand** Kragh, p. 50.
46 **"Now when we see this beautiful order . . ."** quoted in Kragh p. 53.
46 **"to imagine anything more laughable . . ."** Montaigne (ed. Screech), p. 502.
46–7 **"Nothing could be more obvious . . ."** Boorstin, p. 294.
47 **"Was it still possible to believe . . ."** Kocher, p. 146.
47 **"How astonishing if . . ."** Danielson, p. 112.
48 **"No wonder, then . . ."** Danielson, p. 115.
48 **"But, of course, there has never been . . ."** quoted in Kragh, p. 43.
48 **"the alteration of the frame . . ."** I. Bernard Cohen, *The Birth of a New Physics*, p. 51.

3. TYCHO BRAHE AND THOMAS DIGGES

50 **"This majestical roof fretted with golden fire . . ."** *Hamlet* (2.2.301).
50 **A lump of white-dwarf matter . . .** Seeds, p. 269.
52 **"Amazed, and as if astonished and stupefied . . ."** quoted in Gingerich, "Tycho Brahe and the Nova of 1572," p. 7.
53 **"I noticed that a new and unusual star . . ."** quoted in Danielson, p. 129.

53 **"the greatest wonder . . ."** quoted in Sobel, pp. 202–3.

53 **"not some kind of comet or fiery meteor . . ."** Danielson, p. 131.

55 **"locked up in strange tongues"** quoted in Harkness, p. 107.

55 **"with more judgement . . ."** quoted in Harkness, p. 107.

55 **. . . among seventy-five men given a reprieve** Usher, *Shakespeare and the Dawn of Modern Science*, pp. 308–9.

55 **"glasses concave and convex"** quoted in McLean, p. 149.

56 **"Despite their seeming certitude . . ."** Panek, p. 30.

56 **"many people were making . . ."** Dunn, p. 23.

57 **he and Dee used the instrument together** Pumfrey, "'Your astronomers and ours . . . ,'" p. 32.

57 **Pumfrey presents a compelling case** Pumfrey, "'Your astronomers and ours . . .,'" page

57 **"fit for an Umberto Eco novel"** Pumfrey, "'Your astronomers and ours . . . ,'" p. 35.

58 **"mend their bad ways . . ."** quoted in Pumfrey, "'Your astronomers and ours . . . ,'" p. 38.

58 **"Both showed them light . . ."** quoted in Fowler, p. 77.

58 **"your astronomers and ours . . ."** quoted in Pumfrey, "'Your astronomers and ours . . . ,'" p. 42.

59 **"rare and supernaturall"** quoted in Pumfrey, "'Your astronomers and ours . . . ,'" p. 50.

59 **the metaphor of choice** See Ferris, p. 71; Sobel, p. 202; Danielson, p. 128.

60 **"you can live peacefully . . ."** quoted in Falk, "The Rise and Fall of Tycho Brahe," p. 54.

61 **"This was truly a microcosm . . ."** quoted in Falk, "The Rise and Fall of Tycho Brahe," p. 55.

61 **". . . the sheer bulk of the observations"** quoted in Falk, "The Rise and Fall of Tycho Brahe," p. 57.

62 **"the second Ptolemy"** quoted in Kragh, p. 51.

62 **"divinely guided under . . ."** quoted in Margolis, p. 48.

62 **"which authors have invented . . ."** quoted in Hale, p. 570.

64 **"The question of celestial matter . . ."** quoted in Kragh, p. 53.

64 **". . . than for etiquette"** quoted in Falk, "The Rise and Fall of Tycho Brahe," pp. 56–57.

65 **Digges spoke of "that devine Copernicus . . ."** The first quote is from http://www-history.mcs.st-and.ac.uk/Biographies/Digges.html; the second is from Danielson, p. 133.

65 **"turning every twenty-four hours . . ."** quoted in Danielson, p. 133.

65 **"as Mathematicall principles"** These Digges quotations are from Heninger, p. 49.

66 **may well have conducted . . .** Rosen, *Copernicus and the Scientific Revolution*, p. 164.

66 **"A Perfect Description of . . ."** This modernized version of Digges's book chapter is from Danielson, p. 135.

66 **"nearly every writer . . ."** Johnson, *Astronomical Thought*, p. 180.

67 **"infinitely up . . . in altitude"** quoted in Heninger, p. 51.

68 **". . . Englishman of the Renaissance"** Johnson, *Astronomical Thought*, p. 165.

68 **"Even the most casual observation . . ."** Johnson, *Astronomical Thought*, p. 175.

68 **"seems highly likely"** Gribbin, pp. 16–17.

68–9 **". . . the very court of celestial angels"** quoted in Gingerich, *The Book Nobody Read*, p. 119.

69 **By Owen Gingerich's estimate** Gingerich, *The Book Nobody Read*, p. 121.

4. THE SHADOW OF COPERNICUS AND THE DAWN OF SCIENCE

70 **"These earthly godfathers of heaven's lights . . ."** *Love's Labour's Lost* (1.1.88).

70–1 **"These were very popular . . ."** Boris Jardine, author interview, June 27, 2012.

72 **"would only need a crystal ball . . ."** Nigel Jones, "The Arch Conjuror of England."

73 **"more than Herculean labours . . ."** quoted in Russell, p. 191.

73 **". . . and genuine demonstrations"** quoted in Russell, p. 192.

73 **"The whole frame of Gods Creatures . . ."** quoted in Heninger, p. 8.

74 **the largest private library . . .** Johnson, *Astronomical Thought*, p. 139.

74 **"Dee knew everyone . . ."** Nigel Jones, "The Arch Conjuror of England."

74 **"Anyone who was anyone . . ."** Cormack, p. 516.

74 **"It is possible that the actors . . ."** Ackroyd, p. 423.

74 **Dee himself played an important role . . .** Russell, p. 192.

74 **"became the focal point . . ."** McLean, p. 134.

75 **"diligent study and reading . . ."** quoted in Harkness, p. 110.

75 **". . . ever written by an Englishman"** McLean, p. 138.

75 **"may wonderfully helpe . . ."** quoted in Dunn, p. 15.

76 **". . . The Globe therefore becomes . . ."** McLean, p. 142.

76 **"new works, strange engines . . ."** quoted in Harkness, p. 113.

77 **"MASTER: How is it that Copernicus . . ."** quoted in Stimson, p. 43; I have modernized the spelling.

77 **"be not abused . . ."** quoted in Johnson, *Astronomical Thought*, p. 130; I have modernized the spelling.

78 **as Stephen Pumfrey notes . . .** Pumfrey, "Harriot's Maps of the Moon."

79 **". . . of his radical cosmology"** Pumfrey, "Harriot's Maps of the Moon," p. 166.

79 **"those vast and multitudinous lights . . ."** quoted in Johnson, *Astronomical Thought*, p. 217.

80 **"physiologia nova"** quoted in Cohen, *Revolution in Science*, p. 133.

80 **"The Magnetic Force . . ."** quoted in Kocher, p. 181.

80 **"like a community of souls . . ."** Russell, p. 207.

80 **"demonstrated by many arguments . . ."** quoted in Cohen, *Revolution in Science*, pp. 133–34.

80 **"contains the seeds . . ."** Cohen, *Revolution in Science*, p. 135.

82 **"Remarkable that this ass . . ."** quoted in Rowland, p. 77.

82 **"Dante's certainties about . . ."** Rowland, p. 106.
82 **When crossing the English Channel . . .** See Rowland, p. 139.
83 **As Giovanni Aquilecchia has noted** Aquilecchia, p. 9.
84 **". . . which rather did run round"** quoted in Gatti, *Essays on Giordano Bruno*, p. 23.
84 **"For he [Copernicus] had . . ."** quoted in Gatti, *Essays on Giordano Bruno*, p. 65.
84 **"ordained by the gods . . ."** quoted in Greenblatt, *The Swerve*, p. 238.
84 **"There is a single general space . . ."** quoted in McLean, p. 147.
84 **"one of an infinite number . . ."** quoted in Stimson, p. 51.
85 **"Suppose now that all space . . ."** quoted in Decker, p. 603.
85 **. . . an eternal past** Rowland, p. 165.
85 **after the group had visited Oxford . . .** Gatti, *Giordano Bruno: Renaissance Philosopher*, p. 44.
86 **"a whole city . . ."** quoted in Gatti, *Essays on Giordano Bruno*, p. 141.
86 **"England can brag of having . . ."** quoted in Rowland, p. 152.
86 **"superior to all the kings . . ."** quoted in Gatti *Essays on Giordano Bruno*, p. 142.
87 **"from infinity is born . . ."** quoted in Jacob, p. 31.
88 **". . . to share their wives"** Rowland, p. 206.
88 **"an all-pervading world-soul"** Rowland, p. 218.
88 **"*That this infinite space* . . ."** quoted in Rowland, p. 219.
88–89 **"a republic of stars . . ."** Rowland, p. 221.
90 **"incubatory period"** Feingold, *The Mathematician's Apprenticeship*, p. 16.

5. THE RISE OF ENGLISH SCIENCE AND THE QUESTION OF THE TUDOR TELESCOPE

91 **"sorrow's eye, glazed with blinding tears . . ."** *Richard II* (2.2.16).
91 **The annual lecture . . .** Johnson, *Astronomical Thought*, p. 199.
92 **"had little scholarly training . . ."** Johnson, *Astronomical Thought*, p. 200.
92 **"a general clearinghouse . . ."** Johnson, *Astronomical Thought*, p. 265.
92 **"all the Gresham Astronomy Professors . . ."** Chapman, "Thomas Harriot: The First Telescopic Astronomer," p. 318.
93 **one out of every ten books . . .** According to the *Short Title Catalogue of English Books, 1475–1640* (London 1926). See Johnson, *Astronomical Thought*, p. 9.
93 **"was not confined to scholars . . ."** Johnson, *Astronomical Thought*, p. 10.
94 **"Every ship that put in . . ."** Harkness, p. 10.
94 **an ostrich egg, money from China . . .** These particular items were mentioned by Lorraine Daston on "Ideas," CBC Radio, May 4, 2012.
94 **"after a more plain manner . . ."** quoted in Harkness, p. 98.
94 **"Three merchants have . . ."** quoted in Harkness, p. 117.
95 **"By the end of Elizabeth's reign . . ."** Harkness, p. 140.
95 **"agreeable to the hypothesis . . ."** quoted in Johnson, *Astronomical Thought*, p. 208; I have modernized the spelling.
95 **In Blagrave's astrolabe . . .** See Johnson, *Astronomical Thought*, pp. 208–9.
95 **". . . of the Copernican theory"** Johnson, *Astronomical Thought*, p. 210.

96 **Gerard Mercator had built . . .** See Findlen, p. 664.
96 **"the ground of all men's affairs"** quoted in Harkness, p. 98.
96 **about five per year** Harkness, p. 104.
97 **football was deemed . . .** Byrne, p. 204.
97 **". . . in the fourth century B.C."** Sharpe, *Early Modern England*, p. 259.
97 **"during the brief period . . ."** Johnson, *Astronomical Thought*, p. 10.
97 **"entirely dependent upon . . ."** Johnson, *Astronomical Thought*, p. 11.
98 **as Paula Findlen notes** Findlen, p. 662.
98 **". . . rather than a manifesto"** Findlen, p. 662.
98 **"we may be sure . . ."** Johnson, *Astronomical Thought*, p. 181.
98 **"and sometimes taught well"** Feingold, *The Mathematician's Apprenticeship*, p. 20.
98 **Edmund Lee, for example . . .** Feingold, *The Mathematician's Apprenticeship*, p. 100.
99 **"Copernicus, the prince . . ."** Feingold, *The Mathematician's Apprenticeship*, p. 47.
99 **William Camden . . .** Feingold, *The Mathematician's Apprenticeship*, p. 101.
99 **John Mansell . . .** Feingold, *The Mathematician's Apprenticeship*, p. 102.
99 **Richard Crakanthorpe . . .** Feingold, *The Mathematician's Apprenticeship*, p. 66.
99 **Sir William Boswell . . .** Feingold, *The Mathematician's Apprenticeship*, p. 79.
99 **"more cultivated in London . . ."** Feingold, *The Mathematician's Apprenticeship*, p. 87.
100 **"[In] England, perhaps . . ."** Johnson, "The Influence of Thomas Digges . . . ," p. 390.
101 **"explanations take the mystery out . . ."** quoted in Ball, p. 100.
101 **". . . of discovering their own errors"** Bacon (ed. Weinberger), p. 24.
101 **"an intricate puzzle"** Ball, p. 107.
101 **"heaven and earth do conspire . . ."** Bacon, *The Advancement of Learning*, p. 37.
101 **"that anticipate the technology . . ."** Cartwright, p. 70.
102 **"This lady was endued . . ."** Bacon, *The Advancement of Learning*, p. 49.
102 **. . . set of musical chimes** Feingold, *The Mathematician's Apprenticeship*, p. 198.
102 **wore a tiny "alarm watch"** Landes, p. 87.
102 **"a quantity of water . . ."** quoted in Feingold, *The Mathematician's Apprenticeship*, 198.
103 **presented James with a clock . . .** Feingold, *The Mathematician's Apprenticeship*, p. 198.
103 **"the latest book by Galileo"** quoted in Maisano, "Shakespeare's Last Act," p. 427.
105 **"a perspective glasse . . ."** quoted in McLean, p. 150.
106 **"As far as we can tell . . ."** Chapman, "Thomas Harriot: The First Telescopic Astronomer," p. 315.
106 **by February 1610 . . .** Johnson, *Astronomical Thought*, p. 228.
106 **Harriot made numerous astronomical observations . . .** See, for example, Chapman, "Thomas Harriot: The First Telescopic Astronomer."

109 **"two perspective trunks . . ."** quoted in McLean, p. 154.
109 **Dee mentions two visits by Harriot . . .** Feingold, *The Mathematician's Apprenticeship*, p. 137.
109 **Bruno's books . . .** Fox, p. 6.
109 **"There can be no question . . ."** Johnson, *Astronomical Thought*, p. 229.
109 **"could well have intersected . . ."** Stephen Greenblatt, author interview, May 1, 2012.
109 **"there is a possibility . . ."** Stephen Greenblatt, author interview, May 1, 2012.
110 **"was quite happy to accumulate . . ."** Chapman, "The Astronomical Work of Thomas Harriot (1560–1621)," pp. 104–5.
111 **". . . not as a public figure"** Chapman, "Thomas Harriot: the first telescopic astronomer," p. 320.
113 **as David Levy argues** See Levy, *Science in Early Modern English Literature*, p. 64.
113 **"a stealthy reminder . . ."** Dawson and Yachnin, p. 184.
113 **"must be confused with . . ."** Ure, p. 71.
113 **"The allusion is to a 'perspective' . . ."** Wilders, pp. 153–154.
114 **"magic crystal permitting . . ."** Braunmuller, p. 212.
114 **"a specially devised . . ."** Brown, p. 80.
114 **"by the late sixteenth century . . ."** Dunn, p. 15.
115 **"things of a marvellous largeness . . ."** quoted in McLean, p. 149.
115 **"seems entirely probable"** Johnson, "The Influence of Thomas Digges . . . ," p. 401.
115 **"cannot conclude . . ."** Levy, *Science in Early Modern English Literature*, p. 67.
115 **"I simply do not believe . . ."** Chapman, "Thomas Harriot: The First Telescopic Astronomer," p. 324.
116 **"There would have been . . ."** Harkness, p. 2.
116 **"the seeds of modern scientific thought"** Jensen, p. 527.
116 **"prologue of modern science"** Feingold, *The Mathematician's Apprenticeship*, p. 7.

6. A BRIEF HISTORY OF WILLIAM SHAKESPEARE

117 **"Who is it that can tell me who I am?"** *King Lear* (1.4.189).
117 **conducts the walk every day** Stratford Town Walk; see www.stratfordtownwalk.co.uk.
119 **Only three-quarters . . .** Forgeng, p. 47.
120 **". . . as a double consciousness"** Greenblatt, *Will in the World*, p. 103.
120 **"To argue that the Shakespeares . . ."** Shapiro, *1599*, p. 148.
120 **"was a fact of daily life"** Jones, *Shakespeare's England*, p. 41.
120 **"His mind and world . . ."** Bate, p. 12.
121 **"One had to make . . ."** Jones, *Shakespeare's England*, p. 39.
121 **". . . in English history"** Jones, *Shakespeare's England*, p. 41.
121 **about four million** Jones, *Shakespeare's England*, p. 34.
121 **About one-third** Forgeng, p. 43.
121 **Those who lived to thirty . . .** Forgeng, p. 68.
121 **"fickle-headed tailors"** quoted in Pritchard, p. 19.

122 **"Women [were made] to . . ."** quoted in Forgeng, p. 40.
122 **"They go to market . . ."** quoted in Forgeng p. 41.
123 **. . . and were not taught Latin** See, for example, Jensen, p. 512.
123 **more than a hundred works . . .** Travitsky, p. 3.
123 **about 160 such institutions** Jensen, p. 512.
124 **"God hath sanctified . . ."** quoted in Pritchard, p. 91.
124 **". . . exceptionally troublesome adult"** Dodd, p. 91.
125 **"ordered her carriage . . ."** quoted in Doran, p. 54.
126 **"very majestic; her face oblong . . ."** quoted in Nicoll, p. 5.
126 **". . . heart and stomach of a king"** quoted in Bate, p. 226; Williams, p. 210.
126 **twenty-seven for men; twenty-four for women** Forgeng, p. 64.
127 **"It may, in fact, have been . . ."** St. John Parker, p. 6.
127 **"This was an age in which . . ."** Laroque, p. 18.
128 **two kilobytes of memory** The Apollo computers also had an additional 32k of read-only storage. See http://downloadsquad.switched.com/2009/07/20/how-powerful-was-the-apollo-11-computer/.
128 **". . . about the nature of genius"** http://literateur.com/interview-with-james-shapiro.
131 **admission was one penny** Chute, p. 58.
131 **"For at one time . . ."** quoted in Pritchard, p. 164.
132 **"great multitudes of people . . ."** quoted in Ackroyd, p. 94.
132 **". . . brimful of curiosities"** quoted in Harkness, pp. 1–2.
132 **"theatre industry and the sex trade . . ."** Bate, p. 47.
133 **some 2,760 books** Kermode, p. 44.
133 **"Scarce a cat can look . . ."** quoted in Chute, p. 65.
133 **"evil disposed men . . ."** quoted in Fitzmaurice, p. 30.
134 **". . . the latest literary trends"** Shapiro, p. 191.
134 **"self-satisfied pork butcher"** quoted in Nicholl, p. 22.
134 **. . . the possession of the Chandos family** See, for example, Bryson pp. 2–4.
135 **"upstart crow . . ."** quoted in Laroque, p. 48.
136 **"the most experienced playgoers . . ."** Shapiro, p. 9.
137 **"ten times . . . a well-paid schoolmaster"** Schoenbaum, p. 212.
137 **". . . a serious living by his pen"** Bate and Thornton, p. 10.
138 **". . . most women of her class"** St. John Parker, p. 10.
139 **a story laced with maybes** The quotations that follow are from Donnelly and Woledge, p. 10; Greenblatt, *Will in the World*, p. 71; Day, p. 6; Shapiro, p. 190; Ackroyd, p. 428.
139 **"We must assume" / "could well have"** Kermode, pp. 44, 17.
140 **a headline on the BBC News website** See http://news.bbc.co.uk/2/hi/uk_news/1007876.stm.
140 **Again, it made headlines** Leake, "Bad Bard: a tax dodger and famine profiteer"; Lawless, "Shakespeare the 'hard-headed businessman' uncovered."
141 **". . . not the apparel or the books"** Bryson, p. 180.
141 **". . . but that they are dull"** Greenblatt, *The Norton Shakespeare*, p. 47.

141 **"better known than contemporaries . . ."** Stephen Greenblatt, author interview, May 1, 2012.
142 **"It seems rather circular to me . . ."** http://literateur.com/interview-with-james-shapiro.
142 **"The longer you muse over . . ."** Stephen Greenblatt, author interview, May 1, 2012.

7. THE SCIENCE OF HAMLET

145 **"More things in heaven and earth . . ."** *Hamlet* (1.5.174).
145 **more than 1,500 lines** Hunt, p. 2.
146 **". . . but idle coinages"** http://shakespearean.org.uk/ham1-haz.htm.
149 **"in the constellation of Cassiopeia . . ."** quoted in Olson, p. 70.
149–50 **"throughout the scene . . ." / "seems to imply . . ."** Spencer, p. 207.
151 **"The Pole Star's usefulness . . ."** Neill, *Othello*, p. 241.
152 **"knew better than his commentators . . ."** Furness, p. 93.
152 **"the brilliant star Capella . . ."** Jenkins, p. 167.
152 **Capella is out and the supernova is in** Thompson and Taylor, p. 151.
153 **some of his fellow actors . . .** Jenkins, p. 190.
156 **multiple versions of the engraving** See Olson, p. 72; Gingerich, "Astronomical Scrapbook," p. 395.
156 **"the most noble and most learned . . ."** Hotson, p. 123.
156 **"it is entirely possible . . ."** Gingerich, "Astronomical Scrapbook," p. 395.
156 **"I have included four copies . . ."** quoted in Olson, p. 72.
156 **"we may be sure that Tycho's portrait . . ."** Gingerich, "Astronomical Scrapbook," p. 395.
157 **"Shakespeare's imagination . . ."** Olson p. 72.
157 **"an authentic touch of Denmark"** Jenkins, p. 423.
157 **"were common among . . . Danish families"** Jenkins, p. 422.
157 **"an admittedly striking . . ."** Marchitello, p. 78.
157 **". . . isn't just a coincidence"** Scott Maisano, author interview, June 4, 2012.
158 **the character of Fluellen . . .** Hotson, pp. 119–122.
159 **"Students have never had . . ."** quoted in Johnson, *Astronomical Thought*, p. 184; I have modernized the spelling.
160 **"a frequent and welcome guest"** Hotson, p. 112.
162 **"An underground car park . . ."** Nicholl, p. 50.
162 **As Leslie Hotson has noted** Hotson, p. 113.
162 **the wreck of the *Sea Venture*** Bate, p. 58.
163 **"Be sure, our Shakespeare . . ."** quoted in Hotson, p. 247.
163 **"little doubt that from 1590 . . ."** Hotson, p. 124
163 **"There is a single general space . . ."** quoted in Janowitz, p. 79.
165 **". . . of so noble a theory"** quoted in Danielson, p. 133.
166 **"The Ptolemaic science . . ."** Shapiro, p. 299.
166 **"the poem's first two lines . . ."** Jenkins, p. 242.
166 **"is a clever epitome . . ."** Spencer, p. 249.

167 **"to have left a mark . . ."** Gatti, *Essays on Giordano Bruno*, p. 142.

168 **"The Bruno–Shakespeare discussion . . ."** Gatti, *Essays on Giordano Bruno*, p. 146.

168 **favorable mention of Bruno** Gatti, *Essays on Giordano Bruno*, p. 143.

168 **"A convincing basis . . ."** Gatti, *Essays on Giordano Bruno*, p. 144.

168 **"Whoever wishes to philosophise . . ."** quoted in Gatti, *Essays on Giordano Bruno*, p. 155.

168 **"a particularly obstinate heretic . . ."** Gatti, *Essays on Giordano Bruno*, p. 155.

169 **"The truth must be pursued . . ."** Gatti, *Essays on Giordano Bruno*, pp. 155–6.

8. READING SHAKESPEARE, AND READING INTO SHAKESPEARE

170 **"I argue that as early as 1601 . . ."** Usher, *Hamlet and the Infinite Universe*.

171 **"It is simply not credible . . ."** Usher, *Shakespeare and the Dawn of Modern Science*, pp. xix–xx.

171 **"hunting through the canon . . ."** Peter Usher, author interview, May 24, 2012.

172 **"Thomas and Hamlet are both . . ."** Usher, *Shakespeare and the Dawn of Modern Science*, pp. 95–96.

173 **"is an excellent fit . . ."** Usher, *Shakespeare and the Dawn of Modern Science*, p. 73.

174 **he notes that another astronomer** Usher, "Shakespeare's Cosmic World View," p. 23.

175 **along with the *Oxfordian*** The journal's website is http://www.shakespeare-oxford.com.

175 **"widely regarded as the poet Shakespeare . . ."** Usher, *Shakespeare and the Dawn of Modern Science*, p. xxiii.

176 **"it has been argued . . ."** Kragh, p. 57.

176 **"The idea of having . . ."** Peter Usher, author interview, May 24, 2012.

176 **"are in such a condition . . ."** quoted in McLean, p. 152.

177 **"there was no persecution . . ."** Chapman, "Thomas Harriot: The First Telescopic Astronomer," p. 318.

179 **". . . the Brunian version of Copernicanism"** Sacerdoti, p. 8.

179 **"If he keeps on like this . . ."** http://lafrusta.homestead.com/rec_shakespeare_antonio_cleopatra.html; with thanks to Marina de Santis for translating the review and portions of the book.

180 **". . . to invent a vengeful bear"** Usher, *Shakespeare and the Dawn of Modern Science*, p. 280.

180 **". . . the planet Jupiter's Great Red Spot"** Usher, "Shakespeare's Support for the New Astronomy," accessed online.

180 **could represent Saturn** Usher, *Shakespeare and the Dawn of Modern Science*, p. 241.

181 **". . . without telescopic aid"** Usher, *Shakespeare and the Dawn of Modern Science*, p. 69.

181 **"In *Hamlet*, the Bard describes . . ."** Usher, "Shakespeare and Elizabethan Telescopy," p. 17.

181 **". . . of resolving such detail"** Usher, *Shakespeare and the Dawn of Modern Science*, p. xxiii.

182 **"strains credulity"** Levy, *Science in Early Modern English Literature*, p. 67.

182 **"Spyglasses had obvious military uses . . ."** Usher, *Shakespeare and the Dawn of Modern Science*, pp. 310–311.

183 **"relatively good, clear images . . ."** Chapman, "Thomas Harriot: The First Telescopic Astronomer," p. 324.

184 **"quadruple groupings"** McAlindon, p. 80.

184 **"unified duality . . ."** McAlindon, p. 48.

184 **"number symbolism co-operates with . . ."** This and the subsequent quotations are from McAlindon, p. 200.

184 **"The explanation, of course . . ."** McAlindon, p. 205.

184 **"the witches' favourite . . ."** McAlindon, p. 205.

185 **"a far more intricate and artful play . . ."** McAlindon, p. 209.

185 **"the language of proportionality . . ."** These quotations are from Raman, "Specifying Unknown Things," p. 209.

185 **"the language of Renaissance arithmetic . . ."** Raman, "Death by Numbers," p. 159.

185 **". . . on the finititude of existence"** These quotations are from Raman, "Death by Numbers," pp. 162, 168, 174.

186 **"was often lax with numbers"** Jenkins, p. 346.

186 **"The meaning generally given . . ."** http://www.shakespeare-online.com/plays/hamlet/examq/six.html.

187 **"Sometimes a cigar . . ."** Unfortunately, Freud may not have actually said it. See, for example, http://quoteinvestigator.com/2011/08/12/just-a-cigar/.

189 **"the bear's disruptive . . ."** Abstract distributed at the seminar but not yet posted online.

9. SHAKESPEARE AND GALILEO

192 **"Does the world go round?"** *Cymbeline* (5.5.232).

193 **"Aristotle declares . . ."** Crew, pp. 64–65.

194 **"Philosophy is written . . ."** Drake, pp. 237–8.

195 **"I have many diverse inventions"** quoted in Reston, p. 85.

195 **"It is really pitiful . . ."** quoted in Reston, p. 54.

195 **"About ten months ago . . ."** Drake, pp. 28–29.

196 **"by means of which . . ."** quoted in Dunn, p. 22.

196 **"to turn a popular carnival toy . . ."** Gingerich, "Mankind's place in the Universe," p. 28.

197 **"On the seventh day of January . . ."** Drake, p. 51.

199 **". . . their revolution about Jupiter"** These quotations are from Drake, pp. 52–53.

200 **"from the creation of the world . . ."** Drake, pp. 50–51.

200 **". . . through so many ages"** Drake, p. 49.

200 **as far away as Peking** Drake, p. 59.

201 **"without any doubt . . ."** quoted in Swerdlow, p. 261.

201 **"With absolute necessity . . ."** Drake, p. 94.
201 **"now we have not just one . . ."** Drake, p. 57.
202 **"made it intellectually respectable . . ."** Gingerich, "Mankind's place in the Universe," p. 28.
202 **"I send herewith unto his majesty . . ."** quoted in Panek, pp. 40–42.
202 **"Me thinks my diligent Galileo . . ."** quoted in Reston, p. 100; I have modernized the spelling.
202 **"*An luna sit habitabilis?*"** Feingold, "Galileo in England," p. 416.
203 **"became seminal to . . ."** Feingold, "Galileo in England," p. 415.
203 **"They that buy such books . . ."** quoted in Feingold, "Galileo in England," p. 416.
203 **a play published in 1618** Feingold, "Galileo in England," p. 416.
203 **"Columbus gave man . . ."** quoted in Panek, p. 41.
203 **"tragical-comical-historical-pastoral"** Bate, *Cymbeline*, p. vii.
204 **"to waste criticism upon . . ."** These quotations are from Bate, *Cymbeline*, pp. xii–xiii.
204 **"narrative grips and compels . . ."** Butler, p. 1.
206 **"has often been questioned . . ."** Warren, *Cymbeline*, p. 54.
207 **"perhaps the head . . ."** Warren, *Cymbeline*, p. 235.
208 **"These ghosts happen . . ."** Usher, "Jupiter and *Cymbeline*," p. 8.
208 **"The book placed on the bosom . . ."** Usher, "Jupiter and *Cymbeline*," p. 8.
208 **"*Cymbeline* has mystical . . ."** Usher, *Shakespeare and the Dawn of Modern Science*, p. 171.
209 **". . . on new discoveries"** Usher, *Shakespeare and the Dawn of Modern Science*, p. 183.
211 **"is replete with zodiacal . . ."** Usher, *Shakespeare and the Dawn of Modern Science*, p. 224.
211 **"is a paean to the glories . . ."** Usher, *Shakespeare and the Dawn of Modern Science*, p. xxii.
211 **"If it seems incongruous . . ."** Maisano, "Shakespeare's Last Act," p. 403.
212 **"Shakespeare must have seen . . ."** quoted in Maisano, "Shakespeare's Last Act," p. 403.
212 **"a backward-looking romance . . ."** Maisano, "Shakespeare's Last Act," p. 411.
212 **". . . undoubtedly Galileo"** Maisano, "Shakespeare's science fictions," accessed online.
212 **"the only such utterance . . ."** Maisano, "Shakespeare's Last Act," p. 413.
213 **"calls our attention . . ."** Maisano, "Shakespeare's Last Act," p. 429.
213 **"seems to have set . . ."** Maisano, "Shakespeare's Last Act," p. 415.
213 **"by the evidence of . . ."** Pitcher, p. lxix.
213 **"in Ptolemaic cosmology . . ."** Butler, *Cymbeline*, p. 220.
214 **"In that year, because of Galileo . . ."** Pitcher, p. lxix.
214 **"as a deliberate and subtle . . ."** Pitcher, p. lxxiii.
215 **"If Galileo's telescope . . ."** Pitcher, p. lxxiii.
215 **"possibly a way of saying . . ."** Pitcher, p. lxxvi.
215 **"a scientific publication . . ."** Pitcher, p. lxxii.

10. THE ALLURE OF ASTROLOGY

218 "Treachers by spherical predominance . . ." *King Lear* (1.2.108).
218 . . . the date of July 11, 1564 Ackroyd, p. 4.
218 "Saturn was passing through . . ." quoted in Brigden, p. 299.
218 "People watched the sky . . ." Olsen, vol. 1, p. 71.
219 "a fringe position" Olsen, vol. 1, p. 60.
220 "an essential aspect . . ." Thomas, *Religion and the Decline of Magic,* p. 338.
220 "had a fair claim to being . . ." Sharpe, *Early Modern England*, p. 307.
220 "man's body, and all other . . ." quoted in Hale, p. 568.
221 "the most part of men . . ." quoted in Kocher, p. 210.
221 "An expert and prudent astrologer . . ." quoted in Thomas, p. 392.
221 "If thou want'st an heir . . ." quoted in Thomas, p. 393.
222 Shakespeare's company consulted . . . Ackroyd, p. 374.
222 an English pamphlet See Hale, p. 567.
222 ". . . both truth and falsehood" Montaigne (ed. Screech), p. 44.
224 "Nowadays among the common . . ." quoted in Thomas, p. 269.
224 "probably the most ambitious" / "coherent and comprehensive" Thomas, pp. 340, 391.
224 "often looked suspiciously . . ." Kocher, p. 201.
225 "If we cannot deny . . ." quoted in Thomas, p. 395.
226 "not to hear about the stars . . ." Kocher, p. 207.
226 "For in those thinges . . ." quoted in Kocher, p. 213.
227 "So fared another clerk . . ." The quotation from *The Miller's Tale,* is lines 349–53; http://www.librarius.com/canttran/milltale/milltale331-387.htm.
228 In the New Cambridge edition These quotations are from Halio, p. 121.
228 "He scorns the platitudes . . ." Bevington, *Shakesepare's Ideas*, p. 168.
231 "part of an endeavour . . ." McAlindon, p. 4.
232 "hardly any would work . . ." quoted in Thomas, p. 355.
232 "lost their reputation . . ." quoted in Thomas, p. 355.
232 "For their observations . . ." quoted in Thomas, p. 398.
232 "The paradox . . ." Thomas, p. 401.
233 "The world could no longer . . ." Thomas, p. 415.

11. MAGIC IN THE AGE OF SHAKESPEARE

234 "Fair is foul, and foul is fair" *Macbeth* (1.1.12).
234 "perhaps the most striking . . ." Braunmuller, p. 118.
234 "The weird sisters rise . . ." quoted in Braunmuller, p. 24.
234 "a soul tortured by . . ." Bradley, p. 276.
234–5 "not of brevity but of speed" Bradley, p. 276.
235 As Terry Eagleton asserts See Dickson, p. 210.
235 "Long misogynistic traditions . . ." / "less resistant to Satan's advances" Edwards, p. 45; Sharpe, *Early Modern England*, p. 312.
235 "it was the women who . . ." Thomas, p. 678.
235 "old woman with a wrinkled face . . ." quoted in Thomas, p. 677.
236 "whether the accused . . ." Olsen, vol. 2, p. 678.

237 **In England, records show** The English and Scottish figures are from Sharpe, "The Debate on Witchcraft," p. 513, and from Edwards, p. 32.
237 **"wit, understanding, or sence"** Olsen, vol. 3, p. 675.
238 **"a thing like a black Dog . . ."** quoted in Edwards, p. 37.
238 **young Edmund admitted** Sharpe, "The Debate on Witchcraft," p. 520.
239 **"to chasten sinful humankind . . ."** quoted in Edwards, p. 39.
239 **". . . salvation and damnation"** Brigden, p. 302.
240 **"the growing preoccupation . . ."** Edwards, p. 47.
240 **"Sorcerers are too common . . ."** quoted in Thomas, p. 209.
240 **"only one branch . . ."** Thomas, p. 210.
241 **endorsed by scientists . . .** Thomas, p. 261.
241 **"Let a man's child . . ."** quoted in Thomas, p. 300.
241 **"both one Master . . ."** quoted in Vaughan and Vaughan, p. 63.
241 **"witches, conjurers, enchanters . . ."** quoted in Thomas, p. 307.
242 **"can only be imagined . . ."** Spiller, pp. 26, 36.
242 **". . . serious magician and carnival illusionist"** Vaughan and Vaughan, p. 63.
243 **"at least possible . . ."** Gatti, *Essays on Giordano Bruno*, p. 163.
244 **". . . science and imposture"** Campbell, p. xviii.
246 **"the possibility of telepathy . . ."** Thomas, p. 266.
248 **"I erect the whole of astronomy . . ."** quoted in Rosen, *Kepler's Somnium*, p. 100.
249 **the first astrophysicist** Gingerich, *The Book Nobody Read*, p. 168.
249 **"implied an end to the . . ."** Cohen, *Revolution in Science*, p. 129.
249 **"soul principle"** Cohen, *Revolution in Science*, p. 132.
250 **"a Renaissance scientific paradox . . ."** Debus, p. 100.
250 **"we could easily assemble . . ."** Cohen, *Revolution in Science*, p. 127.
250 **"foolish little daughter . . ."** quoted in Baumgardt, p. 27.
250 **"the last major astronomer . . ."** Cohen, *Revolution in Science*, p. 127.
250 **"The overwhelming impression . . ."** Sharpe, *Early Modern England*, pp. 308–9.
251 **"is precisely because . . ."** Henry, p. 55.
251 **"The end of our foundation . . ."** quoted in Henry, p. 59.

12. SHAKESPEARE AND MEDICINE

252 **"A body yet distempered . . ."** *Henry IV, Part 2* (3.1.40).
252 **"Behold at Southwark . . ."** http://www.thegarret.org.uk/pdfs/exhibitions/stthomashospitalmedieval.pdf
257 **As Kirstin Olsen notes** See Olsen, p. 473.
257 **more than six hundred editions** Hale, p. 557.
259 **"an angry mob . . ."** Olsen, vol. 1, p. 10.
260 **". . . a blood-letting lancet"** quoted in Hale, p. 543.
261 **". . . usually failed to work"** Olsen, vol. 1, p. 177.
263 **"houses lately infected . . ."** quoted in Ackroyd, p. 477.
263 **"Mortality and anxiety . . ."** Ackroyd, p. 119.

264 **"simply wandered from . . ."** Olsen, vol. 1, p. 398.
265 **"unobtrusive"** Pope, p. 287.
265 **"The number of medical references . . ."** Andrews, vol. 2, p. 98.
266 **"several dignified . . . medical men"** Bate, p. 46.
266 **"Mrs *Hall of Stratford* . . ."** quoted in Bate, p. 44.
267 **"the intimate relationship . . ."** Bate, p. 46.
267 **"Then was a pigeon cut . . ."** Kevin Flude, unpublished article.
267 **". . . at this particular stage"** Flude, author interview, June 22, 2012.
267 **". . . from the reality"** Flude, unpublished article.

13. LIVING IN THE MATERIAL WORLD

268 **"Drawn with a team of little atomi . . ."** *Romeo and Juliet* (1.4.58).
269 **"Fly all around . . ."** Lucretius, pp. 65–66.
269 **". . . risking eternal damnation"** Bertram, p. 171.
269 **"Of her own, by chance . . ."** quoted in Hecht, p. 151.
270 **". . . in Einstein or Freud . . ."** Flow, accessed online.
270 **"There is no master plan . . ."** Greenblatt, *The Swerve*, p. 6.
270 **"For certainty not by design . . ."** Lucretius, p. 32.
271 **". . . the taint of vile religion"** quoted in Hecht, p. 150.
271 **"the notion, intrinsic . . ."** Hyman, p. 3.
271 **"Lucretian thoughts percolated . . ."** Greenblatt, *The Swerve*, p. 220.
271 **Some thirty Latin editions** Palmer, p. 414.
272 **he had a medal struck** Burke, p. 14.
273 **"a thorny undertaking"** quoted in Bakewell, p. 33.
273 **"I turn my gaze inward . . ."** quoted in Bakewell, p. 224.
274 **"What kind of Good . . ."** Montaigne (ed. Screech), p. 653.
274 **"The fury of the mob . . ."** Montaigne (ed. Screech), p. 656.
274 **". . . to roast someone alive"** quoted in Jacob, p. 35.
274 **"monstrous thing" / ". . . darkness of irreligion"** Montaigne (ed. Screech), pp. 498, 500.
274 **"The wise man ought . . ."** quoted in Jacob, p. 37.
274 **". . . as a social ripple"** Jacob, p. 37.
275 **"has convinced [himself] that . . ."** Montaigne (ed. Screech), p. 502.
275 **"atheistic naturalism"** Montaigne (ed. Screech), p. xli.
275 **"anthropological curiosity"** Friedrich, p. 134.
275 **"erroneous conclusion"** Friedrich, p. 137.
276 **goat's blood was probably . . .** Sayce, p. 186.
276 **"points to the scientific method . . ."** Sayce, p. 187.
276 **"[Diagoras] was shown . . ."** Montaigne (ed. Screech), p. 44.
276 **"We have formed a truth . . ."** quoted in Bakewell, p. 129.
276 **If colored glass is used . . .** Montaigne (ed. Screech), p. 676.
277 **"So whoever judges . . ."** quoted in Jacob, p. 36.
277 **". . . inevitably swept away"** Montaigne (ed. Screech), p. 676.
277 **"When I play with my cat . . ."** Montaigne (ed. Screech), p. 505.
277 **"it was the earth that moved . . ."** Montaigne (ed. Florio), p. 514.
278 **"one of the first in France . . ."** Friedrich, p. 140.

278 **"Montaigne shows . . ." / "the theories of Copernicus . . ."** Sayce, pp. 79, 110.

278 **"Is it not more likely . . ."** Montaigne (ed. Screech), pp. 644–5.

278 **"raised a huge question mark . . ."** Bakewell, p. 139.

278 **"Montaigne's preference for details . . ."** Bakewell, p. 275.

279 **"hidden Englishman"** Bakewell, p. 276.

279 **"that hath no kinde of traffike . . ."** quoted in Vaughan and Vaughan, p. 61.

280 **"Montaigne's fingerprints are . . ."** Stephen Greenblatt, author interview, May 1, 2012.

280 **"rhetorical strategy of exploring . . ."** Vaughan and Vaughan, p. 61.

280 **the British Museum has a copy** See, for example, Bell, p. 20.

281 **"held up as *modern* writers . . ."** Bakewell, p. 279.

281 **"seems indebted to the French essayist . . ."** Halio, p. 9.

281 **contains more than a hundred words** Bell, pp. 17, 146.

281 **"skeptic philosopher"** Bell, p. 169.

281 **"Edmond is an adventurer . . ."** Bradley, pp. 249–50.

282 **"I think of Edmond . . ."** Stephen Greenblatt, author interview, May 1, 2012.

282 **"Edmond is the embodiment . . ."** Bate, p. 65.

283 **"dangerously attractive"** Harold Bloom, *King Lear in the 20th Century*, p. 314.

284 **"Lear's resort to natural, rather than divine . . ."** Elton, *King Lear and the Gods*, p. 220.

284 **"Lear aptly summarizes . . ."** Halio, p. 167.

284 **"is like his brother Edmond . . ."** Bevington, *Shakespeare's Ideas*, p. 172.

14. THE DISAPPEARING GODS

285 **"As flies to wanton boys . . ."** *King Lear* (4.1.36).

285 **". . . his multitudinous powers"** Bradley, p. 200.

285 **"for the immediate future . . ."** quoted in Halio, p. 54.

286 **"natural ideas of justice"** quoted in Halio, p. 24.

286 **"horror is piled upon horror . . ."** Wilson, p. 120.

286 **"how much misery . . ."** McAlindon, p. 181.

286 **even very young children** See Paul Bloom, "The Moral Life of Babies."

287 **as one psychologist puts it** This summary of Lerner's work is from Furnham, p. 795.

287 **"the BJW seems to provide . . ."** Furnham, p. 796.

288 **"threatens deeply held beliefs . . ."** Feinberg and Willer, p. 34.

288 **"a battleground over which . . ."** Bevington, *Shakespeare's Ideas*, p. 169.

288 **"essentially moral, dictating altruism . . ."** McAlindon, p. 173.

288 **". . . in which the wicked prosper"** quoted in Bradley, p. 252.

288 **"pose disturbing questions . . ."** Bevington, *Shakespeare's Ideas*, p. 171.

289 **"a gesture, however inadequate . . ."** Greenblatt, *Will in the World*, p. 180.

289 **"What is truly frightening . . ."** Bevington, *Shakespeare's Ideas*, pp. 169–70.

289 **"the stage panache . . ."** Bate, p. 308.

290 **"an intrinsically modern disposition" / "roughly contemporaneous . . ."** Hyman, pp. xviii, 2.

290 **"the notion of a worldview . . ."** Hyman, p. 4.

290 **A similar piece of legislation** The laws are quoted in Berman, pp. 48–49.
290 **"must have flourished in . . ."** Bertram, p. 167
291 **". . . were deeply upsetting"** Greenblatt, *The Swerve*, p. 252.
291 **"to exorcise atomism"** Greenblatt, *The Swerve*, p. 250.
291 **"whether human souls . . ."** quoted in Allen, p. 56.
291 **"great many in England . . ."** quoted in Buckley, p. 30; I have modernized the spelling.
291 **"Others grant God to be . . ."** quoted in Buckley, pp. 65–66.
292 **". . . no simple morality play"** Brigden, p. 308.
292 **"That which nourishes me . . ."** from Brigden, p. 309.
293 **"that Moses was but a Jugler . . ."** quoted in Buckley, p. 130.
293 **"no evidence of religious incredulity"** Buckley, p. 146.
293 **"Not inferior to any of the former . . ."** quoted in Buckley, p. 91.
293 **"Marlowe is able to show more . . ."** quoted in Marlowe, p. xii.
294 **"a temper of mind . . ."** Buckley, p. 129.
294 **". . . and followed Machiavelli"** Bertram, p. 170.
294 **a French poet . . . named Théophile de Viau** Grayling, p. 121.
294 **"taken to be expressive of . . ."** Grayling, p. 122.
295 **"religion as terrified sadism . . ."** Mallin, p. 9.
295 **"Aaron arranges things . . ."** Eric Mallin, author interview, June 14, 2013.
296 **". . . particularly in his tragedies"** Eric Mallin, author interview, June 14, 2013.
296 **"their presence is . . ." / ". . . devastatingly silent"** Halio, p. 15; Greenblatt, *Will in the World*, p. 357.
296 **"with the death of the good . . ."** Elton, *King Lear and the Gods*, p. 337.
296 **"The gods perdie doe reckon . . ."** quoted in Halio, p. 207.
296 **"Lear's disillusionment . . ."** Elton, *King Lear and the Gods*, p. 230.
296 **"must have at least evoked . . ."** McAlindon, p. 195.
296 **"Human beings are here left . . ."** McAlindon, p. 195.
296 **"essentially a godless document . . ."** Eric Mallin, author interview, June 14, 2013.
297 **"pagan atheist and libertine naturalist"** Harold Bloom, *King Lear/Bloom's Shakespeare Through the Ages*, p. 317.
297 **"fraught with danger . . ."** Elton, *King Lear and the Gods*, p. 337.
297 **"For Shakespeare, in the matter of . . ."** quoted in Mallin, p. 6.
298 **". . . strikes me as pretty modern"** Eric Mallin, author interview, June 14, 2013.
298 **"entirely secular . . ." / "He is simply saying . . ."** McGinn, pp. 15, 185–6.
298 **"justice is entirely man-made"** Colin McGinn, author interview, March 12, 2012.
299 **"disturbingly amoral"** Orgel, p. 41.
299 **"seems at once Catholic, Protestant . . ." / "William Shakespeare was . . ."** Greenblattt, *Will in the World*, pp. 103, 113.
299 **". . . faith and scepticism"** Bate, p. 12.
300 **". . . but similar to a clock"** quoted in Shapin, p. 33.
300 **". . . a great piece of clock-work"** quoted in Shapin, p. 34.
301 **"no difference between . . ."** quoted in Shapin, p. 32.

301 **"it is not less natural for a clock . . ."** quoted in Shapin, p. 32.
301 **"nothing but a statue . . ."** quoted in Grayling, p. 158.
301 **". . . configuration of its wheels"** quoted in Maisano, "Descartes avec Milton," p. 38.
302 **"depending on the various . . ."** quoted in Grayling, p. 158.
302 **"a vehicle for 'taking the wonder out' . . ."** Shapin, p. 36.
302 **As Scott Maisano points out** Maisano, "Shakespeare's science fictions," pp. 102–3. Note that Maisano develops these arguments further in "Infinite Gesture."
303 **Shakespeare may well have had** Maisano, "Shakespeare's science fictions," p. 80
303 **"Any sufficiently advanced technology . . ."** http://edge.org/response-detail/11150.
304 **". . . forward-looking speculations"** Maisano, "Shakespeare's science fictions," p. vi.
304 **". . . inventing science fiction"** Nuttall, p. 361.
304 **"When I consider the short duration . . ."** Pascal's *Pensées*, accessed online at http://www.gutenberg.org/files/18269/18269-h/18269-h.htm#p_205.
305 **". . . it also seems pointless"** Weinberg, p. 154.

CONCLUSION

306 **"They say miracles are past . . ."** *All's Well That Ends Well* (2.3.1).
306 **"no longer had its origins . . ."** Findlen, p. 669.
306 **Howard Margolis lists . . . the discoveries** Margolis, p. 5.
308 **"you will learn more about . . ."** quoted in Bouwsma, p. 66.
308 **"surpass by far any Latin and Greek . . ."** quoted in Hale, p. 589.
308 **"one of the most vivid . . ."** Shapin, p. 20.
308 **". . . central, immovable earth"** Stimson, p. 50.
308 **"was no amateur of . . ."** Boas Hall, p. 140.
309 **"no sign of [the Copernican] revolution . . ."** McAlindon, p. 4.
309 **"among Shakespeare's myriad . . ."** Hotson, p. 123.
309 **"Even if Shakespeare had . . ."** Levy, *Starry Night*, p. 69; see also Maisano, "Shakespeare's Last Act," p. 404.
309 **". . . the new heliocentric astronomy"** Bate, p. 62.
309 **". . . by the Copernican revolution"** Shapiro, p. 299.
309 **"laments for a dying world . . ."** Byard, pp. 122–3.
309 **"And freely men confesse . . ."** quoted in Byard, p. 123.
310 **"an intelligent student of science" / "The discoveries of Galileo . . ."** Johnson, *Astronomical Thought*, p. 243.
310 **Donne had connections . . .** Byard, p. 122.
310 **"For of Meridians, and Parallels . . ."** quoted in Byard, p. 123.
310 **"Only the Earth doth stand . . ."** quoted in Cheney, p. 189.
311 **"a proficient and intimate acquaintance . . ."** Orchard, p. 52.
311 **"By the end of the 17th century . . ."** Byard, p. 129.
312 **applied . . . physics to the Last Judgment** See Allen, p. 101.
313 **"a late-medieval worldview . . ."** Stephen Greenblatt, author interview, May 1, 2012.

313 **"is actually surprisingly alert . . ."** Stephen Greenblatt, author interview, May 1, 2012.

313 **"is neither ignorant nor indifferent . . ."** Scott Maisano, author interview, June 4, 2012.

313 **"the privileging of scientism . . ."** Mazzio, p. 1.

314 **". . . of science over literature"** quoted in Mazzio, p. 18.

314 **". . . a Shackleton or an Einstein"** Wilson, p. 124.

Bibliography

SHAKESPEARE EDITIONS CITED

NOTE: Quotations from Shakespeare (along with the cited line numbers) are taken from the editions marked with a ★. In all other cases, they are from *The Arden Shakespeare: The Complete Works, Revised Edition*, edited by Richard Proudfoot, Ann Thompson, and David Scott Kastan. London: Methuen Drama, 2001.

Bate, Jonathan, ed. *Cymbeline*. The RSC Shakespeare. New York: Modern Library, 2011.

★ Bevington, David, ed. *Henry IV, Part 2*. The Oxford Shakespeare. Oxford: Clarendon Press, 1998.

★ Bevington, David, ed. *Troilus and Cressida*. The Arden Shakespeare. London: Thomson Learning, 2006.

★ Braunmuller, A. R., ed. *Macbeth*. Updated edition. The New Cambridge Shakespeare. Cambridge: Cambridge University Press, 2008.

Brooke, Nicholas, ed. *William Shakespeare: The Tragedy of Macbeth*. Oxford: Oxford University Press, 1990.

Butler, Martin, ed. *Cymbeline*. The New Cambridge Shakespeare. Cambridge: Cambridge University Press, 2005.

★ Daniell, David, ed. *Julius Caesar*. The Arden Shakespeare. London: Thomson Learning, 1998.

Dawson, Anthony B., and Paul Yachnin, eds. *William Shakespeare: Richard II*. Oxford: Oxford University Press, 2011.

Edwards, Philip, ed. *Hamlet, Prince of Denmark*. The New Cambridge Shakespeare. Cambridge: Cambridge University Press, 2003.

Furness, Horace Howard, ed. *Othello: The New Variorum Edition*. New York: Dover, 2000. Original publication 1886.

★ Halio, Jay L., ed. *King Lear*. Updated Edition. The New Cambridge Shakespeare. Cambridge: Cambridge University Press, 2007.

Hibbard, G. H., ed. *Hamlet*. The Oxford Shakespeare. Oxford: Clarendon Press, 1987.

Honigmann, E. A. J., ed. *Othello*. The Arden Shakespeare. London: Thomson Learning, 1997.

Hoy, Cyrus, ed. *Hamlet*. A Norton Critical Edition. New York: W. W. Norton, 1992.

Humphreys, Arthur, ed. *Julius Caesar*. The Oxford Shakespeare. Oxford: Clarendon Press, 1984.

★ Jenkins, Harold, ed. *Hamlet*. The Arden Shakespeare. London: Thomson Learning, 1982 (1990 ed).

★ Neill, Michael, ed. *Othello*. The Oxford Shakespeare. Oxford: Oxford University Press, 2006.

★Orgel, Stephen, ed. *The Winter's Tale*. The Oxford Shakespeare. Oxford: Oxford University Press, 2008.

★ Pitcher, John, ed. *Cymbeline by William Shakespeare*. London: Penguin Books, 2005.

Spencer, T. J. B., ed. *Hamlet by William Shakespeare*. London: Penguin Books, 1996.

Thompson, Ann, and Neil Taylor, eds. *Hamlet*. The Arden Shakespeare. London: Thomson Learning, 2006.

Ure, Peter, ed. *King Richard II*. The Arden Shakespeare. London: Routledge, 1991.

★Vaughan, Virginia Mason, and Alden T. Vaughan, eds. *The Tempest*. The Arden Shakespeare. London: Thomson Learning, 1999.

Warren, Roger, ed. *Cymbeline by William Shakespeare*. Oxford World's Classics. Oxford: Oxford University Press, 1998.

★ Wilders, John, ed. *Antony and Cleopatra*. The Arden Shakespeare. London: Thomson Learning, 1995.

OTHER WORKS CITED

Ackroyd, Peter. *London: The Biography.* New York: Anchor Books, 2003.

Ackroyd, Peter. *Shakespeare: The Biography.* New York: Anchor Books, 2006.

Allen, Don Cameron. *Doubt's Boundless Sea: Skepticism and Faith in the Renaissance.* Baltimore: Johns Hopkins Press, 1964.

Aquilecchia, Giovanni. "Giordano Bruno as Philosopher of the Renaissance." In *Giordano Bruno: Philosopher of the Renaissance*, edited by Hilary Gatti. Aldershot, UK: Ashgate, 2002.

Bacon, Francis. *The Advancement of Learning.* New York: Modern Library, 2001.

Bacon, Francis. *New Atlantis and the Great Instauration.* Edited by Jerry Weinberger. Wheeling, IL: Harlan Davidson, 1989.

Bakewell, Sarah. *How to Live: A Life of Montaigne in One Question and Twenty Attempts at an Answer.* London: Vintage Books, 2011.

Ball, Philip. *Curiosity: How Science Became Interested in Everything.* Chicago: University of Chicago Press, 2013.

Bate, Jonathan, and Dora Thornton. *Shakespeare's Britain.* London: British Museum Press, 2012.

Bate, Jonathan. *Soul of the Age: A Biography of the Mind of William Shakespeare.* New York: Random House, 2009. Citations are to the 2010 edition.

Baumgardt, Carola. *Johannes Kepler: Life and Letters.* New York: Philosophical Library, 1951.

Bell, Millicent. *Shakespeare's Tragic Skepticism.* New Haven, CT: Yale University Press, 2002.

Berman, David. *A History of Atheism in Britain: From Hobbes to Russell.* New York: Croom Helm, 1988.

Bertram, Benjamin. *The Time Is Out of Joint: Skepticism in Shakespeare's England.* Newark, DE: University of Delaware Press, 2004.

Bevington, David. *Shakespeare's Ideas: More Things in Heaven and Earth.* Chichester, UK: Wiley-Blackwell, 2008.

Bloom, Harold. *King Lear.* Bloom's Shakespeare through the Ages. New York: Bloom's Literary Criticism, 2008.

Bloom, Paul. "The Moral Life of Babies." *New York Times*, May 5, 2010. Accessed online at http://www.nytimes.com/2010/05/09/magazine/09babies-t.html?pagewanted=all.

Boas Hall, Marie. "Scientific Thought." In *Shakespeare in His Own Age*, edited by Allardyce Nicoll. Cambridge: Cambridge University Press, 1964.

Boorstin, Daniel J. *The Discoverers: A History of Man's Search to Know His World and Himself.* New York: Vintage Books, 1985.

Bouwsma, William James. *The Waning of the Renaissance, 1550–1640.* New Haven, CT: Yale University Press, 2002.

Bradley, A. C. *Shakespearean Tragedy.* Greenwich, CT: Fawcett Publications, n.d. First published 1904.

Brecht, Bertolt. *Life of Galileo.* Translated by John Willett. London: Methuen, 1980.

Bryson, Bill. *Shakespeare.* London: Harper Perennial, 2008.

Buckley, George T. *Atheism in the English Renaissance.* Chicago: University of Chicago Press, 1932.

Burke, Peter. *Montaigne.* Oxford: Oxford University Press, 1981.

Burns, William E. *The Scientific Revolution: An Encyclopedia.* Santa Barbara, CA: ABC-CLIO, 2001.

Byard, Margaret M. "Poetic Responses to the Copernican Revolution." *Scientific American*, June 1977, 121–29.

Campbell, Gordon, ed. *Ben Jonson: The Alchemist and Other Plays.* Oxford: Oxford University Press, 1995.

Cartwright, John. "Science and Literature in the Elizabethan Renaissance." In *Literature and Science: Social Impact and Interaction* by John H. Cartwright and Brian Baker. Santa Barbara, CA: ABC-CLIO, 2005.

Chapman, Allan. "The Astronomical Work of Thomas Harriot (1560–1621)." *Quarterly Journal of the Royal Astronomical Society* 36 (1995): 97–107.

Chapman, Allan. "Thomas Harriot: The First Telescopic Astronomer." *Journal of the British Astronomical Association* 118, no. 6 (2008): 315–25.

Cheney, Patrick. *Reading Sixteenth-Century Poetry.* Malden, MA: Wiley-Blackwell, 2011.

Chute, Marchette. *Shakespeare of London*. New York: E. P. Dutton, 1949.

Cohen, I. Bernard. *Birth of a New Physics*. New York: W. W. Norton, 1985.

Cohen, I. Bernard. *Revolution in Science*. Cambridge, MA: Harvard University Press, 1985.

Cormack, Lesley B. "Science and Technology." Chapter 28 in *A Companion to Tudor Britain*, edited by Robert Tittler and Norman Jones. New York: Blackwell Publishing, 2004.

Crew, Henry, and Alfonso de Salivo, eds. *Galileo Galilei: Dialogues Concerning Two New Sciences*. New York: Dover, 1954.

Danielson, Dennis. *The Book of the Cosmos: Imagining the Universe from Heraclitus to Hawking*. Cambridge, MA: Perseus Publishing, 2000.

David, Ariel. "Heavens Big Enough for Both God and Aliens, Says Vatican Astronomer." *Globe and Mail*, May 14, 2008, A3.

Davies, Norman. *God's Playground: A History of Poland in Two Volumes*. Oxford: Oxford University Press, 2005.

Day, Malcolm. *Shakespeare's London*. London: Batsford, 2011.

Dear, Peter. "Miracles, Experiments, and the Ordinary Course of Nature." *ISIS* 81 (1990): 663–83.

Dear, Peter. *Revolutionizing the Sciences: European Knowledge and Its Ambitions, 1500–1700*. 2nd ed. Princeton: Princeton University Press, 2009.

Debus, Allen G. *Man and Nature in the Renaissance*. Cambridge: Cambridge University Press, 1999.

Decker, Kevin S. "The Open System and Its Enemies: Bruno, the Idea of Infinity, and Speculation in Early Modern Philosophy of Science." *American Catholic Philosophical Quarterly* 74, no. 4 (2000).

DeWitt, Richard. *Worldviews: An Introduction to the History and Philosophy of Science*. Malden, MA: Blackwell Publishing, 2004.

Dickson, Andrew. *The Rough Guide to Shakespeare*. 2nd ed. London: Rough Guides, 2009.

Dodd, A. H. *Life in Elizabethan England*. London: Batsford, 1961.

Donnelly, Ann, and Elizabeth Woledge. *Shakespeare: Work, Life and Times.* Stratford, UK: Shakespeare Birthplace Trust/Jigsaw Design and Publishing, 2010.

Doran, Susan. "The Queen." In *The Elizabethan World*, edited by Susan Doran and Norman Jones. London: Routledge, 2011.

Drake, Stillman, trans. *Discoveries and Opinions of Galileo.* New York: Anchor Books, 1957.

Dunn, Richard. *The Telescope: A Short History.* London: National Maritime Museum, 2009.

Edwards, Kathryn A. "Witchcraft in Tudor England and Scotland." In *A Companion to Tudor Literature*, edited by Kent Cartwright, 31–48. Chichester, UK: Wiley-Blackwell, 2010.

Elton, William R. *King Lear and the Gods.* San Marino, CA: Huntington Library, 1968.

Elton, W. R. "Shakespeare and the Thought of His Age." In *A New Companion to Shakespeare Studies*, edited by Kenneth Muir and S. Schoenbaum. Cambridge: Cambridge University Press, 1971.

Falk, Dan. "The Rise and Fall of Tycho Brahe." *Astronomy*, December 2003, 52–57.

Falk, Dan. *In Search of Time: The Science of a Curious Dimension.* New York: St. Martin's Press, 2008.

Falk, Dan. *Universe on a T-Shirt: The Quest for the Theory of Everything.* New York: Arcade Publishing, 2004.

Feinberg, Matthew, and Robb Willer. "Apocalypse Soon? Dire Messages Reduce Belief in Global Warming by Contradicting Just-World Beliefs." *Psychological Science* 22, no. 1 (2011): 34–38.

Feingold, Mordechai. "Galileo in England: The First Phase." In *Novità celesti e crisi del sapere*, edited by P. Galluzzi. Florence: Giunti Barbèra, 1984.

Feingold, Mordechai. *The Mathematicians' Apprenticeship: Science, Universities and Society in England, 1560–1640.* Cambridge: Cambridge University Press, 1984.

Findlen, Paula. "The Sun at the Center of the World." In *The Renaissance World*, edited by John Jeffries Martin. New York: Routledge, 2007.

Fitzmaurice, James, ed. *Major Women Writers of Seventeenth-Century England.* Ann Arbor, MI: University of Michigan Press, 2000.

Flow, Christian. "Swerves." *Harvard Magazine*, July–August 2011. Accessed online at http://harvardmagazine.com/2011/07/swerves.

Flude, Kevin. "The Hospital, the Bard, and the Son-in-law." Unpublished manuscript.

Flude, Kevin, and Paul Herbert. *The Old Operating Theatre, Museum, and Herb Garret: Museum Guide*. London: The Old Operating Theatre, Museum, and Herb Garret, 1995.

Friedrich, Hugo. *Montaigne*. Berkeley, CA: University of California Press, 1991.

Forgeng, Jeffrey L. *Daily Life in Elizabethan England*. Santa Barbara, CA: Greenwood Press, 2010.

Fowler, Alastair. *Time's Purpled Masquers: Stars and the Afterlife in Renaissance English Literature*. Oxford: Clarendon Press, 1996.

Fox, Robert. *Thomas Harriot: An Elizabethan Man of Science*. Aldershot, UK: Ashgate, 2000.

Furnham, Adrian. "Belief in a Just World: Research Progress over the Past Decade." *Personality and Individual Differences* 34, no. 5 (2003): 795–817.

Gatti, Hilary. *Essays on Giordano Bruno*. Princeton, NJ: Princeton University Press, 2011.

Gatti, Hilary. *Giordano Bruno: Philosopher of the Renaissance*. Aldershot, UK: Ashgate, 2002.

Gingerich, Owen. "Astronomical Scrapbook." *Sky & Telescope*, May 1981. 394–5.

Gingerich, Owen. *The Book Nobody Read: Chasing the Revolutions of Nicolaus Copernicus*. New York: Penguin Books, 2005.

Gingerich, Owen. "Mankind's Place in the Universe." *Nature* 457 (1 January 2009): 28–29.

Gingerich, Owen. "Tycho Brahe and the Nova of 1572." *1604–2004: Supernovae as Cosmological Lighthouses*. ASP Conference Series 342 (2005).

Grayling, A. C. *Descartes*. London: Pocket Books, 2006.

Greenblatt, Stephen. *Will in the World*. New York: W. W. Norton, 2004. Citations are to the 2005 edition.

Greenblatt, Stephen. *The Norton Shakespeare*. 2nd ed. New York: W. W. Norton, 2008.

Greenblatt, Stephen. *The Swerve: How the World Became Modern*. New York: W. W. Norton, 2011.

Gribbin, John. *Science: A History, 1543–2001*. New York: BCA, 2002.

Guthrie, Douglas. "The Medical and Scientific Exploits of King James IV of Scotland." *British Medical Journal*, May 30, 1953, 1191–1193.

Haidt, Jonathan. *The Righteous Mind: Why Good People Are Divided by Politics and Religion*. New York: Random House, 2012.

Hafer, Carolyn L., and Laurent Bègue. "Experimental Research on Just-World Theory: Problems, Developments, and Future Challenges." *Psychological Bulletin* 131, no. 1 (2005): 128–167.

Hale, John. *The Civilization of Europe in the Renaissance*. London: Harper Collins, 1993.

Harkness, Deborah. *The Jewel House: Elizabethan London and the Scientific Revolution*. New Haven, CT: Yale University Press, 2007.

Hawkes, Nigel. "Astronomer discovers cast of stars hidden in Hamlet." *Times* (London), January 14, 1997.

Hecht, Jennifer Michael. *Doubt: A History*. New York: HarperOne, 2004.

Heilbron, John L. *The Sun in the Church: Cathedrals as Solar Observatories*. Cambridge, MA: Harvard University Press, 1999.

Heninger, S. K., Jr. *The Cosmographical Glass*. San Marino, CA: Huntington Library, 1977.

Henry, John. *The Scientific Revolution and the Origin of Modern Science*. 2nd ed. Houndmills, UK: Palgrave, 2002.

Hotson, Leslie. *I, William Shakespeare Do Appoint Thomas Russell, Esquire . . .* London: Jonathan Cape, 1937.

Hunt, Marvin W. *Looking for Hamlet*. New York: Palgrave MacMillan, 2007.

Hyman, Gavin. *A Short History of Atheism*. London: I. B. Tauris, 2010.

Jacob, James R. *The Scientific Revolution: Aspirations and Achievements, 1500–1700*. Amherst, NY: Humanity Books, 1999.

Janowitz, Henry. "Some Evidence on Shakespeare's Knowledge of the Copernican Revolution and the 'New Philosophy.'" *The Shakespeare Newsletter* (Fall 2001): 79–80.

Jensen, Freyja Cox. "Intellectual Developments." In *The Elizabethan World*, edited by Susan Doran and Norman Jones. London: Routledge, 2011.

Johnson, Francis R. *Astronomical Thought in Renaissance England: A Study of the English Scientific Writings from 1500 to 1645*. New York: Octagon Books, 1968.

Johnson, Francis R. "The Influence of Thomas Digges on the Progress of Modern Astronomy in Sixteenth-Century England." *Osiris* 1 (January 1936): 390–410.

Jones, Nigel. "The Arch Conjuror of England: John Dee by Glyn Parry." *Daily Telegraph*, March 6, 2012 (accessed online).

Jones, Norman. "Shakespeare's England." In *A Companion to Shakespeare*, edited by David Scott Kastan. Malden, MA: Blackwell, 1999.

Kermode, Frank. *The Age of Shakespeare*. New York: Modern Library, 2004. Citations are to the 2005 edition.

Kocher, Paul H. *Science and Religion in Elizabethan England*. San Marino, CA: Huntington Library, 1953.

Kragh, Helge S. *Conceptions of Cosmos: From Myths to the Accelerating Universe: A History of Cosmology*. Oxford: Oxford University Press, 2007.

Landes, David S. *Revolution in Time: Clocks and the Making of the Modern World*. Cambridge, MA: Harvard University Press, 1983.

Laroque, Francois. *The Age of Shakespeare*. New York: Harry N. Abrams, 1993.

Lawless, Jill. "Shakespeare the 'Hard-headed Businessman' Uncovered." *Independent*, April 1, 2013 (accessed online).

Leake, Jonathan. "Bad Bard: A Tax Dodger and Famine Profiteer." *Sunday Times*, March 31, 2013 (accessed online).

Lerner, Melvin J. *The Belief in a Just World: A Fundamental Delusion*. New York: Plenum Press, 1980.

Levy, David H. *The Sky in Early Modern English Literature: A Study of Allusions to Celestial Events in Elizabethan and Jacobean Writing, 1572–1620*. New York: Springer, 2011.

Levy, David H. *Starry Night: Astronomers and Poets Read the Sky*. Amherst, NY: Prometheus Books, 2001.

Lucretius. *On the Nature of the Universe.* Translated by Ronald Melville. Oxford: Oxford University Press, 2008.

Maisano, Scott. "Descartes avec Milton." In *The Automaton in English Renaissance Literature*, edited by Wendy Beth Hyman. Farnham, UK: Ashgate, 2011.

Maisano, Scott. "Infinite Gesture: Automata and the Emotions in Descartes and Shakespeare." In *Genesis Redux*, edited by Jessica Riskin, 63–84. Chicago: University of Chicago Press, 2007.

Maisano, Scott. "Shakespeare's Last Act: The Starry Messenger and the Galilean Book in Cymbeline." *Configurations* 12, no. 3 (Fall 2004): 401–434.

Maisano, Scott. "Shakespeare's Science Fictions: The Future History of the Late Romances." PhD thesis, Indiana University, August 2004.

Marchitello, Howard. *The Text in the Machine: Science and Literature in the Age of Shakespeare and Galileo.* Oxford: Oxford University Press, 2011.

Margolis, Howard. *It Started with Copernicus: How Turning the World Inside Out Led to the Scientific Revolution.* New York: McGraw-Hill, 2002.

Marlowe, Christopher. *The Complete Plays.* Edited by Frank Romany and Robert Lindsey. London: Penguin Books, 2003.

Martin, Randall, ed. *Women Writers in Renaissance England.* London: Longman, 1997.

Mazzio, Carla. "Shakespeare and Science, c. 1600." *South Central Review* 26 (Winter and Spring 2009): 1–23.

McAlindon, Thomas. *Shakespeare's Tragic Cosmos.* Cambridge: Cambridge University Press, 1991.

McGinn, Colin. *Shakespeare's Philosophy.* New York: Harper Perennial, 2006.

McLean, Antonia. *Humanism and the Rise of Science in Tudor England.* New York: Neale Watson Academic Publications, 1972.

Meadows, A. J. *The High Firmament: A Survey of Astronomy in English Literature.* Leicester, UK: Leicester University Press, 1969.

Milton, John. *Paradise Lost.* Edited by Scott Elledge. New York: W. W. Norton, 1975. Citations are to the 1993 edition.

Montaigne, Michel de. *The Complete Essays.* Edited by M. A. Screech. London: Penguin Books, 2003.

Montaigne, Michel de. *The Essays of Montaigne: The John Florio Translation*. London: Modern Library, 1933.

Nicholl, Charles. *The Lodger: Shakespeare on Silver Street*. New York: Allen Lane, 2007.

Nicoll, Allardyce. *The Elizabethans*. Cambridge: Cambridge University Press, 1957.

Numbers, Ronald L., ed. *Galileo Goes to Jail and Other Myths about Science and Religion*. Cambridge, MA: Harvard University Press, 2009.

Nuttall, A. D. *Shakespeare the Thinker*. New Haven, CT: Yale University Press, 2007.

Olsen, Kirstin. *All Things Shakespeare: An Encyclopedia of Shakespeare's World*. 3 vols. Westport, CT: Greenwood Press, 2002.

Olson, Donald W., Marilyn S. Olson, and Russell L. Doescher. "The Stars of Hamlet." *Sky & Telescope*, November 1998, 68–73.

Orchard, Thomas N. *The Astronomy of Milton's "Paradise Lost."* Fairford, UK: Echo Press, 2010.

Palmer, Ada. "Reading Lucretius in the Renaissance." *Journal of the History of Ideas* 73, no. 3 (July 2012): 395–416.

Panek, Richard. *Seeing and Believing: How the Telescope Opened our Eyes and Minds to the Heavens*. New York: Penguin Books, 1998. Citations are to the 1999 edition.

Pope, Maurice. "Medicine." In *The Oxford Companion to Shakespeare*, edited by Michael Dobson and Stanley Wells, 283–87. Oxford: Oxford University Press, 2001.

Principe, Lawrence M. *The Scientific Revolution: A Very Short Introduction*. Oxford: Oxford University Press, 2011.

Pritchard, R. E., ed. *Shakespeare's England*. Phoenix Mill, UK: Sutton Publishing, 2000.

Pumfrey, Stephen. "Harriot's Maps of the Moon: New Interpretations." *Notes and Records of the Royal Society* 63 (2009): 163–68.

Pumfrey, Stephen. "'Your Astronomers and Ours Differ Exceedingly': The Controversy over the 'New Star' of 1572 in the Light of a Newly Discovered Text by Thomas Digges." *British Journal for the History of Science* 44, no. 1 (2011): 29–60.

Raman, Shankar. "Death by Numbers: Counting and Accounting in *The Winter's Tale*." In *Alternative Shakespeares 3*, edited by Diana E. Henderson. London: Routledge, 2008.

Raman, Shankar. "Specifying Unknown Things: The Algebra of *The Merchant of Venice*." In *Making Publics in Early Modern Europe*, edited by Bronwen Wilson and Paul Yachnin. New York: Routledge, 2010.

Reston, James. *Galileo: A Life*. New York: Harper Collins, 1994.

Ridley, Jasper. *The Tudor Age*. London: Robinson, 2002.

Rosen, Edward. *Copernicus and the Scientific Revolution*. Malabar, FL: Robert E. Krieger, 1984.

Rosen, Edward, trans. *Kepler's Somnium: The Dream, or Posthumous Work on Lunar Astronomy*. Madison, WI: University of Wisconsin Press, 1967.

Ross, James Bruce, and Mary Martin McLaughlin, eds. *The Portable Renaissance Reader*. New York: Penguin Books, 1988.

Rowland, Ingrid. *Giordano Bruno: Philosopher/Heretic*. New York: Farrar, Straus and Giroux, 2008.

Russell, John L. "The Copernican System in Great Britain." In *The Reception of Copernicus's Heliocentric Theory*, edited by Jerzy Dobrzycki. Boston: D. Reidel Publishing Company, 1973.

Sacerdoti, Gilberto. *Nuovo cielo, nuova terra: La rivelazione copernicana di "Antonio e Cleopatra" di Shakespeare*. Bologna: Società editrice il Mulino, 1990.

Sayce, R. A. *The Essays of Montaigne: A Critical Exploration*. London: Weidenfeld and Nicholson, 1972.

Schoenbaum, Samuel. *Shakespeare: A Compact Documentary Life*. Oxford: Oxford University Press, 1987.

Seeds, Michael A. *Foundations of Astronomy*. Belmont, CA: Wadsworth, 1999.

Shapin, Steven. *The Scientific Revolution*. Chicago: University of Chicago Press, 1996.

Shapiro, James. *A Year in the Life of William Shakespeare*. New York: Harper Perennial, 2005.

Sharpe, J. A. *Early Modern England: A Social History 1550–1760*. London: Edward Arnold, 1987.

Sharpe, James. "The Debate on Witchcraft." In *A New Companion to English Renaissance Literature and Culture* vol. 2, edited by Michael Hattaway. Malden, MA: Wiley-Blackwell, 2010.

Sobel, Dava. *A More Perfect Heaven: How Copernicus Revolutionized the Cosmos*. New York: Walker, 2011.

Spiller, Elizabeth. "Shakespeare and the Making of Early Modern Science." *South Central Review* 26 (Winter and Spring 2009): 24–41.

St. John Parker, Michael. *Shakespeare*. Andover: Pitkin Publishing, 2010.

Stimson, Dorothy. *The Gradual Acceptance of the Copernican Theory*. Gloucester, MA: Peter Smith, 1972.

Swerdlow, Noel M. "Galileo's Discoveries with the Telescope and Their Evidence for the Copernican Theory." In *The Cambridge Companion to Galileo*, edited by Peter Machammer. Cambridge: Cambridge University Press, 1999.

Thomas, Keith. *Religion and the Decline of Magic*. London: Penguin Books, 1991.

Travitsky, Betty, ed. *The Paradise of Women: Writings by Englishwomen of the Renaissance*. Westport, CT: Greenwood Press, 1981.

Usher, Peter. "Hamlet and Infinite Universe." *Research Penn State* 18, no. 3 (September 1997). Accessed online at http://www.rps.psu.edu/sep97/hamlet.html.

Usher, Peter. "Jupiter and *Cymbeline*." *The Shakespeare Newsletter*, Spring 2003, 7–12.

Usher, Peter. *Shakespeare and the Dawn of Modern Science*. Amherst, NY: Cambria Press, 2010.

Usher, Peter. "Shakespeare and Elizabethan Telescopy." *Journal of the Royal Astronomical Society of Canada* (February 2009): 16–18.

Usher, Peter. "Shakespeare's Cosmic World View." *Mercury* 26, no. 1 (January–February 1997): 20–23.

Usher, Peter. "Shakespeare's Support for the New Astronomy." *The Oxfordian* 5 (2002): 132–46. Accessed online at http://www.shakespearedigges.org/ox2.htm.

Weinberg, Steven. *The First Three Minutes: A Modern View of the Origin of the Universe*. New York: Basic Books, 1977. Citations are to the 1988 updated edition.

Westfall, Richard S. "The Scientific Revolution Reasserted." In *Rethinking the Scientific Revolution*, edited by Margaret J. Osler. Cambridge: Cambridge University Press, 2000.

Williams, Neville. "The Tudors." In *The Lives of the Kings and Queens of England*, edited by Antonia Fraser. London: Weidenfeld and Nicholson, 1993.

Wilson, John Dover. *The Essential Shakespeare*. Cambridge: Cambridge University Press, 1964.

Index